John Wellesley Russell

**An elementary treatise on pure geometry with numerous examples**

John Wellesley Russell
**An elementary treatise on pure geometry with numerous examples**
ISBN/EAN: 9783337278441
Printed in Europe, USA, Canada, Australia, Japan
Cover: Foto ©berggeist007 / pixelio.de

More available books at **www.hansebooks.com**

AN

ELEMENTARY TREATISE

ON

# PURE GEOMETRY

*WITH NUMEROUS EXAMPLES*

BY

JOHN WELLESLEY RUSSELL, M.A.

FORMERLY FELLOW OF MERTON COLLEGE
MATHEMATICAL LECTURER OF BALLIOL AND ST. JOHN'S COLLEGES, OXFORD

Oxford
AT THE CLARENDON PRESS
1893

# PREFACE

In this treatise, the author has attempted to bring together all the well-known theorems and examples connected with Harmonics, Anharmonics, Involution, Projection (including Homology), and Reciprocation. In order to avoid the difficulty of framing a general geometrical theory of Imaginary Points and Lines, the Principle of Continuity is appealed to. The properties of Circular Points and Circular Lines are then discussed, and applied to the theory of the Foci of Conics.

The examples at the ends of the articles are intended to be solved by the help of the article to which they are appended. Among these examples will be found many interesting theorems which were not considered important enough to be included in the text. At the end of the book there is, besides, a large number of Miscellaneous Examples. Of these, the first part is taken mainly from examination papers of the University of Oxford. Scattered throughout the book will be found examples taken from that admirable collection of problems called *Mathematical Questions and Solutions from the 'Educational Times.'* For permission to make use of these, I am indebted to the kindness of the able editor, Mr. W. J. C. Miller, B.A., Registrar of the General Medical Council.

The book has been read both in MS. and in proof by my old pupil, Mr. A. E. Jolliffe, B.A., Fellow of Corpus Christi College, and formerly Scholar of Balliol College, Oxford, whose valuable suggestions I have made free use of. To him I am also indebted for the second part of the Miscellaneous Examples. I am glad of this opportunity of acknowledging my great obligations to my former tutor, the late Professor H. J. S. Smith. My first lessons in Pure Geometry were learnt from his lectures; and many of the proofs in this book are derived from the same source.

I have assumed that the reader has passed through the ordinary curriculum in Geometry before attempting to read the present subject; viz. Euclid, some Appendix to Euclid, and Geometrical Conics.

I have not found it convenient to keep rigidly to any single notation. But, ordinarily, points have been denoted by $A, B, C, \ldots$, lines by $a, b, c, \ldots$, and planes and conics by $\alpha, \beta, \gamma, \ldots$.

The following abbreviations have been used—

A straight line has been called a *line*, and a curved line has been called a *curve*.

The point of intersection of two lines has been called the *meet* of the lines.

The line joining two points has been called the *join* of the points.

The meet of the lines $AB$ and $CD$ has been denoted by $(AB\,;\,CD)$.

To avoid the frequent use of the phrase 'with respect to' or 'with regard to,' the word 'for' has been substituted.

The abbreviation 'r. h.' has sometimes been used for 'rectangular hyperbola.'

The single word 'director' has been used to include the 'director circle' of a central conic and the 'directrix' of a parabola.

# *Preface.*

The angle between the lines $a$ and $b$ has been denoted by $\angle ab$ and the sine of this angle by $\sin ab$.

The length of the perpendicular from the point $A$ on the line $b$ has been denoted by $(A, b)$.

I have ventured to use the word 'mate' to mean 'the point (or line) corresponding.' I have avoided using the word 'conjugate' except in its legitimate sense in connection with the theory of polars.

I shall be glad to receive, from any of my readers, corrections, or suggestions for the improvement of the book; interesting theorems and examples which are not already included will also be welcomed.

J. W. RUSSELL.

*February,* 1893.

# CONTENTS

## CHAPTER I.
### FORMULAE CONNECTING SEGMENTS OF THE SAME LINE.

|  | PAGE |
|---|---|
| General properties of points and lines | 2 |
| Menelaus's theorem. Ceva's theorem | 7 |

## CHAPTER II.
### HARMONIC RANGES AND PENCILS.

| | |
|---|---|
| Polar of a point for two lines. Pole of a line for two points | 18 |
| Locus of $P$, given $1/OP = \Sigma a/OA$ | 19 |
| Complete quadrilateral. Complete quadrangle | 19 |

## CHAPTER III.
### HARMONIC PROPERTIES OF A CIRCLE.

| | |
|---|---|
| Imaginary points and tangents | 23 |
| Inverse points. Orthogonal circles | 24 |
| Harmonic chord of two circles | 25 |
| Salmon's theorem | 27 |
| Self-conjugate triangle—polar circle | 31 |
| Inscribed quadrangle—circumscribed quadrilateral | 33 |

## CHAPTER IV.

### PROJECTION.

|   | PAGE |
|---|---|
| Line at infinity | 37 |
| The eight vertices of two quadrangles | 40 |
| Homologous triangles | 42 |

## CHAPTER V.

### HARMONIC PROPERTIES OF A CONIC.

| | |
|---|---|
| Two lines or two points form a conic | 47 |
| Self-conjugate triangle | 50 |
| — of parabola, given a circumscribing triangle | 53 |
| Centre—diameter—conjugate diameter—condition for circle—asymptotes—rectangular hyperbola—principal axes | 53 |

## CHAPTER VI.

### CARNOT'S THEOREM.

| | |
|---|---|
| Newton's theorem—equations of conics | 61 |
| Hyperbola—conventional conjugate diameter—asymptotic properties | 63 |
| Rectangular hyperbola—passes through orthocentre | 65 |

## CHAPTER VII.

### FOCI OF A CONIC.

| | |
|---|---|
| Confocal conics. Focal projection | 70 |

## CHAPTER VIII.

### RECIPROCATION.

| | |
|---|---|
| Focal properties of conics. Envelopes | 88 |
| Poncelet's theorem | 94 |

## CHAPTER IX.

### ANHARMONIC OR CROSS RATIO.

| | |
|---|---|
| Harmonic range | 100 |
| Cross ratio unchanged by projection | 103 |
| To project the figure $ABCD$ into the figure $A'B'C'D'$ | 103 |

## Contents.     xi

|  | PAGE |
|---|---|
| Homographic ranges and pencils | 104 |
| Ranges and pencils in perspective | 106 |
| Triangle inscribed in one triangle and circumscribed to another | 108 |
| Projective ranges and pencils | 110 |

### CHAPTER X.

#### VANISHING POINTS OF TWO HOMOGRAPHIC RANGES.

Locus of vertex of projection of two homologous triangles . . 113
Formula $kxx' + lx + mx' + n = 0$ . . . . . . . . 114
Equation of a line and of a point . . . . . . . 116
Common points of two homographic ranges . . . . . 118
Ranges formed by the mates of a point in two homographic ranges . . . . . . . . . . . . 119
Determine $X$, given $AX . A'X \div BX . B'X$ . . . . . 120
Similar ranges—superposable ranges . . . . . . 120
Common rays of two homographic pencils . . . . . 121
Transversal cutting two homographic pencils in superposable ranges . . . . . . . . . . . . . 121
Two homographic ranges subtend superposable pencils . . 123

### CHAPTER XI.

#### ANHARMONIC PROPERTIES OF POINTS ON A CONIC.

Harmonic points on a conic . . . . . . . . 126
$xy = $ constant in hyperbola . . . . . . . . 126
Tangents at meets of four-point conics . . . . . . 127
Pappus's theorem—$a . \gamma = k . \beta . \delta$ . . . . . . . 127
Constant intercept by chords on asymptote of hyperbola . . 128
Locus of meets of corresponding rays of two homographic pencils 128
One conic can be drawn through five points. Two conics meet in four points. Conics through four points have a common self-conjugate triangle . . . . . . . . . . 130
One rectangular hyperbola can be drawn through four points . 131
Construct a conic, given two pairs of conjugate diameters and a point . . . . . . . . . . . . 131
A conic—its own reciprocal . . . . . . . . 131
To trisect a circular arc . . . . . . . . . 133
Locus of centre of circle, given two pairs of conjugate lines . . 134
Locus of foci of conics touching a parallelogram . . . 134
The projection of a conic is a conic . . . . . . 134

## CHAPTER XII.

#### ANHARMONIC PROPERTIES OF TANGENTS OF A CONIC.

|   | PAGE |
|---|---|
| Product of intercepts of a tangent on parallel tangents | 136 |
| Harmonic tangents of a conic | 136 |
| $p \cdot r = k \cdot q \cdot s$ | 137 |
| Envelope of lines joining corresponding points of two homographic ranges. | 139 |
| One conic can be drawn touching five lines. Two conics have four common tangents. Conics touching four lines have a common self-conjugate triangle | 140 |
| Construct a conic, given two pairs of conjugate diameters and a tangent | 141 |
| The eight vertices of two quadrangles | 141 |
| Tangents of a parabola divide any tangent similarly | 142 |
| Envelope of axes of conics having double contact | 143 |

## CHAPTER XIII.

#### POLES AND POLARS. RECIPROCATION.

| Poles homographic with polars. Range homographic with reciprocal pencil | 144 |
|---|---|
| Locus of fourth harmonics of a line for a conic on concurrent radii | 145 |
| Envelope of join of conjugate points on two lines. | 145 |
| Envelope of perpendicular (or oblique) from a point on its polar (which passes through a fixed point) | 145 |
| Reciprocal of conic—of pole and polar | 146 |
| Any two conics are reciprocal | 147 |
| Reciprocation of $AB:BC$ | 148 |

## CHAPTER XIV.

#### PROPERTIES OF TWO TRIANGLES.

| Gaskin's theorem | 151 |
|---|---|
| Centre of circle circumscribed to a triangle self-conjugate for a parabola is on the directrix | 152 |
| Centre of circle touching a triangle self-conjugate for a rectangular hyperbola is on the r. h. | 152 |
| Given a self-conjugate triangle and a point on the director | 153 |
| Pole and polar of a triangle for a conic. Hesse's theorem | 154 |

## Contents.

### CHAPTER XV.
#### PASCAL'S THEOREM AND BRIANCHON'S THEOREM.

|   |   |
|---|---|
| Conjugate points on a line through a given point | 158 |
| Steiner's theorem | 161 |
| Conjugate lines through a point on a given line | 161 |

### CHAPTER XVI.
#### HOMOGRAPHIC RANGES ON A CONIC.

|   |   |
|---|---|
| Homographic sets of tangents | 164 |
| Envelope of joins of corresponding points | 165 |
| Construction of common points (or rays) of two homographic ranges on a line (or pencils at a point) | 165 |

### CHAPTER XVII.
#### RANGES IN INVOLUTION.

|   |   |
|---|---|
| Two homographic ranges can be placed in involution | 172 |
| Ranges formed by the mates of a point in two homographic ranges | 174 |
| Involution of coaxal circles | 175 |
| Determine $C$, given $CA \cdot CA' \div CB \cdot CB'$ | 177 |

### CHAPTER XVIII.
#### PENCILS IN INVOLUTION.

|   |   |
|---|---|
| Two homographic pencils can be placed in involution | 181 |
| Properties of an Involution Range obtained by Projection | 185 |

### CHAPTER XIX.
#### INVOLUTION OF CONJUGATE POINTS AND LINES.

|   |   |
|---|---|
| Conjugate diameters. Asymptotes. Principal Axes | 188 |
| Feet of normals from a point to a conic. Locus of meet with conjugate diameter of perpendicular from a point on a diameter. Locus of point such that the perpendicular from it on its polar passes through a fixed point. Feet of obliques | 189 |
| Common chords. Two conics have only one common self-conjugate triangle. Exception. Common apexes | 190 |
| Homothetic figures—homothetic conics | 192 |

## CHAPTER XX.

### INVOLUTION RANGE ON A CONIC.

|  | PAGE |
|---|---|
| Set of pairs of tangents in involution | 197 |
| Join of feet of perpendiculars from a point on a pair of rays of an involution pencil | 198 |
| Common points (or lines) of two involutions | 199 |
| The orthogonal pair of an involution pencil | 200 |
| Frégier point. Frégier line | 201 |
| Construction of double points (or lines) of an involution | 201 |

## CHAPTER XXI.

### INVOLUTION OF A QUADRANGLE.

| | |
|---|---|
| Construction for mate of a point in an involution | 203 |
| Hesse's theorem | 204 |
| Involution of four-point conics. (Desargues's theorem) | 204 |
| Polygon of $2n$ sides inscribed in a conic, so that each side shall pass through one of $2n$ collinear points | 206 |
| Problem on pole of triangle | 206 |
| Rectangular hyperbola about a triangle passes through orthocentre; and conversely | 209 |

## CHAPTER XXII.

### POLE-LOCUS AND CENTRE-LOCUS.

| | |
|---|---|
| Conjugate points for a system of four-point conics | 210 |
| Pole-locus (or eleven-point conic) is also the locus of conjugate points | 211 |
| The polars of two points for a system of four-point conics form homographic pencils | 212 |

## CHAPTER XXIII.

### INVOLUTION OF A QUADRILATERAL.

| | |
|---|---|
| Involution of four-tangent conics | 216 |
| The circles on the diagonals of a quadrilateral and the directors of inscribed conics form a coaxal system. Locus of middle points of diagonals and of centres of inscribed conics is the diameter of the quadrilateral | 219 |
| The polar circle of a triangle about a conic is orthogonal to the director | 220 |

## Contents.

|  | PAGE |
|---|---|
| Locus of centre of rectangular hyperbola inscribed in a given triangle is the polar circle | 220 |
| Steiner's theorem. Gaskin's theorem | 220 |
| Two rectangular hyperbolas can be drawn touching four lines | 221 |
| Locus of poles of a given line | 221 |
| Polar-envelope | 222 |

### CHAPTER XXIV.

#### CONSTRUCTIONS OF THE FIRST DEGREE.

### CHAPTER XXV.

#### CONSTRUCTIONS OF THE SECOND DEGREE.

| | |
|---|---|
| Two conics cannot have two common self-conjugate triangles | 235 |
| If a variable conic through $ABCD$ meet fixed lines through $A$ and $B$ in $P$ and $Q$, then $PQ$ and $CD$ meet in a fixed point | 237 |

### CHAPTER XXVI.

#### METHOD OF TRIAL AND ERROR.

| | |
|---|---|
| Solution of certain algebraical equations | 242 |
| To inscribe a polygon in a conic, so that each side shall pass through a given point | 243 |

### CHAPTER XXVII.

#### IMAGINARY POINTS AND LINES.

| | |
|---|---|
| The Principle of Continuity | 245 |

### CHAPTER XXVIII.

#### CIRCULAR POINTS AND CIRCULAR LINES.

| | |
|---|---|
| Concentric circles have double contact | 252 |
| The circle about a triangle self-conjugate for a rectangular hyperbola passes through the centre. Gaskin's theorem | 253 |
| Axes of conics through four concyclic points. Director circle | 253 |
| Coaxal circles | 254 |
| Foci of a conic. The circle about a triangle about a parabola passes through the focus. Confocal conics | 255 |

## CHAPTER XXIX.

### PROJECTION, REAL AND IMAGINARY.

|   | PAGE |
|---|---|
| Homologous triangles | 264 |
| Pole-locus touches sixteen conics | 265 |
| Common chords and common apexes | 266 |
| Harmonic envelope and harmonic locus of two conics | 267 |
| Poncelet's theorem | 268 |
| Envelope of join of corresponding points of two homographic ranges on a conic. Conics having double contact. Envelope of last side of polygon inscribed in a conic, so that each side but one shall pass through a fixed point | 269 |

## CHAPTER XXX.

### GENERALISATION BY PROJECTION.

Generalisation by Reciprocation . . . . . . . 279

## CHAPTER XXXI.

### HOMOLOGY.

| Locus of vertex of projection of two figures in perspective | 283 |
|---|---|
| Coaxal figures and copolar figures | 285 |
| Given a parallelogram, construct a parallel to a given line through a given point | 288 |

MISCELLANEOUS EXAMPLES . . . . . . . . 299

# TEXT-BOOK OF PURE GEOMETRY.

## CHAPTER I.

FORMULAE CONNECTING SEGMENTS OF THE SAME LINE.

**1.** ONE of the differences between Modern Geometry and the Geometry of Euclid is that a length in Modern Geometry has a sign as well as a magnitude. Lengths measured on a line in one direction are considered positive and those measured in the opposite direction are considered negative. Thus if $AB$, i.e. the segment extending from $A$ to $B$, be considered positive, then $BA$, i.e. the segment extending from $B$ to $A$, must be considered negative. Also $AB$ and $BA$ differ only in sign. Hence we obtain the first formula, viz. $AB = -BA$.

Notice that by allowing lengths to have a sign as well as a magnitude, we are enabled to utilise the formulae of Algebra in geometrical investigations. In making use of Algebra it is generally best to reduce all the segments we employ to the same origin. This is done in the following way.

Take any segment $AB$ on a line and also any origin $O$. Then $AB = OB - OA$. This is obviously true in the above figure, and it is true for any figure. For

$$OB - OA = OB + AO = AO + OB = AB;$$

for $AO + OB$ means that the point travels from $A$ to $O$ and then from $O$ to $B$, and thus the point has gone from $A$ to $B$.

The fundamental formulae then are

(1) $AB = -BA$;  (2) $AB = OB - OA$.

In the above discussion the lengths have been taken on a line. But this is not necessary; the lengths might have been taken on any curve.

It is generally convenient to use an abridged form of the formula $AB = OB - OA$, viz. $AB = b - a$, where $a = OA$ and $b = OB$.

**2.** $A, B, C, D$ are any four collinear points; show that
$$AB \cdot CD + AC \cdot DB + AD \cdot BC = 0.$$

Take $A$ as origin, then $CD = AD - AC = d - c$, and so on. Hence
$$AB \cdot CD + AC \cdot DB + AD \cdot BC = b(d-c) + c(b-d) + d(c-b)$$
$$= bd - bc + cb - cd + dc - db = 0.$$

**Ex. 1.** $A, B, C, D, O$ are any five points in a plane; show that
$$AOB \cdot COD + AOC \cdot DOB + AOD \cdot BOC = 0,$$
where $AOB$ denotes the area of the triangle $AOB$.

Let a line meet $OA, OB, OC, OD$ in $A', B', C', D'$. Then
$$AOB = \tfrac{1}{2} \cdot OA \cdot OB \sin AOB.$$
Hence the given relation is true if
$$\Sigma \{\sin AOB \cdot \sin COD\} = 0,$$
i.e. if
$$\Sigma \{\sin A'OB' \cdot \sin C'OD'\} = 0.$$
But $p \cdot A'B' = OA' \cdot OB' \sin A'OB'$, where $p$ is the perpendicular from $O$ on $A'B'C'D'$. Hence the given relation is true if
$$A'B' \cdot C'D' + A'C' \cdot D'B' + A'D' \cdot B'C' = 0.$$

**Ex. 2.** If $OA, OB, OC, OD$ be any four lines meeting in a point, show that
$$\sin AOB \cdot \sin COD + \sin AOC \cdot \sin DOB + \sin AOD \cdot \sin BOC = 0.$$

**Ex. 3.** Show also that
$$\cos AOB \cdot \sin COD + \cos AOC \cdot \sin DOB + \cos AOD \cdot \sin BOC = 0,$$
and
$$\cos AOB \cdot \cos COD - \cos AOC \cdot \cos DOB - \sin AOD \cdot \sin BOC = 0.$$

For Ex. 2 is true for $OA'$ where $A'OA$ is a right angle, and also for $OA'$ and $OD'$ where $A'OA$ and $D'OD$ are right angles.

**Ex. 4.** From Ex. 2 deduce Ptolemy's Theorem connecting four points on a circle.

Take $O$ also on the circle. Then $AB = 2 \cdot R \cdot \sin AOB$.

**Ex. 5.** Show also that the relation of Ex. 2 holds if each angle involved be multiplied by the same quantity.

For $AOB = VOB - VOA$, if $OV$ be the initial line. Now take $VOB' = h \cdot VOB$, $VOC' = h \cdot VOC$, and so on. Then $A'OB' = h \cdot AOB$, &c.; and the theorem is true for $A'OB'$, &c.

**Ex. 6.** *If $A$, $B$, $C$ be the angles of a triangle and $A'$, $B'$, $C'$ be the angles which the sides $BC$, $CA$, $AB$ make with any line, then*
$$\sin A \cdot \sin A' + \sin B \cdot \sin B' + \sin C \cdot \sin C' = 0.$$
Draw parallels through any point.

**Ex. 7.** *$OL$, $OM$, $ON$ are any three lines through $O$ and $PL$, $PM$, $PN$ make equal angles with $OL$, $OM$, $ON$ in the same way, show that*
$$PL \cdot \sin MON + PM \cdot \sin NOL + PN \cdot \sin LOM = 0.$$

**3.** $A$, $B$, $C$ *are any three collinear points, and $P$ is any other point; show that*
$$PA^2 \cdot BC + PB^2 \cdot CA + PC^2 \cdot AB + BC \cdot CA \cdot AB = 0.$$
Drop the perpendicular $PO$ from $P$ on $ABC$.
Then $PA^2 \cdot BC = (OA^2 + OP^2) BC = (a^2 + p^2)(c-b)$.
Hence $\Sigma (PA^2 \cdot BC) = \Sigma a^2 (c-b) + p^2 \Sigma (c-b) = \Sigma a^2 (c-b)$
$= a^2 c - a^2 b + b^2 a - b^2 c + c^2 b - c^2 a = -(c-b)(a-c)(b-a)$
$= -BC \cdot CA \cdot AB.$

**Ex. 1.** *If $A$, $B$, $C$ be three collinear points and $a$, $b$, $c$ be the tangents from $A$, $B$, $C$ to a given circle, then*
$$a^2 \cdot BC + b^2 \cdot CA + c^2 \cdot AB + BC \cdot CA \cdot AB = 0.$$

**Ex. 2.** *If $P$ be any point on the base $AB$ of the triangle $ABC$, then*
$$AP \cdot CB^2 - BP \cdot CA^2 = AB \cdot (CP^2 - AP \cdot BP).$$

**Ex. 3.** *If $A$, $B$, $C$, $D$ be four points on a circle and $P$ any point whatever, show that*
$$\triangle BCD \cdot AP^2 - \triangle CDA \cdot BP^2 + \triangle DAB \cdot CP^2 - \triangle ABC \cdot DP^2 = 0,$$
*disregarding signs.*
Let $AC$, $BD$ meet in $O$ inside the circle.
Then $\triangle BCD \propto BD \cdot CO$ and $BO \cdot OD = CO \cdot OA$.

**Ex. 4.** *If $VA$, $VB$, $VC$, $VD$ be any four lines through $V$, then*
$$\frac{\sin BVD \cdot \sin CVD}{\sin BVA \cdot \sin CVA} + \frac{\sin CVD \cdot \sin AVD}{\sin CVB \cdot \sin AVB} + \frac{\sin AVD \cdot \sin BVD}{\sin AVC \cdot \sin BVC} = 1.$$
Draw a parallel to $VD$.

**Ex. 5.** *If $A$, $B$, $C$ be the angles of a triangle and $A'$, $B'$, $C'$ the angles which the sides $BC$, $CA$, $AB$ make with any line, then*
$$\frac{\sin B' \cdot \sin C'}{\sin B \cdot \sin C} + \frac{\sin C' \cdot \sin A'}{\sin C \cdot \sin A} + \frac{\sin A' \cdot \sin B'}{\sin A \cdot \sin B} = 1.$$
Draw parallels through any point.

**Ex. 6.** *If $OA$, $OB$, $OC$ be any three lines through $O$ and $PA$, $PB$, $PC$ be the three perpendiculars from any point $P$ on $OA$, $OB$, $OC$, then*
$$\Sigma \{PB \cdot PC \sin BOC\} = -PO^2 \sin BOC \cdot \sin COA \cdot \sin AOB.$$

**Ex. 7.** *Through the vertices $A$, $B$, $C$ of a triangle are drawn the parallels $AX$, $BY$, $CZ$ to meet the sides $BC$, $CA$, $AB$ in $X$, $Y$, $Z$, show that*
$$\frac{BX \cdot CX}{AX^2} + \frac{CY \cdot AY}{BY^2} + \frac{AZ \cdot BZ}{CZ^2} = 1.$$

**4. Ex. 1.** *If $O, A, B$ be any three collinear points, then*
$$OA^2 + OB^2 = AB^2 + 2 \cdot OA \cdot OB.$$

**Ex. 2.** *If from any point $P$ there be drawn the perpendicular $PQ$ on the line $AB$, then*
$$PA^2 - PB^2 = AB^2 + 2 \cdot AB \cdot BQ.$$

**Ex. 3.** *If $ABCDE \ldots XY$ be any number of collinear points, show that*
$$AB + BC + CD + \ldots + XY + YA = 0.$$

**Ex. 4.** *If $\lambda$ denote the ratio $OA : OB$ and $\lambda'$ the ratio $OA' : OB'$, $OABA'B'$ being collinear points, show that*
$$BB' \cdot \lambda \cdot \lambda' + A'B \cdot \lambda + B'A \cdot \lambda' + AA' = 0.$$

**Ex. 5.** *If $O, A, B, C, D$ be any five points on a line, then*
$$\frac{AC}{AD} + \frac{BC}{BD} = \left(\frac{OB}{AB} - \frac{OD}{AD}\right) + \left(\frac{OB}{AB} - \frac{OC}{AC}\right).$$

**Ex. 6.** *If $A, B, C, D, O, O'$ be any six points on a line, and if*
$$OA : O'A = a, \quad OB : O'B = \beta, \quad OC : O'C = \gamma, \quad OD : O'D = \delta,$$
show that
$$\frac{AC}{AD} + \frac{BC}{BD} = \frac{\gamma - a}{\delta - a} + \frac{\gamma - \beta}{\delta - \beta}.$$

**Ex. 7.** *If $VA, VB, VC, VD, VO$ be any five lines meeting in a point, show that*
$$\frac{\sin AVC}{\sin AVD} + \frac{\sin BVC}{\sin BVD} = \left(\frac{\sin OVB}{\sin AVB} - \frac{\sin OVD}{\sin AVD}\right) + \left(\frac{\sin OVB}{\sin AVB} - \frac{\sin OVC}{\sin AVC}\right).$$

**Ex. 8.** *If $VA, VB, VC, VD$ be any four lines meeting in a point, show that*
$$\frac{\sin AVC}{\sin AVD} + \frac{\sin BVC}{\sin BVD} = \frac{\cot AVB - \cot AVD}{\cot AVB - \cot AVC}.$$

Let $VX$ be the initial line and in Ex. 7 take $VO$ 90° behind $VA$. Then $\sin OVB = \sin(XVB - XVO) = \sin(XVB - XVA + 90°)$
$= \sin(90° + AVB) = \cos AVB;$ and so on.

**Ex. 9.** *If $VA, VB, VC, VD, VO, VO'$ be any six lines meeting in a point, and if $a = \sin OVA \div \sin O'VA$, and so on, show that*
$$\frac{\sin AVC}{\sin AVD} + \frac{\sin BVC}{\sin BVD} = \frac{\gamma - a}{\delta - a} + \frac{\gamma - \beta}{\delta - \beta}.$$

**Ex. 10.** *If $VA, VB, VC, VD, VO$ be any five lines meeting in a point, show that*
$$\frac{\sin AVC}{\sin AVD} + \frac{\sin BVC}{\sin BVD} = \frac{\tan OVA - \tan OVC}{\tan OVA - \tan OVD} + \frac{\tan OVB - \tan OVC}{\tan OVB - \tan OVD}.$$
In Ex. 9 take $VO$ and $VO'$ at right angles.

**Ex. 11.** *Three lines $OAA'$, $OBB'$, $OCC'$ are cut by two lines $ABC$, $A'B'C'$, show that*
$$\frac{OA}{OC} + \frac{AB}{BC} = \frac{OA'}{OC'} + \frac{A'B'}{B'C'}$$
and
$$\frac{AA' \cdot BC}{OA'} + \frac{BB' \cdot CA}{OB'} + \frac{CC' \cdot AB}{OC'} = 0.$$

**Ex. 12.** *If the polygon $abcd \ldots$ be inscribed in the polygon $ABCD \ldots$, so that $a$ is on $AB$, $b$ on $BC$, and so on, and $O$ be any point in the plane, then the continued product of such ratios as*
$$\sin AOa / \sin aOB \div Aa/aB \quad \text{is unity.}$$

**Ex. 13.** *If $D, E, F$ be any three points on the sides $BC, CA, AB$ of a triangle, show that*
$$\frac{DB \cdot EC \cdot FA}{DC \cdot EA \cdot FB} = \frac{\sin DAB \cdot \sin EBC \cdot \sin FCA}{\sin DAC \cdot \sin EBA \cdot \sin FCB}.$$

**Ex. 14.** *If the sides $DE, EF, FD$ of one triangle pass through the vertices $C, A, B$ of another triangle, show that*
$$\frac{AF \cdot BD \cdot CE}{FB \cdot DC \cdot EA} = \frac{\sin FAC \cdot \sin DBA \cdot \sin ECB}{\sin FAB \cdot \sin DBC \cdot \sin ECA}.$$

**5.** *If $C$ be the middle point of $AB$, then whatever origin $O$ be chosen, we have $OC = \frac{1}{2}(OA + OB)$.*

For $OC = OA + AC = OA + \frac{1}{2} AB = OA + \frac{1}{2}(OB - OA)$
$= \frac{1}{2}(OA + OB).$

As we have used general formulae throughout this proof, the formula holds for every relative position of the points $O, A$ and $B$.

**Ex. 1.** *If $C$ be the middle point of $AB$, and $O$ be any point on the line $ACB$, show that*

(i) $OA \cdot OB = OC^2 - AC^2$;
(ii) $OA^2 + OB^2 = CA^2 + CB^2 + 2 \cdot OC^2$;
(iii) $OA^2 - OB^2 = 2 \cdot AB \cdot CO.$

**Ex. 2.** *If $AA', BB', CC'$ be collinear segments whose middle points are $\alpha, \beta, \gamma$, and if $P$ be a variable point on the line, show that*
$$PA \cdot PA' \cdot \beta\gamma + PB \cdot PB' \cdot \gamma\alpha + PC \cdot PC' \cdot \alpha\beta \text{ is constant.}$$

Take $O$ as origin. Then $2 \cdot \beta\gamma = 2 \cdot O\gamma - 2 \cdot O\beta = c + c' - b - b'$. Twice the given expression is
$$(a-p)(a'-p)(c+c'-b-b') + \ldots + \ldots,$$
which is equal to $aa'(c+c'-b-b') + \ldots + \ldots$.

**Ex. 3.** *If $P$ be the middle point of the segment $AA'$ and $Q$ be the middle point of the segment $BB'$ (on the same line as $AA'$), show that*
$$2 \cdot PQ = AB' + A'B = AB + A'B'$$
and $2 \cdot PQ \cdot AA' = AB \cdot AB' - A'B \cdot A'B'.$

**Ex. 4.** *If $AX \cdot AY = BX \cdot BY$ and $A$ and $B$ do not coincide, show that $AB$ and $XY$ have the same bisector.*

**Ex. 5.** *If on the line $AB$ the point $G$ be taken such that $a \cdot GA + b \cdot GB = 0$, $a$ and $b$ being any numbers, positive or negative, then, $O$ being also on $AB$,*
$$a \cdot OA + b \cdot OB = (a+b) \cdot OG$$
and $a \cdot OA^2 + b \cdot OB^2 = a \cdot GA^2 + b \cdot GB^2 + (a+b) \cdot GO^2.$

**Ex. 6.** *If on the line $ABCD \ldots$ a point $G$ be taken such that*
$$GA + GB + GC + \ldots = 0,$$
*and $O$ be any other point on the line, then,*
$$OA + OB + OC + \ldots = n \cdot OG$$
$$OA^2 + OB^2 + OC^2 + \ldots = GA^2 + GB^2 + GC^2 + \ldots + n \cdot GO^2,$$
*$n$ being the number of the points $ABCD \ldots$.*

**Ex. 7.** *If $GABC$... and $G'A'B'C'$... be points so situated on the same line that $GA + GB + GC + ... = 0$, and also $G'A' + G'B' + G'C' + ... = 0$, then*
$$n \cdot GG' = AA' + BB' + CC' + ...,$$
*where $n$ is the number of the points $ABC$... and also of the points $A'B'C'$....*

**Ex. 8.** *If there be $n$ of the points $ABC$... and $n'$ of the points $A'B'C'$..., then*
$$n \cdot n' \cdot GG' = \Sigma (AA' + AB' + AC' + ...).$$

**6.** The following is an interesting application of Algebra to Geometry.

*If $A, B, C, D, P, Q$ be any six collinear points, then*
$$\frac{AP.AQ}{AB.AC.AD} + \frac{BP.BQ}{BC.BD.BA} + \frac{CP.CQ}{CD.CA.CB} + \frac{DP.DQ}{DA.DB.DC} = 0.$$

Put $X$ for $A$, and reduce the resulting equation to any origin, after getting rid of the denominators. We shall have an equation of the second order in $x$ to determine $X$.

Put $x = b$, i.e. $X = B$, and we get an identity.

Hence $x = b$ is one solution of this equation.

Similarly $x = c$, and $x = d$ are solutions.

Hence the equation of the second order has three solutions; and hence is an identity.

*If $A, B, C, P, Q$ be any five collinear points, then*
$$\frac{AP.AQ}{AB.AC} + \frac{BP.BQ}{BC.BA} + \frac{CP.CQ}{CA.CB} = 1.$$

Multiply the identity just proved by $AD$ throughout and let $D$ be at infinity.

Then $AD = AB + BD$, $\therefore$ $AD/BD = AB/BD + 1$.

But when $D$ is at infinity $AB/BD = 0$.

Hence $AD/BD = 1$. Similarly $AD/CD = 1$.

So $DP/DB = 1$ and $DQ/DC = 1$.

Hence we obtain the result enunciated.

*If $A, B, C, D, P$ be any five collinear points, then*
$$\frac{AP}{AB.AC.AD} + \frac{BP}{BC.BD.BA} + \frac{CP}{CD.CA.CB} + \frac{DP}{DA.DB.DC} = 0.$$

In the first identity take $Q$ at infinity, then since
$$BQ/AQ = 1, \quad CQ/AQ = 1, \quad DQ/AQ = 1,$$
the required result follows.

**Ex. 1.** *Show that the first result is true for n points $A, B, \ldots$ and $(n-2)$ points $P, Q, \ldots$.*

**Ex. 2.** *Show that the second result is true for n points $A, B, \ldots$ and $(n-1)$ points $P, Q, \ldots$.*

**Ex. 3.** *Show that the third result is true for n points $A, B, \ldots$ and $(n-2-m)$ points $P, Q, \ldots$; where m may be 0, 1, 2, 3, $\ldots (n-2)$.*

**Ex. 4.** *Enunciate the theorems obtained from Ex. 2 and Ex. 3 by taking the points $P, Q, \ldots$ all coincident; and show that the theorems still hold when $P$ is outside the line, provided the index of $AP$ is even.*

Use $AP^2 = Ap^2 + pP^2$, and the Binomial Theorem.

## Menelaus's Theorem.

**7.** *If any transversal meet the sides $BC, CA, AB$ of a triangle in $D, E, F$; then*
$$AF \cdot BD \cdot CE = -FB \cdot DC \cdot EA.$$

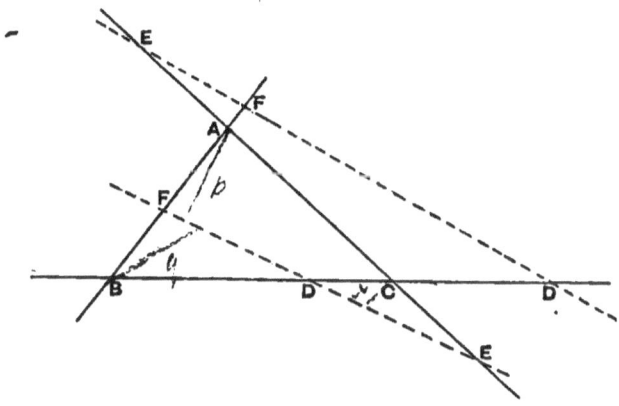

The transversal must cut all the sides externally, or two sides internally and one externally; for as a point proceeds along the transversal from infinity, at a point where the transversal cuts a side internally, the point enters the triangle and at the point where the point leaves the triangle, the transversal must cut another side internally. Hence of the ratios $AF : FB$, $BD : DC$, $CE : EA$, one is negative and the other two are either both positive or both negative. Hence the sign of the formula is correct.

To prove that the formula is numerically correct, drop the perpendiculars $p, q, r$ from $A, B, C$ on the transversal. Then $AF/FB = p/q$, and $BD/DC = q/r$, and $CE/EA = r/p$.

Hence, multiplying, we see that the formula is true numerically.

Conversely, *if three points $D$, $E$, $F$, taken on the sides $BC$, $CA$, $AB$ of a triangle, satisfy the relation*
$$AF \cdot BD \cdot CE = -FB \cdot DC \cdot EA,$$
then $D$, $E$, $F$ are collinear.

For, if not, let $DE$ cut $AB$ in $F'$. Then since $D$, $E$, $F'$ are collinear, we have
$$AF' \cdot BD \cdot CE = -F'B \cdot DC \cdot EA.$$
But by hypothesis we have
$$AF \cdot BD \cdot CE = -FB \cdot DC \cdot EA.$$
Dividing we get $AF' : F'B :: AF : FB$; hence
$$AF' + F'B : AF + FB :: AF' : AF,$$
i.e. $AF' = AF$, i.e. $F'$ coincides with $F$. Hence $D$, $E$, $F$ are collinear.

**Ex. 1.** *Show that the above relation is equivalent to*
$$\sin ACF \cdot \sin BAD \cdot \sin CBE = -\sin FCB \cdot \sin DAC \cdot \sin EBA.$$
For $AF : FB = \triangle ACF : \triangle FCB$
$$= \tfrac{1}{2} AC \cdot CF \sin ACF : \tfrac{1}{2} FC \cdot CB \sin FCB.$$

**Ex. 2.** *The tangents to a circle at the vertices of an inscribed triangle meet the opposite sides in collinear points.*

**Ex. 3.** *A line meets $BC$, $CA$, $AB$ in $D$, $E$, $F$. $P$, $Q$, $R$ bisect $EF$, $FD$, $DE$. $AP$, $BQ$, $CR$ meet $BC$, $CA$, $AB$ in $X$, $Y$, $Z$. Show that $X$, $Y$, $Z$ are collinear.*
For $BX : CX :: BA \sin FAP : CA \sin PAE :: BA \cdot EA : CA \cdot FA$.

**Ex. 4.** *A line meets $BC$, $CA$, $AB$ in $X$, $Y$, $Z$, and $O$ is any point; show that*
$$\sin BOX \cdot \sin COY \cdot \sin AOZ = \sin COX \cdot \sin AOY \cdot \sin BOZ.$$

**Ex. 5.** *If any transversal cut the sides $AB$, $BC$, $CD$, $DE$, ... of any polygon in the points $a$, $b$, $c$, $d$, ..., show that the continued product of the ratios*
$$Aa/Ba, \; Bb/Cb, \; Cc/Dc, \; Dd/Ed, \ldots \text{ is unity.}$$
Let $AC$ cut the transversal in $\gamma$, $AD$ in $\delta$, and so on,
then $\quad Aa/Ba \times Bb/Cb \times C\gamma/A\gamma = 1$
and $\quad A\gamma/C\gamma \times Cc/Dc \times D\delta/A\delta = 1$, and so on.
Multiplying up and cancelling, we get the theorem.

**Ex. 6.** *A transversal meets the sides of a polygon $ABCD$ ... in $\alpha$, $\beta$, $\gamma$, ... and meets any lines through the vertices $A$, $B$, $C$, ... in $a$, $b$, $c$, ...; show that the continued product of such ratios as*
$$\sin a B b / \sin b B \beta + a b / b \beta \text{ is unity.}$$

**Ex. 7.** *If on the four lines $AB$, $BC$, $CD$, $DA$ there be taken four points $a$, $b$, $c$, $d$ such that* $Aa \cdot Bb \cdot Cc \cdot Dd = aB \cdot bC \cdot cD \cdot dA$,
*show that $ab$ and $cd$ meet on $AC$ and $ad$ and $bc$ meet on $BD$.*

I.] *of the same Line.* 9

Apply Menelaus's Theorem to $ABD$ and $ad$ and to $BCD$ and $bc$; multiply, and divide by the given relation; and we see that $ad$ and $bc$ meet $BD$ in the same point; similarly for $AC$.

**Ex. 8.** *If $ad$ and $bc$ meet on $BD$, then $ab$ and $cd$ meet on $AC$, and the above relation holds.*

**Ex. 9.** *If $AB$ and $CD$ meet in $E$ and $AD$ and $BC$ in $F$, and if $Edb$ cut $AD$ in $d$ and $BC$ in $b$, and if $Fac$ cut $AB$ in $a$ and $CD$ in $c$, then*
$$Aa \cdot Bb \cdot Cc \cdot Dd = aB \cdot bC \cdot cD \cdot dA.$$
Use the theorem $\sin A / \sin B = a/b$.

**Ex. 10.** *Between $ABCD$, $abcd$ there holds also the following relation*
$$\sin ab\, B \cdot \sin bc\, C \cdot \sin cd\, D \cdot \sin da\, A = \sin B bc \cdot \sin C cd \cdot \sin D da \cdot \sin A ab.$$

**Ex. 11.** *If the lines $AB$, $BC$, $CD$, $DA$, which are not in the same plane, be met by any plane in $a$, $b$, $c$, $d$; then the relation of Ex. 7 holds; and if this relation hold, the four points are in one plane.*

For the planes $ABD$, $CBD$, $abcd$ meet in a point, i.e. $ad$ and $bc$ meet on $BD$.

**Ex. 12.** *If the sides of the triangle $ABC$ which is inscribed in a circle be cut by any transversal in $D$, $E$, $F$, show that the product of the tangents from $D$, $E$, $F$ to the circle is numerically equal to $AF \cdot BD \cdot CE$.*

**Ex. 13.** *Construct geometrically the ratio $a/b + c/d$.*

**Ex. 14.** *The bisectors of the supplements of the angles of a triangle meet the opposite sides in collinear points.*

**Ex. 15.** *The bisectors of two angles of a triangle and the bisector of the supplement of the third angle meet the opposite sides in collinear points.*

### Ceva's Theorem.

**8.** *If the lines joining any point to the vertices $A$, $B$, $C$ of a triangle meet the opposite sides in $D$, $E$, $F$; then*
$$AF \cdot BD \cdot CE = FB \cdot DC \cdot EA.$$

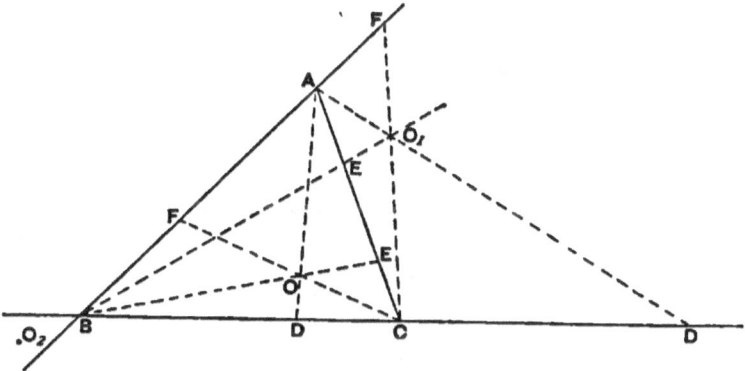

To verify the sign of the formula. $O$ the point in which

$AD$, $BE$, $CF$ meet must be *either* inside the triangle, in which case each of the ratios $AF:FB$ and $BD:DC$ and $CE:EA$ is positive, *or* as at $O_1$ or $O_2$, in which cases two of the ratios are negative and one positive. Hence the sign of the formula is correct.

To prove the formula numerically, we have
$$AF:FB :: \triangle ACF : \triangle FCB :: \triangle AOF : \triangle FOB$$
$$:: \triangle ACF - \triangle AOF : \triangle FCB - \triangle FOB$$
$$:: \triangle AOC : \triangle BOC.$$

Similarly $\quad BD:DC :: \triangle BOA : \triangle AOC$
and $\qquad CE:EA :: \triangle COB : \triangle AOB.$

Hence, multiplying, we see that the formula is true numerically.

Conversely, *if three points $D$, $E$, $F$, taken on the sides $BC$, $CA$, $AB$ of a triangle, satisfy the relation*
$$AF.BD.CE = FB.DC.EA,$$
*then $AD$, $BE$, $CF$ are concurrent.*

For, if not, let $AD$, $BE$ cut in $O$; and let $CO$ cut $AB$ in $F'$. Then since $AD$, $BE$, $CF'$ are concurrent, we have
$$AF'.BD.CE = F'B.DC.EA.$$

But by hypothesis we have
$$AF.BD.CE = FB.DC.EA.$$

Dividing, we get $AF':F'B :: AF:FB$. Hence $F$ and $F'$ coincide, i.e. $AD$, $BE$, $CF$ are concurrent.

**Ex. 1.** *In the figure, show that*
$$\frac{OD}{AD} + \frac{OE}{BE} + \frac{OF}{CF} = 1.$$

**Ex. 2.** *$AO$ meets $BC$ in $D$, $BO$ meets $CA$ in $E$, $CO$ meets $AB$ in $F$. $GH$ is equal and parallel to $BC$ and passes through $A$. $BC$ meets $GO$ in $L$ and $HO$ in $K$. Similarly segments like $KL$ are formed on $CA$ and $AB$. Show that the product of these segments is* $\qquad AF.BD.CE.$

**Ex. 3.** *Show that the necessary and sufficient condition that $Aa$, $Bb$, $Cc$ should meet in a point is*
$$\sin a AB . \sin b BC . \sin c CA = \sin CA a . \sin AB b . \sin BC c.$$

**Ex. 4.** *If the lines $Aa$, $Bb$, $Cc$, $Dd$, ... drawn through the vertices of a plane polygon $ABCD$... in the same plane meet in a point, then the continued product of such ratios as $\sin a AB : \sin AB b$ is unity.*

**Ex. 5.** *If the lines joining a fixed point O to the opposite vertices of a polygon of an odd number of sides meet the sides $AB, BC, CD, DE, \ldots$ in the points $a, b, c, d, \ldots$, show that the continued product of such ratios as $Aa/aB$ is unity.*

For $Aa/aB = AO \cdot aO \sin AOa/aO \cdot BO \sin aOB$.

**Ex. 6.** *The lines joining the centres of the escribed circles of a triangle to the middle points of the corresponding sides of the triangle are concurrent.*

**Ex. 7.** *$AO$ meets $BC$ in $P$, $BO$ meets $CA$ in $Q$, $CO$ meets $AB$ in $R$; $PU$ meets $QR$ in $X$, $QU$ meets $RP$ in $Y$, $RU$ meets $PQ$ in $Z$; show that $AX, BY, CZ$ are concurrent.*

**Ex. 8.** *Through the vertices of a triangle are drawn parallels to the reflexions of the opposite sides in any line; show that these parallels meet in a point.*

For the angle between the reflexions of two lines is equal to the angle between the lines.

**Ex. 9.** *$A', B', C'$ are the reflexions of $A, B, C$ in a given line through $O$. $OA', OB', OC'$ meet $BC, CA, AB$ in $D, E, F$. Show that $D, E, F$ are collinear.*

For $BD : DC :: BO \sin BOA' : CO \sin A'OC$.

**Ex. 10.** *A circle meets $BC$ in $D, D'$, $CA$ in $E, E'$, and $AB$ in $F, F'$. Show that if $AD, BE, CF$ meet in a point, so do $AD', BE', CF'$.*

**Ex. 11.** *A line meets $BC, CA, AB$ in $P, Q, R$ and $AO, BO, CO$ in $X, Y, Z, O$ being any point. Show that*

$$QX \cdot RY \cdot PZ = RX \cdot PY \cdot QZ.$$

**Ex. 12.** *$AA', BB', CC'$ meet in a point, show that the meets of $BC, B'C'$, of $CA, C'A'$ and of $AB, A'B'$ are collinear; and conversely, if the meets are collinear, the joins are concurrent.* (See also IV. 11.)

Let $p, p'$ be the perpendiculars from $A$ on $A'B', A'C'$ and $q, q'$ those from $B$ on $B'C', B'A'$ and $r, r'$ those from $C$ on $C'A', C'B'$. Then

$$\sin B'A'A : \sin AA'C' :: p : p' \text{ and } p' : r :: AY : CY,$$

if $AC$ meet $A'C'$ in $Y$.

**Ex. 13.** *The lines from the vertices of a triangle to the points of contact of any circle touching the sides of the triangle are concurrent.*

# CHAPTER II.

### HARMONIC RANGES AND PENCILS.

**1.** *A range* or *row* is a set of points on the same line, called the *axis* or *base* of the range.

*A pencil* is a set of lines, called *rays*, passing through the same point, called the *vertex* or *centre* of the pencil.

If $A$, $B$, $A'$, $B'$ are collinear points such that
$$AB:BA'::AB':A'B'$$
or (which is the same thing) such that
$$AB/BA' = -AB'/B'A',$$
then $(ABA'B')$ is called a *harmonic range*. $A$, $A'$ and $B$, $B'$ are called *harmonic pairs* of points; and $A$, $A'$ are said to *correspond* and $B$, $B'$ are said to correspond. Also $A$ is said to be the *fourth harmonic* of $A'$ (and $A'$ of $A$) for $B$ and $B'$; so $B$ is said to be the fourth harmonic of $B'$ (and $B'$ of $B$) for $A$ and $A'$. Also $AA'$ and $BB'$ are called *harmonic segments* and are said to divide one another harmonically. The briefest and clearest way of stating the harmonic relation is to say that $(AA', BB')$ is harmonic. The relation may be stated in words thus—each pair of harmonic points divides the segment joining the other pair in the same ratio internally and externally.

$$\underline{\quad A \qquad\qquad B \quad\; A' \qquad\qquad\qquad B'\quad}$$

For $\qquad BA:AB' = -BA':A'B'.$

**Ex. 1.** *The centres of similitude of two circles divide the segment joining the centres of the circles harmonically.*

**Ex. 2.** *The internal and external bisectors of the vertical angle of a triangle cut the base harmonically.*

CH. II.]  *Harmonic Ranges and Pencils.*  13

**Ex. 3.** $(BC, XX')$, $(CA, YY')$, $(AB, ZZ')$ are harmonic ranges; show that if $AX$, $BY$, $CZ$ are concurrent, then $X'Y'Z'$ are collinear, and that if $X'Y'Z'$ are collinear, then $AX$, $BY$, $CZ$ are concurrent.
Use the theorems of Ceva and Menelaus.

**2.** If $(AA', BB')$ be harmonic, then

$$\frac{2}{AA'} = \frac{1}{AB} + \frac{1}{AB'}, \qquad \frac{2}{BB'} = \frac{1}{BA} + \frac{1}{BA'},$$

$$\frac{2}{A'A} = \frac{1}{A'B} + \frac{1}{A'B'}, \qquad \frac{2}{B'B} = \frac{1}{B'A} + \frac{1}{B'A'}.$$

Taking any one of these formulae, say

$$\frac{2}{A'A} = \frac{1}{A'B} + \frac{1}{A'B'},$$

choose $A'$ as origin in the defining relation

$$AB : BA' :: AB' : A'B'$$

and use abridged notation. Then $AB \cdot A'B' = BA' \cdot AB'$ gives us

$$(b-a)b' = (-b)(b'-a) \quad \text{or} \quad bb' - ab' + bb' - ab = 0,$$

$$\text{or} \quad 2bb' = ab + ab' \quad \text{or} \quad \frac{2}{a} = \frac{1}{b} + \frac{1}{b'}.$$

*Conversely, if any one of these relations is true, then $(AA', BB')$ is harmonic.*

For retracing our steps we see that

$$AB : BA' :: AB' : A'B'.$$

**Ex. 1.** *If $\beta$ bisect $BB'$, then $AB \cdot AB' = AA' \cdot A\beta$.*

**Ex. 2.** *$AD$, $BE$, $CF$ are the perpendiculars on $BC$, $CA$, $AB$, and $(BC, DP)$, $(CA, EQ)$ and $(AB, FR)$ are harmonic; show that $PQR$ is the radical axis of the circum-circle and the nine-point circle of $ABC$.*

**Ex. 3.** *If $(AA', BB')$ be harmonic, and $P$ be any point on the line $AB'$, show that*

$$2 \cdot \frac{PA'}{AA'} = \frac{PB}{AB} + \frac{PB'}{AB'}.$$

Put $PA' = PA + AA'$, &c.

**Ex. 4.** *If $EF$ divide both $AA'$ and $BB'$ harmonically, then*

$$AB \cdot B'E \cdot EA' = -A'B' \cdot BE \cdot EA.$$

For we have $1/EA - 1/EB = 1/EB' - 1/EA'$.

If we call $AA'$ the harmonic mean between $AB$ and $AB'$ and so on, Ex. 1 shows us that *the G. M. between two lengths is equal to the G. M. between the A. M. and the H. M.*

For $A\beta$ is the A. M. between $AB$ and $AB'$.

14   *Harmonic Ranges and Pencils.*   [CH.

**3.** *If $(AA', BB')$ be harmonic, then $aA^2 = aB \cdot aB'$; and conversely, if $aA^2 = aB \cdot aB'$, then $(AA', BB')$ is harmonic, $a$ being the middle point of $AA'$.*

For taking $a$ as the origin in the defining relation
$$AB : BA' :: AB' : A'B',$$
we have $\quad (b-a)(b'-a') = (a'-b)(b'-a).$

But $a' = -a$, hence $(b-a)(b'+a) = (-a-b)(b'-a)$,

i.e. $\quad bb' + ba - ab' - a^2 + ab' - a^2 + bb' - ba = 0,$

i.e. $\quad bb' = a^2,\quad$ i.e. $\quad aA^2 = aB \cdot aB'.$

The converse follows by retracing our steps.

**Ex. 1.** *Show that the middle point of either of two harmonic segments is outside the other segment.*

**Ex. 2.** *The chord of contact of tangents from $A$ to any circle cuts the diameter $BB'$ through $A$ in the fourth harmonic of $A$ with respect to $BB'$.*

For $OB^2 = OP^2 = OA \cdot OA'$ by similar triangles, $P$ being one of the points of contact and $O$ the centre.

**Ex. 3.** *Deduce a construction for the fourth harmonic of $A'$ with respect to $BB'$ when $A'$ is between $B$ and $B'$.*

**Ex. 4.** *Deduce the connexion between the A. M., G. M., and H. M. of $AB$ and $AB'$.*

**Ex. 5.** *Deduce the formula $2/AA' = 1/AB + 1/AB'$.*

We have $\quad AO \cdot AA' = AP^2 = AB \cdot AB'.$

Hence the result follows from $2 \cdot AO = AB + AB'$.

**Ex. 6.** *Do Ex. 4 and Ex. 5, interchanging $A$ and $A'$.*

**Ex. 7.** *Show that if $(AA', BB')$ be harmonic and $\alpha$ bisect $AA'$ and $\beta$ bisect $BB'$, then*
$$2 \cdot \alpha B = (\sqrt{A\beta} \pm \sqrt{A'\beta})^2.$$

**Ex. 8.** *Also $AB^2 : A'B^2 :: \beta A : \beta A'$*

For $\beta A : \beta A' = \beta A^2 : \beta A \cdot \beta A' = \beta A^2 : \beta B^2.$

**Ex. 9.** *Given two segments $AB$, $CD$ upon the same line, construct a segment $XY$ which shall divide both $AB$ and $CD$ harmonically.*

Take any point $P$ not on the given line. Through $ABP$ and $CDP$ construct circles cutting again in $Q$. Let $PQ$ cut $ABCD$ in $O$. From $O$ draw tangents to the circles. With $O$ as centre and any one of these tangents as radius, describe a circle. This circle will cut the given line in the required points $X$ and $Y$. For
$$OX^2 = OY^2 = OP \cdot OQ = OA \cdot OB = OC \cdot OD.$$

**4.** *To find the relation between four harmonic points and a fifth point on the same line.*

Let $(AA', BB')$ be harmonic, and take the fifth point $P$ as origin. Then by definition $AB/BA' = -AB'/B'A'.$

But $AB = PB - PA = b - a$, &c. Hence
$$(b-a)(a'-b') + (a'-b)(b'-a) = 0$$
or $2aa' + 2bb' = (a+a')(b+b')$,

i.e. $2 \cdot PA \cdot PA' + 2 \cdot PB \cdot PB' = (PA + PA')(PB + PB')$.

Conversely, *if this relation hold, $(AA', BB')$ is harmonic*.
For reasoning backwards we deduce the relation
$$AB/BA' = -AB'/B'A'.$$

If $(AA', BB')$ be harmonic, and $a$ bisect $AA'$ and $\beta$ bisect $BB'$, then $PA \cdot PA' + PB \cdot PB' = 2 \cdot Pa \cdot P\beta$.

For $PA + PA' = 2 \cdot Pa$ and $PB + PB' = 2 \cdot P\beta$.

Note that every relation of the second order connecting harmonic points must be identical with the relation of this article. Hence the following relations can be proved.

**Ex. 1.** $2 \cdot AB' \cdot BA' = 2 \cdot AB \cdot A'B' = AA' \cdot BB'$.
**Ex. 2.** $AB \cdot AB' + A'B \cdot A'B' = A'A^2$.
**Ex. 3.** $A'A^2 + B'B^2 = (AB + A'B')^2 = 4 \cdot a\beta^2$.
**Ex. 4.** $PA \cdot A'B' + PA' \cdot AB + PB \cdot B'A + PB' \cdot BA' = 0$.
**Ex. 5.** $AB^2 = 2 \cdot aB \cdot A\beta$.
**Ex. 6.** $BA : BA' :: \beta B : A'\beta$.
**Ex. 7.** $PA \cdot PA' - PB^2 + 2 \cdot aB \cdot P\beta$.
**Ex. 8.** *If P and Q be arbitrary points, then*
$PA \cdot QB' \cdot A'B + PA' \cdot QB \cdot AB' + PB \cdot QA \cdot B'A' + PB' \cdot QA' \cdot BA = 0$.
Take $P$ as origin and put $QB' = b' - x$, &c.
**Ex. 9.** $PB \cdot PB' \cdot AA' + PA^2 \cdot A'\beta + A'P^2 \cdot \beta A = 0$.

This is a relation of the third order, which vanishes when $A$ coincides with $A'$. Hence we guess that it is the product of $(a-a')$ into the harmonic relation.

**5.** If $B$, $B'$ divide $AA'$ in the same ratio internally and externally, then by definition $(AA', BB')$ is a harmonic range. Now suppose this ratio is one of equality, then $B$ becomes the internal bisector of the segment $AA'$, i.e. $B$ is the middle point of $AA'$; also $B'$ becomes the external bisector of the segment $AA'$, i.e. a point such that $AB' = A'B'$, $B'$ being outside $AA'$. But
$$AB'/A'B' = (AA' + A'B')/A'B' = AA'/A'B' + 1;$$
and this can only be 1 when $AA' = 0$ or $A'B' = \infty$. Hence,

assuming that $A$ and $A'$ do not coincide, we must have $A'B' = \infty$, i.e. $B'$ must be at infinity. Also if $B'$ is at infinity, then $AB'/A'B' = 1$ as above. Hence $AB' = A'B'$, i.e. $B'$ at infinity bisects $AA'$ externally. Hence the two theorems—

*The point at infinity on any line bisects externally every segment on this line.*

*Every segment is divided harmonically by its middle point and the point at infinity on the line, or, in other words, by its internal and external bisectors.*

**6.** *If any two points of a harmonic range coincide, then a third point coincides with them and the fourth may be anywhere on the line.*

Suppose $AA'$ coincide. Then $B$ lying between $A$ and $A'$ must coincide with them. So for $BB'$.

Suppose $AB$ coincide. Then $AB = 0$; hence, from the defining relation $AB \cdot A'B' = BA' \cdot AB'$, we conclude that $BA' = 0$ or $AB' = 0$, i.e. $ABA'$ coincide or $ABB'$. So for $AB'$, $A'B$, $A'B'$.

Again, if $ABA'$ coincide, then $AB = 0$ and $BA' = 0$; hence the relation $AB \cdot A'B' = BA' \cdot AB'$ is satisfied wherever $B'$ is. So for $BA'B'$, &c.

**7.** A pencil of four concurrent rays is called a *harmonic pencil* if every transversal cuts it in a harmonic range.

Harmonic pencils exist for—

*If a pencil be obtained by joining any point to the points of a harmonic range, then every transversal cuts this pencil in a harmonic range.*

Let $(AA', BB')$ be a harmonic range and $V$ any point. Join $V$ to $AA'BB'$, and let any transversal cut the joining lines in $aa'bb'$.

Then $ab : ba' = \triangle aVb : \triangle bVa'$
$= Va \cdot Vb \sin aVb : Vb \cdot Va' \sin bVa'$.

Hence $\dfrac{ab}{ba'} \div \dfrac{ab'}{b'a'} = \dfrac{\sin aVb}{\sin bVa'} \div \dfrac{\sin aVb'}{\sin b'Va'}$.

Now $aVb' = AVB'$; but for the transversal $a\beta'$ we should have $aV\beta' = 180° - AVB'$.

So in all cases $aVb'$ is either equal to or supplemental to $AVB'$; hence in all cases $\sin aVb' = \sin AVB'$. So for the other angles.

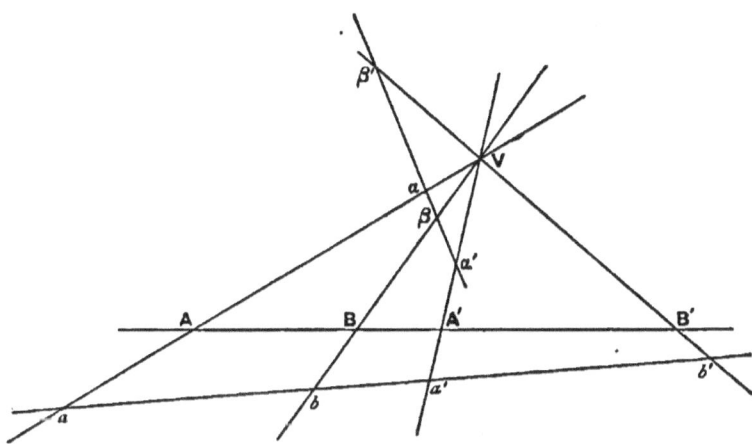

Hence
$$\frac{ab}{ba'} \div \frac{ab'}{b'a'} = \frac{\sin AVB}{\sin BVA'} \div \frac{\sin AVB'}{\sin B'VA'}$$
$$= \frac{AB}{BA'} \div \frac{AB'}{B'A'} \quad \text{by similar reasoning}$$
$$= -1 \quad \text{by definition.}$$

Hence $ab/ba' \div ab'/b'a' = -1$; hence $(aa', bb')$ is a harmonic range.

We denote the pencil subtended by $ABA'B'$ at $V$ by $V(ABA'B')$; and we may briefly state the above theorem thus—if $(AA', BB')$ is a harmonic range, then $V(AA', BB')$ is a harmonic pencil (or more briefly still—is harmonic).

**Ex. 1.** *If $B$ bisect $AA'$ and $V\Omega$ be drawn parallel to $AA'$, then the pencil $V(AA', B\Omega)$ is harmonic.*

For $(AA', B\Omega)$ is harmonic, $\Omega$ being the point at infinity on $AA'$.

**Ex. 2.** *If a transversal be drawn parallel to the ray $VB$ of the harmonic pencil $V(AA', BB')$ meeting the other rays in $ab'a'$, then $b'$ bisects $aa'$.*

For $b$ is at infinity.

**Ex. 3.** *The internal and external bisectors of an angle form with the rays of the angle a harmonic pencil.*
Draw a parallel to one of the bisectors; or use Eu. VI. 3 and A.

**Ex. 4.** *If a pair of corresponding rays of a harmonic pencil be perpendicular, they are the bisectors of the angles between the other pair of rays.*

**Ex. 5.** *If $V(AA', BB')$ be harmonic, prove that*
$$2 \cot AVA' = \cot AVB + \cot AVB'.$$
Take a transversal perpendicular to $VA$.

**Ex. 6.** *Also if $Va$ bisect the angle $AVA'$, then*
$$\tan^2 a\, VA = \tan a\, VB \cdot \tan a\, VB'.$$
Take a transversal perpendicular to $Va$.

**Ex. 7.** $2 \sin AVB' \cdot \sin BVA' = 2 \sin AVB \cdot \sin A'VB'$
$$= \sin AVA' \cdot \sin BVB'.$$

**Ex. 8.** $2 \cdot \dfrac{\sin PVA'}{\sin AVA'} = \dfrac{\sin PVB}{\sin AVB} + \dfrac{\sin PVB'}{\sin AVB'}$

where $VP$ is an arbitrary line through $V$.
Also deduce Ex. 5.

**8.** *The polar of a point $O$ for two lines $BA$ and $BC$ is the fourth harmonic of $BO$ for $BA$ and $BC$.*

*The pole of a line $LM$ for two points $A$, $B$ is the fourth harmonic of the meet of $LM$ and $AB$ for $A$ and $B$.*

*If through $O$ there be drawn the transversal $OPQ$ cutting $BA$ in $P$ and $BC$ in $Q$, then the locus of $R$, the fourth harmonic of $O$ for $P$ and $Q$, is the polar of $O$ for $BA$ and $BC$.*

For the pencil $B(OPRQ)$ is harmonic.

If the two lines $BA$, $BC$ be parallel, i.e. if $B$ be at infinity, the theory still holds, if we consider $B$ to be the limit of a finite point.

To construct the polar of $O$ for $\Omega A$, $\Omega C$ where $\Omega$ is at infinity, draw any transversal $OPQ$ meeting $\Omega A$ in $P$ and $\Omega C$ in $Q$, and take $R$ so that $(OPQR)$ is harmonic, and through $R$ draw a parallel $\Omega R$ to $\Omega A$ and $\Omega C$; then $\Omega R$ is the polar of $O$ for the parallels $\Omega A$, $\Omega C$.

**Ex. 1.** *The polars of any point for the three pairs of sides of a triangle meet the opposite sides in three collinear points.*
Let $AO$, $BO$, $CO$ meet the opposite sides in $P$, $Q$, $R$, and let the polars of $O$ meet these sides in $P'$, $Q'$, $R'$.
Then $BP/PC = -BP'/P'C$, and so on
Now use the theorems of Menelaus and Ceva.

**Ex. 2.** *The poles of any line for the pairs of vertices of a triangle connect concurrently with the opposite vertices.*

II.] *Harmonic Ranges and Pencils.* 19

**Ex. 3.** *The poles of any line for the pairs of points BC and CA are collinear with the meet of the line and AB.*

**Ex. 4.** *The polars of O for BA, BC and for CB, CA meet on AO.*

**9.** *Through a given point O is drawn a line meeting two fixed lines in P and Q, and on OPQ is taken the point R such that $1/OR = 1/OP + 1/OQ$; find the locus of R.*

Take the polar $n$ of $O$ for the two given lines, and let $OPQ$ meet this line in $R$. Then we know that
$$2/OR = 1/OP + 1/OQ.$$
Now draw parallel to $n$ and half-way between $O$ and $n$ the line $n'$ cutting $OPQ$ in $R'$.
Then $OR' = OR/2$,
 i.e. $2/OR = 1/OR'$.
Hence $1/OR' = 1/OP + 1/OQ$;
hence $n'$ is the required locus.

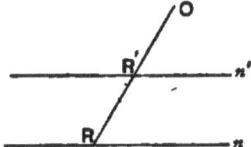

**Ex. 1.** *A transversal through the fixed point O meets fixed lines in A, B, C, ... and on OA is taken a point P such that $1/OP = 1/OA + 1/OB + 1/OC + ...$; find the locus of P.*
Replace $1/OA + 1/OB$ by $1/OL$, and so on.

**Ex. 2.** *A transversal through the fixed point O meets fixed lines in A, B, C,...; find the direction of the transversal when $\Sigma 1/OA$ is* (i) *a maximum* (ii) *a minimum.*
Perpendicular and parallel to the locus of $P$.

**Ex. 3.** *A transversal through the fixed point O meets fixed lines in A, B, C, ... and on OA is taken a point P such that $1/OP = a/OA + b/OB + c/OC + ...$, where a, b, c, ... are any multipliers; find the locus of P. Also find the direction of the transversal when $\Sigma a/OA$ is* (i) *a max.*, (ii) *a min.*
Whatever $a, b, c, ...$ are, we can, by taking the integer $k$ large enough, make $ka, kb, kc, ...$ all integers. Hence
$$k/OP = a'/OA + b'/OB + c'/OC + ...$$
where $k, a', b', c', ...$ are all integers. Now by Ex. 1 find the locus of $Q$ such that $1/OQ = (1/OA + ... a'$ times$) + (1/OB + ... b'$ times$) + ...$ and draw a parallel through $P$ to the locus of $Q$ such that $OP = k \cdot OQ$. This parallel is the required locus.

**10.** A *complete quadrilateral* is formed by four lines called the *sides* which meet in six points called the *vertices* of the quadrilateral. These six points can be joined by three other lines called the *diagonals*. The diagonals are also called the *harmonic lines* of the quadrilateral and the harmonic lines form the sides of the *harmonic triangle*. These names are derived from the following property—called *the harmonic property of a complete quadrilateral.*

*Each diagonal of a complete quadrilateral is divided harmonically by the other two diagonals.*

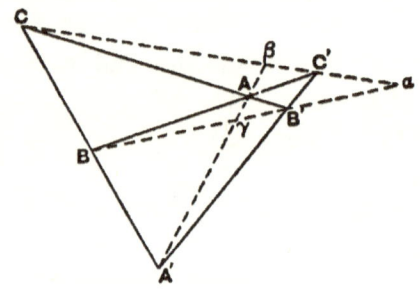

Let the four sides of the complete quadrilateral meet in the three pairs of opposite vertices $AA'$, $BB'$, $CC'$. Then $AA'$, $BB'$, $CC'$, or $\beta\gamma$, $\gamma a$, $a\beta$ are the harmonic lines. We have to show that the ranges $(AA', \beta\gamma)$, $(BB', \gamma a)$, $(CC', a\beta)$ are harmonic.

To prove that $(AA', \beta\gamma)$ is harmonic consider the triangle whose vertices are $AA'$ and any other of the vertices, say $AA'C$. Since $B\gamma B'$ are collinear, we have
$$CB \cdot A'\gamma \cdot AB' = CB' \cdot A\gamma \cdot A'B.$$
Also since $AB$, $A'B'$, $C\beta$ are concurrent, we have
$$CB \cdot A'\beta \cdot AB' = -CB' \cdot A\beta \cdot A'B.$$
Dividing we get $A'\gamma/A\gamma = -A'\beta/A\beta$; hence $(AA', \beta\gamma)$ is harmonic. Similarly $(BB', \gamma a)$ and $(CC', a\beta)$ are harmonic.

**11.** *Using a ruler only, construct the fourth harmonic of a given point for two given points.*

To construct the fourth harmonic of $\gamma$ for $B$ and $B'$. On any line through $\gamma$ take two points $A$ and $A'$. Let $A'B$, $AB'$ cut in $C$ and $AB$, $A'B'$ in $C'$. Then $CC'$ cuts $BB'$ in the required point $a$. For $BB'$ is a diagonal of the complete quadrilateral formed by $AB$, $AB'$, $A'B$, $A'B'$; hence $(BB', \gamma a)$ is harmonic.

**Ex. 1.** *$AO$, $BO$, $CO$ meet $BC$, $CA$, $AB$ in $P$, $Q$, $R$; $QR$, $RP$, $PQ$ meet $BC$, $CA$, $AB$ in $X$, $Y$, $Z$. Show that $(BC, PX)$, $(CA, QY)$, $(AB, RZ)$ are harmonic ranges, and that $XYZ$ are collinear.*

**Ex. 2.** *If a transversal meet $BC$, $CA$, $AB$ in $X$, $Y$, $Z$, and the join of $A$ to the meet of $BY$ and $CZ$ cut $BC$ in $P$; show that $(BC, PX)$ is harmonic, and that the three lines formed like $AP$ are concurrent.*

## Harmonic Ranges and Pencils.

**12.** A *complete quadrangle* is formed by four points called the *vertices* which are joined by six lines called the *sides* of the quadrangle. These six lines meet in three other points called *the harmonic points* of the quadrangle; and the harmonic points are the vertices of the *harmonic triangle*. Some writers give the name diagonal-points to the harmonic points.

The following is *the harmonic property of a complete quadrangle*.

*The angle at each harmonic point is divided harmonically by the joins to the other harmonic points.*

Let $ABCD$ be the four points forming the quadrangle. Then $U$, $V$, $W$ are the harmonic points of the quadrangle; and we have to show that the pencils

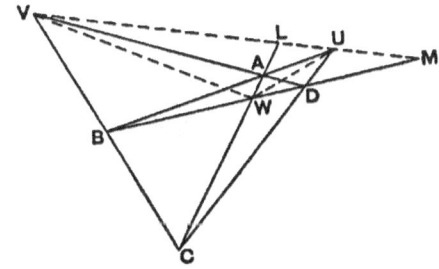

$$U(AD,\ VW),\quad V(BA,\ WU),\quad W(CD,\ UV)$$

are harmonic.

To show that the pencil $W(CD, UV)$ is harmonic, it is sufficient to show that the range $(LM, UV)$ is harmonic, $L$ being the meet of $AC$ and $UV$, and $M$ of $BD$ and $UV$. Consider the triangle formed by $UV$ and any vertex, say $UVC$.

Then because $BDM$ are collinear, we have
$$CB\,.\,VM\,.\,UD = CD\,.\,UM\,.\,VB.$$
Also because $UB$, $VD$, $CL$ are concurrent, we have
$$CB\,.\,VL\,.\,UD = -CD\,.\,UL\,.\,VB.$$
Hence dividing we get $VM/UM = -VL/UL$. Hence $(UV, LM)$ is harmonic, i.e. $W(CD, UV)$ is harmonic. Similarly $U(AD, VW)$ and $V(BA, WU)$ are harmonic.

**13.** *Using a ruler only, construct the fourth harmonic of a given line for two given lines.*

To construct the fourth harmonic of $VU$ for $VA$ and $VB$.

Through any point $U$ on $VU$ draw any two lines $VAB$ and $VDC$, cutting $VA$ in $A$ and $D$, and $VB$ in $B$ and $C$. Then if $AC$ and $BD$ meet in $W$, $VW$ is the required line. For $U$, $V$, $W$ are the harmonic points of the quadrangle $A$, $B$, $C$, $D$. Hence $V(BA, WU)$ is harmonic.

**Ex. 1.** *Through one of the harmonic points of a complete quadrangle is drawn the line parallel to the join of the other two harmonic points; show that two of the segments cut off between opposite sides of the quadrangle are bisected at the harmonic point.*

**Ex. 2.** *Through $V$, one of the harmonic points of a quadrangle, is drawn a line parallel to one side and meeting the opposite side in $P$ and the join of the other harmonic points in $Q$, show that $VP = PQ$.*

**Ex. 3.** *In the figure of the quadrilateral in § 10, show that $A\alpha$, $A'\alpha$, $B\beta$, $B'\beta$, $C\gamma$, $C'\gamma$ form the six sides of a quadrangle.*

We have to show that the six lines pass three by three through four points. Consider $\alpha A$, $\beta B'$, $\gamma C'$. Since $A'B'C'$ are collinear and $(\beta\gamma, AA')$ is harmonic, $\alpha A$, $\beta B'$, $\gamma C'$ are concurrent. Similarly $\alpha A$, $\beta B$, $\gamma C$ are concurrent, also $\alpha A'$, $\beta B$, $\gamma C'$, and also $\alpha A'$, $\beta B'$, $\gamma C$.

**Ex. 4.** *In the figure of the quadrangle in § 12, the sides of the triangle $UVW$ meet the sides in six new points which are the vertices of a quadrilateral.*

# CHAPTER III.

### HARMONIC PROPERTIES OF A CIRCLE.

**1.** *Every line meets a circle in two points, real, coincident or imaginary.*

For take any line $l$ cutting a circle in the points $A$ and $B$. Now move $l$ parallel to itself away from the centre of the circle. Then $A$ and $B$ approach, and ultimately coincide when $l$ touches the circle. But when $l$ moves still further from the centre, the points $A$ and $B$ become invisible; yet, for the sake of continuity, we say that they still exist, but are invisible or *imaginary*. (See also XXVII.)

**2.** *From every point can be drawn to a circle two tangents, real, coincident or imaginary.*

For take any point $T$ outside the circle, and let $TP$ and $TQ$ be the tangents from $T$ to the circle. Now let $T$ approach the centre $O$ of the circle along $OT$. Then $TP$ and $TQ$ approach, and ultimately coincide when $T$ reaches the circumference. But when $T$ moves still further towards $O$, the tangents $TP$ and $TQ$ become invisible; yet, for the sake of continuity, we say that they still exist, but are invisible or *imaginary*. (See also XXVII.)

**3.** Two points which divide any diameter of a circle harmonically are said to be *inverse points* for this circle.

If $O$ be the centre and $r$ the radius of the circle, then inverse points $B$, $B'$ must lie on the same radius of the circle and be such that $OB \cdot OB' = r^2$.

**Ex. 1.** *The inverse of any point at infinity for a circle is the centre of the circle; and conversely, the inverse of the centre is any point at infinity.*

**Ex. 2.** *Every two points and their inverses for a circle lie on a circle.*

**Ex. 3.** *Given a pair of inverse points for a circle, the circle must be one of a certain system of coaxal circles.*

**Ex. 4.** *If four points $(AA', BB')$ be harmonic, so are the four inverse points $(aa', bb')$ for any circle.*

For $Oa = \dfrac{r^2}{OA}$, $Ob = \dfrac{r^2}{OB}$; hence $ab = -\dfrac{r^2 \cdot AB}{OA \cdot OB}$.

**Ex. 5.** *If $BB'$ be a pair of inverse points on the diameter $AA'$ of a circle, and if $P$ be any point on the circle; then $PA$, $PA'$ bisect the angle $BPB'$, and the ratio $PB : PB'$ is independent of the position of $P$.*

**Ex. 6.** *Also if perpendiculars to $AA'$ at $AA'$ $BB'$ meet any tangent to the circle in $aa'$ $bb'$, show that $Oa$ and $Oa'$ bisect the angle $bOb'$, $O$ being the centre, and that the ratio $Ob : Ob'$ is independent of the position of the tangent.*

**4.** Two *circles* are said to be *orthogonal* when the tangents to the circles at each point of intersection are at right angles.

**Ex. 1.** *If two circles are orthogonal at one of their meets, they are orthogonal at the other.*

**Ex. 2.** *If the orthogonal circles $\alpha$ and $\beta$ whose centres are $A$ and $B$ meet in $P$, show that $AP$ touches $\beta$ and $BP$ touches $\alpha$.*

**Ex. 3.** *The radii of two circles are $a$ and $b$ and the distance between their centres is $\delta$; show that the necessary and sufficient condition that the circles should be orthogonal is $a^2 + b^2 = \delta^2$.*

**5.** *Every circle which passes through a pair of points inverse for a circle is orthogonal to this circle; and conversely, every circle orthogonal to a circle cuts every diameter of this circle in a pair of inverse points.*

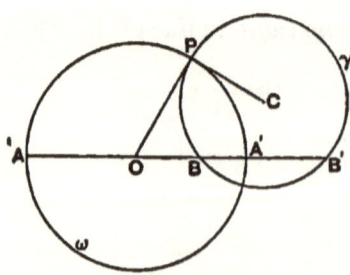

First, let the circle $\gamma$ pass through the inverse points $BB'$ of the circle $\omega$. Let $P$ be one of the meets of $\omega$ and $\gamma$. Then $OB \cdot OB' = OP^2$. Hence $OP$ touches $\gamma$. Hence $OPC$ is a right angle. Hence $CP$ touches $\omega$. Hence the tangents $OP$ and $CP$ are at right angles, i.e. the two circles are orthogonal.

Second, let the two circles $\omega$ and $\gamma$ be orthogonal.

III.]   *Harmonic Properties of a Circle.*   25

Through the centre $O$ of $\omega$ draw the diameter $AA'$ cutting $\gamma$ in $BB'$. Then since the circles are orthogonal, $OPC$ is a right angle; hence $OP$ touches $\gamma$. Hence $OB \cdot OB' = OP^2$. Hence $B$ and $B'$ are inverse points for $\omega$.

**Ex. 1.** *If a circle a divide one diameter of the circle $\beta$ harmonically, it divides every diameter of $\beta$ harmonically.*

**Ex. 2.** *On the diagonals of a complete quadrilateral as diameters are drawn three circles; show that each of these cuts orthogonally the circle about the harmonic triangle.*

**Ex. 3.** *Through two given points draw a circle to cut a given segment harmonically.*

The circle cuts the circle on the segment as diameter orthogonally.

**6.** *A line cuts two circles in the points $PP'$ and $QQ'$, so that $(PP', QQ')$ is harmonic; show that the product of the perpendiculars from the centres of the circles on the line is constant.*

Let $A$ be the centre and $a$ the radius of one circle, and $B$ and $b$ those of the other circle. Let $AX = p$ and $BY = q$ be the perpendiculars from $A$ and $B$ on the line. Then $X$ bisects $PP'$, $Y$ bisects $QQ'$, and since $(PP', QQ')$ is harmonic, we have

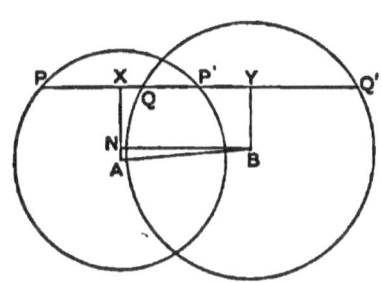

$$XP^2 = XQ \cdot XQ'.$$

Draw $BN$ perpendicular to $AX$. Denote $AB$ by $\delta$.

Now $2pq = p^2 + q^2 - (p-q)^2 = a^2 - PX^2 + b^2 - QY^2 - AN^2$
$= a^2 + b^2 - PX^2 - QY^2 - \delta^2 + XY^2 = a^2 + b^2 - \delta^2.$

For
$$XY^2 - PX^2 - QY^2 = (XY + QY)(XY - QY) - XP^2$$
$$= XQ' \cdot XQ - XP^2 = 0.$$

Hence $pq$ is constant.

**Ex. 1.** *If a line cut two orthogonal circles harmonically, it must pass through one of the centres.*

For $p = 0$ or $q = 0$.

**Ex. 2.** *If a line $l$ cut one circle in the points $PP'$ and another circle in the points $QQ'$, which are such that $(PP', QQ')$ is harmonic; show that the envelope of $l$ is a conic whose foci are the centres of the circles. Show also that if the* *circles meet in C and D, the envelope touches the four tangents of the circles at C and D.*

Since $pq$ is constant, the first follows by Geometrical Conics. Also if $l$ become the tangent at $C$, then $PP'$ and $Q$ coincide at $C$; hence $(PP', QQ')$ is harmonic.

**Ex. 3.** *The locus of the middle points of $PP'$ and $QQ'$ is the coaxal circle whose centre bisects $AB$.*

For the locus of $X$ and $Y$ is the auxiliary circle. Also each meet of the circles is on the locus; for the tangent to either circle at a meet is divided harmonically.

**Ex. 4.** *If $R$ be any point on a circle, $A$ and $B$ fixed points on a diameter and equidistant from the centre, the envelope of a line which cuts harmonically the two circles with $A$, $B$ as centres and $AR$, $BR$ as radii is independent of the position of $R$ on the circle.*

Its foci are $A$ and $B$. Also

$$2b^2 = AR^2 + BR^2 - AB^2 = 2 OR^2 + 2 OA^2 - 4 OA^2.$$

Hence $b^2 = OR^2 - OA^2$, which is constant.

**7.** *Through a point $U$ is drawn a variable chord $PP'$ of a circle and on $PP'$ is taken the point $R$ such that $(UR, PP')$ is harmonic; to show that the locus of $R$ is a line.*

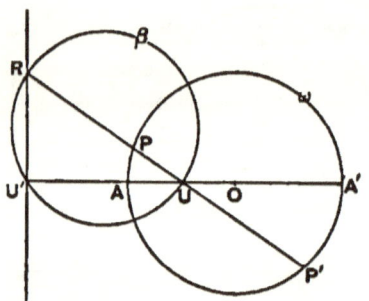

Take $O$ the centre of the given circle $\omega$. Let $OU$ cut $\omega$ in $AA'$. From any position of $R$ drop a perpendicular $RU'$ to $UO$. On $RU$ as diameter describe the circle $\beta$ passing through $U'$. Now since $(RU, PP')$ is harmonic, $PP'$ are inverse points for $\beta$. Hence $\omega$ and $\beta$ are orthogonal. Hence $UU'$ are inverse for $\omega$. Hence $U'$ is a fixed point. Hence the locus of $R$ is a fixed line, viz. the perpendicular to $OU$ through the inverse of $U$ for the given circle.

The locus of $R$ is called the *polar* of $U$ for the circle. We may briefly define the polar of a point for a circle as the locus of the fourth harmonics of the point for the circle. Also if $RU'$ is given, $U$ is called its *pole* for the circle, and $U$ and $RU'$ are said to be *pole and polar* for the circle.

**8.** *If $U$ be outside the circle, the polar of $U$ for the circle is the chord of contact of tangents from $U$ to the circle.*

## Harmonic Properties of a Circle.

For take the chord $UPP'$ very near the tangent $UT$. Then when $PP'$ coincide, $R$, being between them, coincides with them; i.e. one position of $R$ is at $T$. So another position of $R$ is at $T'$. Hence $TT'$ is the polar.

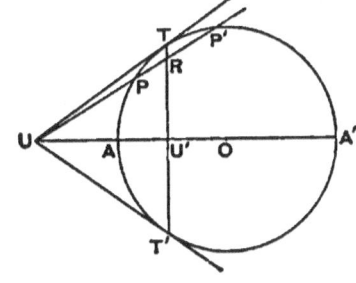

*The polar of the centre of the circle is the line at infinity.* (See IV. 3.)

For if $U$ coincide with $O$, then $PP'$ is bisected at $U$. Hence $R$ is at infinity.

*The pole of the line at infinity for a circle is the centre of the circle.*

For if $R$ be always at infinity, $PP'$ is always bisected at $U$, i.e. $U$ is the centre of the circle.

*The polar of a point on the circle is the tangent at the point.*

For suppose $U$ to approach $A$, then since $OU . OU' = OA^2$, we see that $U'$ also approaches $A$. Hence when $U$ is at $A$, $U'$ is at $A$; and the polar of $U$, being the perpendicular to $OU$ through $U'$, is the tangent at $U$.

Similarly, *the pole of a tangent to a circle is the point of contact.*

**9. Salmon's theorem.**—*If $P$ and $Q$ be any two points and if $PM$ be the perpendicular from $P$ on the polar of $Q$ for any circle, and if $QN$ be the perpendicular from $Q$ on the polar of $P$ for the same circle, then $OP/PM = OQ/QN$, $O$ being the centre of the circle.*

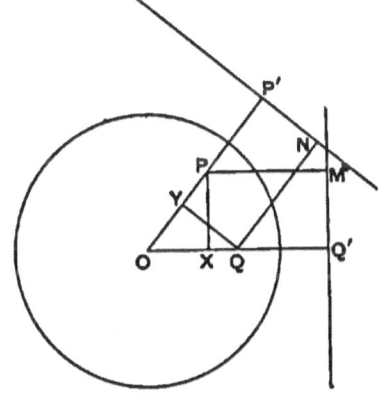

From $P$ drop $PX$ perpendicular to $OQ$ and from $Q$ drop $QY$ perpendicular to $OP$. Then $P'$ being the inverse point

of $P$, and $Q'$ the inverse point of $Q$, we have
$$OP \cdot OP' = OQ \cdot OQ'.$$
Also since the angles at $X$ and $Y$ are right, we have
$$OY \cdot OP = OX \cdot OQ,$$
$$\therefore\ OP/OQ = OQ'/OP' = OX/OY = (OQ' - OX) \div (OP' - OY)$$
$$= XQ'/YP' = PM/QN.$$
Hence $OP/PM = OQ/QN$.

We may enunciate this theorem more briefly thus—*If $p$, $q$ be the polars of $P$, $Q$ for a circle whose centre is $O$, then*
$$OP/(P, q) = OQ/(Q, p).$$

**Ex.** *If $a$, $b$, $p$ be the polars of the points $A$, $B$, $P$ for a circle whose centre is $O$, show that*
$$\frac{(P, a)}{(P, b)} + \frac{(A, p)}{(B, p)} = \frac{(O, a)}{(O, b)} = \frac{(B, a)}{(A, b)}.$$
For $OA \cdot (O, a) = OB \cdot (O, b) = r^2$.

**10.** *If the polar of $P$ pass through $Q$, then the polar of $Q$ passes through $P$.*

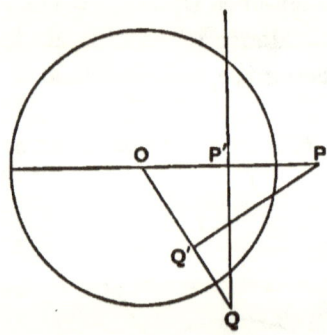

If the polar of $P$ pass through $Q$, then, $P'$ being the inverse of $P$, $P'Q$ is perpendicular to $OP$. Take $Q'$ the inverse of $Q$. Then
$$OP \cdot OP' = OQ \cdot OQ'.$$
Hence $PP'QQ'$ are concyclic. Hence $OQ'P = OP'Q$ is a right angle. Hence $PQ'$ is the polar of $Q$, i.e. the polar of $Q$ passes through $P$.

The points $P$ and $Q$ are called *conjugate points* for the circle. We may define two conjugate points for a circle to be such that the polar of each for the circle passes through the other.

Note that if $PQ$ cut the circle in real points $RR'$, then, since the polar of $P$ passes through $Q$, we see that $(PQ, RR')$ is harmonic; and hence the polar of $Q$ passes through $P$.

*The pole of the join of $P$ and $Q$ is the meet of the polars of $P$ and $Q$.*

III.] *Harmonic Properties of a Circle.* 29

For if the polars of $P$ and $Q$ meet in $R$, then, since the polars of $P$ and $Q$ pass through $R$, therefore the polar of $R$ passes through $P$ and $Q$.

**11.** *On every line there is an infinite number of pairs of conjugate points for a given circle; and each of these pairs is harmonic with the pair of points in which the line meets the circle.*

On the line take any point $P$, and let the polar of $P$ meet the line in $P'$. Then $P$ and $P'$ are conjugate points; for the polar of $P$ passes through $P'$. Also if $PP'$ meet the circle in $RR'$, then $(PP', RR')$ is harmonic; for $P'$ is on the polar of $P$.

Conversely, *every two points which are harmonic with a pair of points on a circle are conjugate for the circle.*

**12.** *If the line $p$ contain the pole of the line $q$, then $q$ contains the pole of $p$.*

Let $P$ be the pole of $p$ and $Q$ of $q$. We are given that $p$ contains $Q$, i.e. that the polar of $P$ passes through $Q$. Hence the polar of $Q$ passes through $P$, i.e. $q$ passes through $P$, i.e. $q$ contains the pole of $p$.

The lines $p$ and $q$ are called *conjugate lines* for the circle. We may define two conjugate lines for a circle to be such that each contains the pole of the other.

*Through every point can be drawn an infinite number of pairs of lines which are conjugate for the circle, and each of these is harmonic with the pair of tangents from the point.*

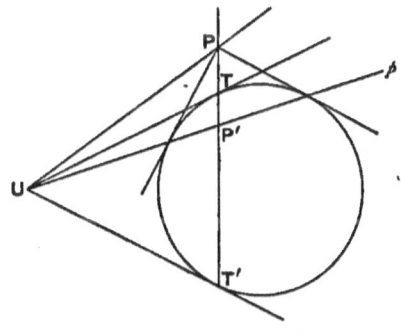

For take any line $p$ through the given point $U$ and join $U$ to the pole $P$ of $p$. Then $p$ and $UP$ are conjugate lines, for $UP$ contains the pole of $p$.

Draw the tangents $UT$ and $UT'$ from $U$, and let the polar $TT'$ of $U$ meet $p$ in $P'$. $TT'$ meets $UP$ in $P$ since $U$ is on the polar of $P$. Now the range $(PP', TT')$ is harmonic, for $P'$ is on the polar of $P$; hence the pencil $U(PP', TT')$ is harmonic, i.e. the conjugate lines $p$ and $UP$ are harmonic with the tangents from $U$.

Conversely, *every pair of lines which are harmonic with the pair of tangents from a point to a circle are conjugate for the circle.*

For let $UQ$ and $UQ'$ be harmonic with the tangents $UT$, $UT'$ from $U$. Let $UQ$ and $UQ'$ cut the polar $TT'$ of $U$ in $P$ and $P'$. Since $U(QQ', TT')$ is harmonic, hence $(PP', TT')$ is harmonic. Hence $UP'$ is the polar of $P$; for the polar of $P$ passes through $P'$ since $(PP', TT')$ is harmonic, and passes through $U$ since $P$ is on the polar of $U$. Hence since the pole of $UP'$ lies on $UP$, we see that $UP$ and $UP'$ are conjugate lines.

**Ex. 1.** *Find the locus of all the points conjugate to a given point.*

**Ex. 2.** *All the lines conjugate to a given line are concurrent.*

**Ex. 3.** *When two points are conjugate, so are their polars; and when two lines are conjugate, so are their poles.*

**Ex. 4.** *A point can be found conjugate to each of two given points; and a line can be found conjugate to each of two given lines.*

**Ex. 5.** *If the circle a be orthogonal to the circle β, then the ends of any diameter of a are conjugate for β.*

**Ex. 6.** *The circle on the segment PQ joining any pair of conjugate points for a circle as diameter is orthogonal to the given circle.*

For $PO$ cuts the new circle in the inverse of $P$.

**Ex. 7.** *If $B'C'$ be the polar of $A$, $C'A'$ of $B$ and $A'B'$ of $C$; then $BC$ is the polar of $A'$, $CA$ of $B'$ and $AB$ of $C'$.*

**Ex. 8.** *Reciprocal triangles are homologous.*

That is, if $A$ is the pole of $B'C'$, $B$ of $C'A'$, $C$ of $A'B'$, then $AA'$, $BB'$, $CC'$ meet in a point. This follows from
$$\sin BAA' : \sin A'AC :: (A', c') : (A', b')$$
and $OA' : (A', c') :: OC' : (C', a')$. (See also XIV. 3.)

**Ex. 9.** *If $P$ and $Q$ be a pair of conjugate points for a circle to which they are external, then*

(i) *$PQ^2$ is equal to the sum of the squares of the tangents from $P$ and $Q$;*

(ii) *$PQ$ is twice the tangent from the middle point of $PQ$;*

(iii) *$PU . UQ$ is equal to the square of the tangent from $U$, $U$ being the foot of the perpendicular from the centre of the circle on $PQ$;*

(iv) *the circle on $PQ$ as diameter is orthogonal to the given circle.*

Take $C$ the middle point of $PQ$ and $R$ the pole of $PQ$.

## Harmonic Properties of a Circle.

Then $RQ$ meets $OP$ perpendicularly in $Y$, say. Hence
$$PQ^2 = OP^2 + OQ^2 - 2OY \cdot OP = OP^2 + OQ^2 - 2r^2 \quad \text{(i)}$$
$$= 2CO^2 + 2CQ^2 - 2r^2 \quad \text{(ii)}$$
and $PU \cdot UQ = UR \cdot UO = UO^2 - r^2$ (iii).

(iv) follows at once from (ii), or because the circle on $PQ$ as diameter passes through $Y$.

**Ex. 10.** *$M$ and $N$ are the projections of a point $P$ on a circle on two perpendicular diameters, $Q$ is the pole of $MN$ for the circle, and $U$ and $V$ are the projections of $Q$ on the diameters. Show that $UV$ touches the circle.*

$UV$ is the polar of $P$.

**13.** Pairs of conjugate lines at the centre of a circle are called pairs of *conjugate diameters* of the circle.

*Every pair of conjugate diameters of a circle is orthogonal.*

Take any diameter $AA'$ of a circle whose centre is $O$. The diameter conjugate to $AA'$ is the line through $O$ conjugate to $AA'$, i.e. is the join of $O$ to the pole of $AA'$. But the tangents at $A$ and $A'$ meet at infinity in $\Omega$, say. Hence $O\Omega$ is the conjugate diameter; hence the diameter conjugate to $AA'$ is parallel to the tangent at $A$, i.e. is perpendicular to $AA'$.

**Ex. 1.** *The pole of a diameter is the point at infinity on any line perpendicular to the diameter; and the polar of any point $\Omega$ at infinity is the diameter perpendicular to any line through $\Omega$.*

**Ex. 2.** *Any two points at infinity which subtend a right angle at the centre are conjugate.*

**14.** A triangle is said to be *self-conjugate* for a circle when every two vertices and every two sides are conjugate for the circle.

Such a triangle is clearly such that each side is the polar of the opposite vertex. Hence the other names—self-reciprocal or self-polar.

*Self-conjugate triangles exist.*

For on the polar of any point $A$ take any point $B$. Then the polar of $B$ passes through $A$ and meets the polar of $A$ in $C$ say. Then $ABC$ is a self-conjugate triangle. For $BC$ is the polar of $A$, $CA$ is the polar of $B$; hence $C$, the meet of $BC$ and $CA$, is the pole of $AB$. Hence $AB$ are conjugate points, and $BC$, $AC$ are conjugate lines. So for other pairs.

**Ex.** *The triangle formed by the line at infinity and any two perpendicular diameters of a circle is self-conjugate for the circle.*

**15.** *There is only one circle for which a given triangle is self-conjugate; and this is real only when the triangle is obtuse-angled.*

Suppose the triangle $ABC$ is self-conjugate for the circle whose centre is $O$. Then since $A$ is the pole of $BC$, it follows that $OA$ is perpendicular to $BC$; so $OB$ is perpendicular to $CA$, and $OC$ to $AB$. Hence $O$ is the orthocentre of $ABC$. Let $OA$ meet $BC$ in $A'$, $OB$ meet $CA$ in $B'$ and $OC$ meet $AB$ in $C'$. Then the square of the radius of the circle must be equal to $OA \cdot OA'$ and to $OB \cdot OB'$ and to $OC \cdot OC'$; and this is possible if $O$ is the orthocentre, for then these products are equal.

Now describe a circle (called the *polar circle* of the triangle) with the orthocentre $O$ as centre and with radius $\rho$, such that $\rho^2 = OA \cdot OA' = OB \cdot OB' = OC \cdot OC'$. Then the triangle $ABC$ is self-conjugate for this circle. For $BC$, being drawn through the inverse point $A'$ of $A$ perpendicular to $OA$, is the polar of $A$; so for $CA$ and $AB$.

Also this circle is imaginary if the triangle is acute-angled; for then $O$ is inside the triangle and hence $\rho^2 (=OA \cdot OA')$ is negative.

**Ex. 1.** *Describe a circle to cut the three sides of a given triangle harmonically. When is this circle real?*

**Ex. 2.** *In any triangle the circles on the sides as diameters are orthogonal to the polar circle.*

**Ex. 3.** *If any three points $X$, $Y$, $Z$ be taken on the sides $BC$, $CA$, $AB$ of a triangle, the circles on $AX$, $BY$, $CZ$ as diameters are orthogonal to the polar circle.*

**Ex. 4.** *The circle on each of the diagonals of a quadrilateral as diameter is orthogonal to the polar circle of each of the four triangles formed by the sides of the quadrilateral.*

**Ex. 5.** *Hence the two sets of circles are coaxal. Hence the middle points of the three diagonals of a quadrilateral are collinear; and the four orthocentres of the four triangles formed by the sides of a quadrilateral are collinear.*

**Ex. 6.** *Every circle cutting two of the circles on the three diagonals of a quadrilateral orthogonally, cuts the third also orthogonally.*

For it cuts two circles of a coaxal system orthogonally.

III.] *Harmonic Properties of a Circle.* 33

**16.** *The harmonic triangle of a quadrangle inscribed in a circle is self-conjugate for the circle.*

Let $UVW$ be the harmonic triangle of the quadrangle $ABCD$ inscribed in a circle. Then $UVW$ is self-conjugate for the circle.

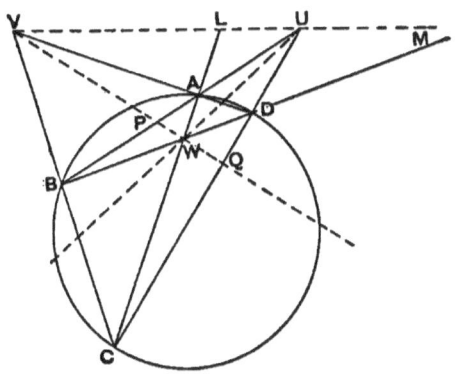

Let $UV$ meet $AC$ in $L$ and $BD$ in $M$. Then since $V(WU, BA)$ is harmonic, hence $(WL, AC)$ and $(WM, BD)$ are harmonic. Hence $L$ and $M$ lie on the polar of $W$, i.e. $UV$ is the polar of $W$. Similarly $VW$ is the polar of $U$, and $WU$ of $V$.

**17.** *With the ruler only, to construct the polar of a given point for a given circle.*

To construct the polar of $V$ for the given circle, draw through $V$ any two chords $AD$ and $BC$ of the circle. Let $BA$, $CD$ meet in $U$, and $AC$, $BD$ meet in $W$. Then by the above theorem $WU$ is the polar of $V$.

**Ex.** *Through $U$ one of the harmonic points of a quadrangle inscribed in a circle is drawn a chord cutting the circle in $aa'$, and the pairs of opposite sides in $bb'$, $cc'$; show that if one of the segments $aa'$, $bb'$, $cc'$ is bisected at $U$, the others are also bisected at $U$.*

Let the transversal cut the opposite side of the harmonic triangle in $X$, then $UX$ divides each segment harmonically.

**18.** *The three diagonals of a quadrilateral circumscribing a circle form a triangle self-conjugate for the circle.*

Let the three diagonals $AA'$, $BB'$, $CC'$ of the quadrilateral

D

$BA$, $AB'$, $B'A'$, $A'B$ circumscribing the circle form the triangle $\alpha\beta\gamma$. Then $\alpha\beta\gamma$ (the harmonic triangle of the quadrilateral) is self-conjugate for the circle.

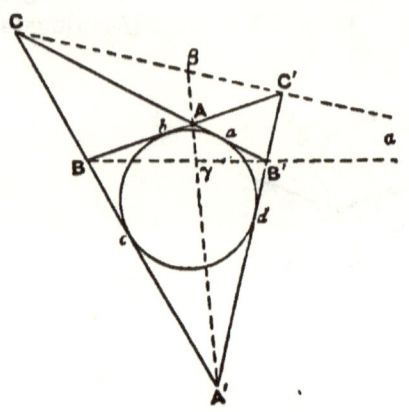

Since $(\beta\gamma, AA')$ is harmonic, hence $C(\beta\gamma, AA')$ and $C'(\beta\gamma, AA')$ are harmonic, i.e. $C\gamma$ is the fourth harmonic of $\alpha\beta$ for the tangents from $C$, and $C'\gamma$ is the fourth harmonic of $\alpha\beta$ for the tangents from $C'$. Hence $\alpha\beta$ is conjugate to $C\gamma$ and to $C'\gamma$, i.e. the pole of $\alpha\beta$ lies on $C\gamma$ and on $C'\gamma$. Hence $\gamma$ is the pole of $\alpha\beta$. Similarly $\alpha$ is the pole of $\beta\gamma$, and $\beta$ of $\gamma\alpha$.

**19.** *With the ruler only, to construct the pole of a given line for a given circle.*

This may be done by the above theorem; but better by finding by § 17 the meet of the polars of two points on the given line.

**Ex.** *The two lines joining the opposite meets of common tangents of two circles which are not centres of similitude cut the line of centres in the limiting points.*

For these points are two vertices of a self-conjugate triangle with respect to both circles.

**20.** *The harmonic triangle of a quadrilateral circumscribed to a circle coincides with the harmonic triangle of the inscribed quadrangle formed by the points of contact.*

In the figure of § 18, let $B'A$, $AB$, $BA'$, $A'B'$ touch the circle in $a$, $b$, $c$, $d$. Comparing with the figure of § 16, we see that we have to prove that $ac$ and $bd$ meet in $\gamma$, that $ba$ and $cd$ meet in $\alpha$, and that $cb$ and $da$ meet in $\beta$. Now $ba$ is the polar of $A$ and $cd$ is the polar of $A'$; hence $ba$ and $cd$ meet in the pole of $AA'$, i.e. in the pole of $\beta\gamma$, i.e. $ba$ and $cd$ pass through $\alpha$. Similarly $ac$ and $bd$ pass through $\gamma$, and $cb$ and $da$ pass through $\beta$.

The theorem is sometimes erroneously stated thus—*Of the two quadrilaterals formed by four tangents to a circle and the points of contact, the four internal diagonals are concurrent and form a harmonic pencil, and the two external diagonals are collinear and divide one another harmonically.*

The former part follows from $\gamma$ being a harmonic point of the quadrangle. The latter part follows from $\beta a$ being a harmonic line of the quadrilateral.

**Ex. 1.** *If the whole figure be symmetrical for $AA'$ and if the angle $ABA'$ be right, show that $ac$, $bd$ bisect the angles between $AA'$ and $BB'$.*

By elementary geometry each of the angles at $\gamma$ is $45°$.

**Ex. 2.** *$AA'$ meets $ab$ in $P$ and $cd$ in $P'$, and so on. Show that the six points $PP'QQ'RR'$ lie three by three on four lines.*

# CHAPTER IV.

### PROJECTION.

**1.** Given a figure $\phi$ in one plane $\pi$ consisting of points $A, B, C, \ldots$ and lines $l, m, n, \ldots$, we can construct another figure $\phi'$ consisting of *corresponding* points $A', B', C', \ldots$ and lines $l', m', n', \ldots$ in the following way. Take any point $V$ (called the *vertex of projection*) and any plane $\pi'$ (called the *plane of projection*). Then $A', B', C', \ldots$ and $l', m', n', \ldots$ are the points and lines in which the plane of projection meets the lines and planes joining the vertex of projection to $A, B, C, \ldots$ and $l, m, n, \ldots$. Each of the figures $\phi$ and $\phi'$ is called the *projection* of the other; and they are said to be *in projection*.

Also each of the points $A$ and $A'$ is said to be the projection of the other; so for the points $B$ and $B'$, $C$ and $C'$, &c., and for the lines $l$ and $l'$, $m$ and $m'$, $n$ and $n'$, &c. The line in which the planes of the figures $\phi$ and $\phi'$ meet may be called the *axis of projection*.

When the vertex of projection is at infinity we get what is called *parallel projection*; in this case all the lines $AA'$, $BB'$, $CC'$, ... are parallel. A particular case of parallel projection is *orthogonal projection*.

The lines $AA'$, $BB'$, $CC'$, ... are called the *rays* of the projection; and projection is sometimes called *radial projection* to distinguish it from orthogonal projection.

Figures in projection are also said to be in perspective in different planes; and then the vertex of projection is called the *centre of perspective*, and the axis of projection is called the *axis of perspective*, and each figure is called

*Projection.*

the *perspective* or *picture* of the other. Note that figures may also be in perspective in the same plane. (See XXXI.)

Some writers use the term *conical projection* or *central projection* or *central perspective* for radial projection.

**2.** *The projection of the join of two points $A$, $B$ is the join of the projections $A'$, $B'$ of the points $A$, $B$.*

*The projection of the meet of the two lines $l$, $m$ is the meet of the projections $l'$, $m'$ of the lines $l$, $m$.*

*The projection of any point on the axis of projection is the point itself.*

*Every line and its projection meet on the axis of projection.*

The proofs of these four theorems are obvious.

*The projection of a tangent to a curve $\gamma$ at a point $A$ is the tangent at $A'$ (the projection of $A$) to the curve $\gamma'$ (the projection of $\gamma$).*

For when the chord $AB$ of $\gamma$ becomes the tangent at $A$ to $\gamma$ by $B$ moving up to $A$, the chord $A'B'$ of $\gamma'$ becomes the tangent at $A'$ to $\gamma'$ by $B'$ moving up to $A'$.

*The projection of a meet (i.e. a common point) of two curves is a meet of the projections of the curves.*

*The projection of a common tangent to two curves is a common tangent to the projections of the curves.*

The proofs of these theorems are obvious.

**3.** The plane through the vertex of projection parallel to the plane of one of two figures in projection meets the plane of the other figure in a line called the *vanishing line* of this plane.

*Each vanishing line is parallel to the axis of projection.*

For the axis of projection and the vanishing line in the plane $\pi$ are the meets of $\pi$ with $\pi'$ and with the plane through $V$ parallel to $\pi'$.

*Every point at infinity in a plane lies on a single line (called the line at infinity).*

Let $A$ be the point at infinity on any line $l$ in the plane $\pi$. Through any point $V$ not in the plane draw a plane $\rho$ parallel to the given plane. Then $\rho$ passes through $A$; for

$\rho$, being parallel to the plane of $l$, meets $l$ at infinity. Similarly $\rho$ passes through every point at infinity in $\pi$. Also every point of intersection of $\pi$ and $\rho$ is at infinity on $\pi$. Hence the points at infinity on $\pi$ are the points of intersection of the two planes $\pi$ and $\rho$. And as two planes when not parallel meet in a line, we may say for the sake of continuity that two parallel planes also meet in a line. Hence the points at infinity in a plane lie on a line.

*The vanishing line in one plane is the projection of the line at infinity in the other plane.*

For the plane joining $V$ to the vanishing line is parallel to the other plane.

*To project a given line to infinity.*

With any vertex of projection, project on to any plane parallel to the plane containing the given line and the vertex of projection. Then the projection of the given line will be the intersection of these two parallel planes and will therefore be entirely at infinity.

**4.** The *vanishing point* of a line is the point in which the line meets the vanishing line of its own plane.

*The angle between the projections of any two lines $l$ and $m$ is the angle which the vanishing points of $l$ and $m$ subtend at the vertex of projection.*

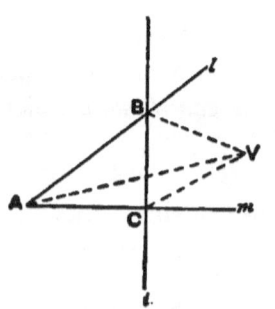

Let $l$ and $m$ meet in $A$, and let $l$ meet the vanishing line $i$ in $B$ and let $m$ meet $i$ in $C$. We have to show that the projection of the angle $BAC$ is equal to $BVC$, $V$ being the vertex of projection. Now the plane of projection $\pi'$ is parallel to the plane $BVC$. Also $A'B'$ is the meet of the plane $AVB$ and $\pi'$. Hence $A'B'$ and $VB$ (being the meets of the plane $AVB$ with the two parallel planes $\pi'$ and $BVC$) are parallel. Similarly $A'C'$ and $VC$ are parallel. Hence $\angle B'A'C' = \angle BVC$.

IV.]  *Projection.*  39

**Ex.** *All angles whose bounding lines have the same vanishing points are projected into equal angles.*

**5.** *To project any two given angles into angles of given magnitudes and at the same time any given line to infinity.*

Let the given angles $ABC$, $DEF$ meet the line which is to be projected to infinity in $AC$, $DF$. Then since $A$, $C$ are the vanishing points of the lines $BA$, $BC$, hence the angle $A'B'C'$ is equal to $AVC$; so $\angle D'E'F' = \angle DVF$. Hence to construct $V$ draw on $AC$ a segment of a circle containing an angle equal to the given angle $A'B'C'$, and on $DF$ and on the same side of it as before describe a segment of a circle containing an angle equal to the given angle $D'E'F'$. Let these segments meet in $V$. Rotate $V$ about $ACDF$ out of the plane of the paper. Then if we project with vertex $V$ on to a plane parallel to the plane $VACDF$, the problem is solved. For the line $ADCF$ will go to infinity. Also $ABC$ will be projected into an angle equal to $AVC$, i.e. into an angle of the required size. So for $DEF$.

The segments may meet in two real points or in one or in none. Hence there may be two real solutions of the problem or one or none.

**Ex.** *In the exceptional case when the vanishing line is parallel to one of the lines of one of the angles, give a construction for the vertex of projection.*
Let $A$ be at infinity. Through $C$ draw a line making with $CF$ the supplement of $A'B'C'$. This will meet the segment on $DF$ in $V$.

**6.** *Given a line $l$ and a triangle $ABC$, to project $l$ to infinity and each of the angles $A$, $B$, $C$ into an angle of given size.*

Suppose we have to project $A$, $B$, $C$ into angles equal to $\alpha$, $\beta$, $\gamma$, where of course $\alpha + \beta + \gamma = 180°$. Let $l$ cut $BC$, $CA$, $AB$ in $P$, $Q$, $R$. Of the points $P$, $Q$, $R$ let $Q$ be the point which lies between the other two. On $RQ$ describe a segment of a circle containing an angle equal to $\alpha$. On $QP$ and on the same side of $l$ describe a segment of a circle containing an angle equal to $\gamma$. These two segments meet in $Q$; hence they meet again in another point, $V$ say. For if the supplements of the segments meet in $V$, then $RVQ + QVP = 180° - \alpha + 180° - \gamma = 180° + \beta > 180°$, which is impossible.

Now rotate $V$ about $l$ out of the plane of the paper. With $V$ as vertex of projection, project on to any plane parallel to $VPQ$; and let $A'B'C'$ be the projection of $ABC$.

We have to prove that $A'=a$, $B'=\beta$, $C'=\gamma$. Through $R$ draw a parallel to $VP$ meeting $VQ$ in $X$. Then $RVX=a$, $VXR=\gamma$, and $XRV=\beta$. Also $A'B'$ is parallel to $VR$, $B'C'$ is parallel to $VP$ and therefore to $RX$, and $C'A'$ is parallel to $VQ$. Hence the sides of the triangles $A'B'C'$ and $VRX$ are parallel. Hence the angles are equal; i.e. $A'=a$, $B'=\beta$, $C'=\gamma$.

**7.** *To project any triangle into a triangle with given angles and sides and any line to infinity.*

Project as above the given triangle $ABC$ into $A'B'C'$ in which $\angle A' = \angle a'$, $\angle B' = \angle b'$, $\angle C' = \angle c'$, $a'b'c'$ being the triangle into which $ABC$ is to be projected. On $VA'$ take a point $P$ such that $VP:VA'::b'c':B'C'$. Through $P$ draw a plane parallel to $A'B'C'$ cutting $VB'$ in $Q$ and $VC'$ in $R$. Then by similar triangles $VP:VA'::QR:B'C'$; hence $QR = b'c'$. So $RP = c'a'$, $PQ = a'b'$. Hence $PQR$ is superposable to $a'b'c'$ and in projection with $ABC$.

Hence we can *project any triangle into an equilateral triangle of any size and any line to infinity.*

**Ex. 1.** *Project any four given points into the angular points of a square of given size.*

Let $ABCD$ (II. 12) be the given points. Project $UV$ to infinity and the angles $VAU$, $LWM$ into right angles. Then in the projected figure $AB$ and $CD$ are parallel, and also $AD$ and $BC$. Also $BAD$ is a right angle and also $AWD$. Hence the figure is a square. We can change its size as before. The construction is always real since the semicircles on $LM$ and $UV$ must meet since $LM$ and $UV$ overlap.

**Ex. 2.** *Project any two homologous triangles (see § 11) simultaneously into equilateral triangles. Is the construction always real?*

**Ex. 3.** *Project any three angles into right angles.*

Let the legs of the angles $A$ and $B$ meet in $L$ and $M$, and let $LM$ cut the legs of $C$ in $DE$; then on $LM$ and $DE$ describe semicircles.

**Ex. 4.** *If two quadrangles have the same harmonic points, then the eight vertices lie on a conic; as a particular case, if any three of the points are collinear, the eight vertices lie on two lines.*

Project one of the sides $UV$ of the harmonic triangle to infinity, and the angles $UAV$ and $UA'V$ into right angles, and the angle $LWM$ into a right angle. The quadrangles are now a square and a rectangle with parallel sides and the same centre; hence the vertices by symmetry lie on a conic whose axes are parallel to the sides. If however $B'$ is on $BD$, clearly this conic degenerates into the common diagonals; so if $B'$ is on $BA$, the conic degenerates into $BA$ and $CD$, and if $B'$ is on $BC$ into $BC$ and $AD$. (See also XII. 7.)

**8.** *In projecting from one plane to another, there are in each plane two points such that every angle at either of them is projected into an equal angle.*

Let the given planes be $\pi$ and $\pi'$. Draw the planes $a$ and $\beta$ bisecting the angles between the planes $\pi$ and $\pi'$. Through the vertex of projection $V$ draw a line perpendicular to $a$ cutting the planes $\pi$ and $\pi'$ in $E$, $E'$, and a line through $V$ perpendicular to $\beta$ cutting the planes $\pi$ and $\pi'$ in $F$, $F'$. Then every angle at $E$ will be projected into an equal angle at $E'$, and every angle at $F$ will be projected into an equal angle at $F'$.

The figure is a section of the solid figure by a plane through $V$ perpendicular to the planes $\pi$ and $\pi'$. Let this plane meet the axis of projection in $K$, and let the legs of any angle at $E$ in $\pi$ meet the axis of projection in $L$, $M$. Then the angle $LEM$ projects into the angle $LE'M$.

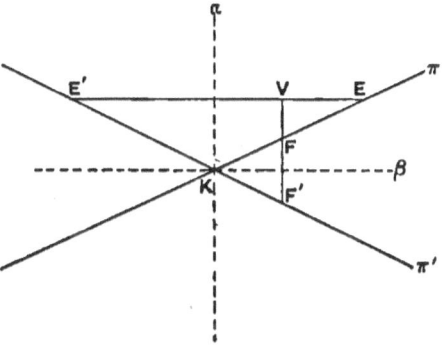

But $EK = E'K$ by construction and $\angle EKL = \angle E'KL = 90°$. Hence the figure $EKLM$ is superposable to the figure $E'KLM$. Hence the angle $LEM$ is equal to the angle $LE'M$, i.e. any angle at $E$ is projected into an equal angle at $E'$. So any angle at $F$ is projected into an equal angle at $F'$.

**9.** *The projection of a harmonic range is a harmonic range.*

For if $A'B'C'D'$ be the projection of the harmonic range $ABCD$, then $V$ and the lines $AB$, $A'B'$ lie in one plane. Hence by II. 7.

*The projection of a harmonic pencil is a harmonic pencil.*

Draw any line cutting the rays of the harmonic pencil $U(ABCD)$ in $a, b, c, d$. Let $U'(A'B'C'D')$ be the projection of the pencil $U(ABCD)$, and $a', b', c', d'$ the projections of $a, b, c, d$. Then $a$ being on $UA$, $a'$ is on $U'A'$, and so on; hence $U'(A'B'C'D')$ is harmonic, if $(a'b'c'd')$ is harmonic. And $(a'b'c'd')$ is harmonic, since $(abcd)$ is harmonic.

**10.** *To prove by Projection the harmonic property of a complete quadrangle.*

In the figure of II. 12, suppose we wish to prove that $V(BA, WU)$ is harmonic. Project $CD$ to infinity. Then $VAWB$ is a parallelogram and $U$ is the point at infinity on $BA$. Let $VW$ cut $BA$ in $O$. Then in the new figure

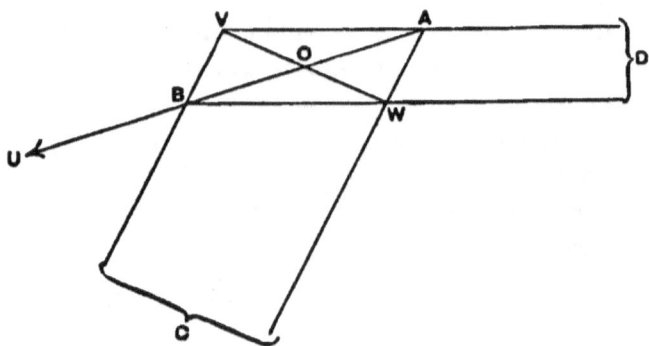

$V(BA, WU)$ is harmonic, for $(BA, OU)$ is harmonic since $BO = OA$ and $U$ is at infinity. It follows that $V(BA, WU)$ is harmonic in the given figure. So $U(AD, VW)$ and $W(CD, UV)$ can be proved to be harmonic.

**Ex.** *Prove by Projection the harmonic property of a complete quadrilateral.*

### Homologous Triangles.

**11.** *Two triangles $ABC$, $A'B'C'$ are said to be homologous* (or *in perspective*) when $AA'$, $BB'$, $CC'$ meet in a point (called the *centre of homology* or centre of perspective) and also $(BC; B'C')$, $(CA; C'A')$, $(AB; A'B')$ lie on a line (called the *axis of homology* or the axis of perspective).

*If two triangles in the same plane be copolar, they are coaxal; and if coaxal, they are copolar.*

(i) Let the two triangles $ABC$, $A'B'C'$ be copolar, i.e. let $AA'$, $BB'$, $CC'$ meet in the point $O$; then they are coaxal, i.e. $(BC; B'C')$, $(CA; C'A')$, $(AB; A'B')$ lie on a line.

Call these three points $X$, $Y$, $Z$. Then we have to show

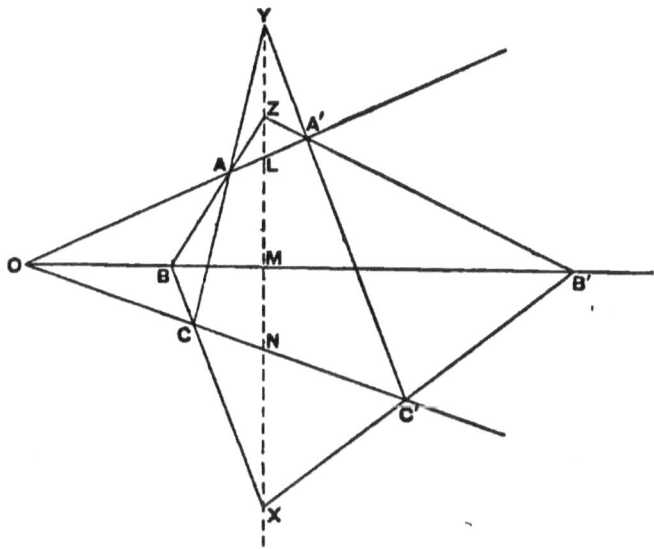

that $YZ$ passes through $X$. Project $YZ$ to infinity. Then in the new figure $AA'$, $BB'$, $CC'$ meet in a point $O$; also $AB$

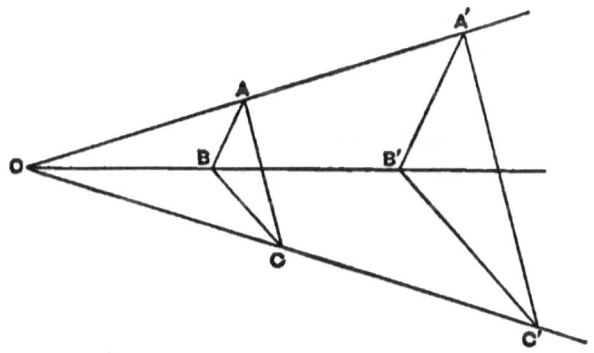

is parallel to $A'B'$ and $AC$ to $A'C'$. Hence
$$OB : OB' :: OA : OA' :: OC : OC'.$$

And since $OB:OB' :: OC:OC'$, $BC$ is parallel to $B'C'$, i.e. $X$ is at infinity, i.e. $X$ lies on $YZ$, i.e. $XYZ$ are collinear. Hence in the original figure $XYZ$ are collinear, i.e. the triangles are coaxal.

(ii) Let the triangles be coaxal, i.e. let $(BC; B'C')$, $(CA; C'A')$, $(AB; A'B')$ be collinear; then they are copolar, i.e. $AA'$, $BB'$, $CC'$ meet in a point.

Project $XYZ$ to infinity. Then in the new figure $BC$ is parallel to $B'C'$, $CA$ to $C'A'$, and $AB$ to $A'B'$. Let $AA'$ and $BB'$ meet in $O$. Then $OB:OB' :: AB:A'B' :: BC:B'C'$; and $\angle OBC = \angle OB'C'$. Hence the triangles $OBC$ and $OB'C'$ are similar. Hence $\angle BOC = \angle B'OC'$. Hence $CC'$ passes through $O$. Hence $AA'$, $BB'$, $CC'$ meet in a point. Hence $AA'$, $BB'$, $CC'$ meet in a point in the original figure.

**12.** If the triangles are not in one plane, the proofs are simpler.

*If two triangles be copolar, they are coaxal.*

(Use the same figure as before, but remember that now the triangles are in different planes.) Since $AB$, $A'B'$ lie in the plane $OAA'BB'$, hence $AB$, $A'B'$ meet in a point on the meet of the planes $ABC$, $A'B'C'$. Similarly $(CA; C'A')$, $(AB; A'B')$ lie on this line, i.e. the triangles are coaxal.

*If two triangles be coaxal, they are copolar.*

The three planes $BCXB'C'$, $CAYC'A'$, $ABZA'B'$ meet in a point; hence their meets $AA'$, $BB'$, $CC'$ pass through this point, i.e. the triangles are copolar.

**Ex. 1.** *Hence (by taking the angle between the planes evanescent) deduce that coaxal triangles in the same plane are copolar; and (by a 'reductio ad absurdum' proof) that copolar triangles are coaxal.*

**Ex. 2.** *If two triangles $ABC$, $A'B'C'$ in the same plane be such that $AA'$, $BB'$, $CC'$ meet in a point $O$; and if on any line through $O$ not in the plane be taken two points $V$, $V'$; show that $VA, V'A'$ meet in a point $A''$, and $VB$, $V'B'$ in a point $B''$, and $VC$, $V'C'$ in a point $C''$; and that the three triangles $ABC$, $A'B'C'$, $A''B''C''$ are such that corresponding sides meet in threes at three points on the same line, viz. the meet of the given plane and the plane of the triangle $A''B''C''$.*

For the triangles $AA'A''$, $BB'B''$ are coaxal (and not in the same plane); hence they are copolar.

This gives us another proof that triangles in the same plane which are copolar are also coaxal.

**Ex. 3.** *The sides $BC$, $B'C'$ of two triangles in the same plane meet in $X$, and $CA$, $C'A'$ meet in $Y$, and $AB$, $A'B'$ meet in $Z$; and $X$, $Y$, $Z$ are collinear. The lines joining $A$, $B$, $C$ to any vertex $V$ not in the plane $ABC$ cut any plane through $X$, $Y$, $Z$ but not through $V$ in $A''$, $B''$, $C''$. Show that $A'A''$, $B'B''$, $C'C''$ meet in a point $V'$ such that $AA'$, $BB'$, $CC'$ meet in the point where $VV'$ cuts the plane of the triangles.*

For $B''A''$ passes through $Z$.

This gives us another proof that triangles in the same plane which are coaxal are also copolar.

**Ex. 4.** *If three triangles $ABC$, $A'B'C'$, $A''B''C''$, which are homologous in pairs, be such that $BC$, $B'C'$, $B''C''$ are concurrent and $CA$, $C'A'$, $C''A''$ and $AB$, $A'B'$, $A''B''$; then the three centres of homology of the triangles taken in pairs are collinear.*

For the triangles $AA'A''$, $BB'B''$ are copolar and therefore coaxal.

**Ex. 5.** *If three triangles $ABC$, $A'B'C'$, $A''B''C''$ be such that $AA'A''$, $BB'B''$, $CC'C''$ are concurrent lines; then the axes of homology of the triangles taken in pairs are concurrent.*

For the triangles whose sides are $AB$, $A'B'$, $A''B''$ and $AC$, $A'C'$, $A''C''$ are coaxal and therefore copolar.

**Ex. 6.** *If the points $A'$, $B'$, $C'$ lie on the lines $BC$, $CA$, $AB$, and if $AA'$, $BB'$, $CC'$ meet in a point, show that the meets of $BC$, $B'C'$, of $CA$, $C'A'$ and of $AB$, $A'B'$ lie on a line which bisects the lines drawn from $A$, $B$, $C$ to $BC$, $CA$, $AB$ parallel to $B'C'$, $C'A'$, $A'B'$.*

The line is the axis of homology of the two triangles. Let $AB$, $A'B'$ meet in $Z$, and $BC$, $B'C'$ in $X$. Bisect $AL$ (parallel to $B'C'$) in $O$. It is sufficient to prove that $AZ.BX.LO = -ZB.XL.OA$. But $LO = OA$; and $AZ:BZ = AC:C'B = LX:XB$.

**Ex. 7.** *The triangles $ABC$, $A'B'C'$ are coaxal; if $(BC; B'C')$ be $X$, $(CA; C'A')$ be $Y$, $(AB; A'B')$ be $Z$, $(BC; B'C)$ be $X'$, $(CA'; C'A)$ be $Y'$, $(AB'; A'B)$ be $Z'$; then $XY'Z'$, $X'YZ'$, $X'Y'Z$ are lines.*

# CHAPTER V.

### HARMONIC PROPERTIES OF A CONIC.

**1.** WE define a *conic section* or briefly a *conic* as the projection of a circle, or in other words, as the plane section of a cone on a circular base. The plane of projection may be called the plane of section.

From the definition of a conic it immediately follows that—

*Every line meets a conic in two points, real, coincident, or imaginary.*

*From every point can be drawn to a conic two tangents, real, coincident, or imaginary.*

For these properties are true for a circle, and therefore for a conic by projection.

**2.** There are three kinds of conics according as the vanishing line meets the circle, touches the circle, or does not meet the circle, or more properly according as the vanishing line meets the circle in real, coincident, or imaginary points.

If the vanishing line meet the circle in two points $P$ and $Q$, then, $V$ being the vertex of projection, the plane of section is parallel to the plane $VPQ$, and therefore cuts the cone on both sides of $V$. Hence we get a conic consisting of two detached portions, extending to infinity in opposite directions, called a *hyperbola*.

If the vanishing line touch the circle, and $TT'$ be the tangent, then the plane of section, being parallel to the plane $VTT'$ which touches the cone, cuts the cone on one side only of $V$. Hence we get a conic consisting of one portion extending to infinity, called a *parabola*.

If the vanishing line does not meet the circle, the plane of section is parallel to a plane through $V$ which does not meet the cone except at the vertex, and therefore cuts the cone in a single closed oval curve, called an *ellipse*.

Since the line at infinity is the projection of the vanishing line, it follows that the line at infinity meets a hyperbola in two points, touches a parabola and does not meet an ellipse, in other words, *the line at infinity meets a hyperbola in two real points, a parabola in two coincident points, and an ellipse in two imaginary points*, or, again, a hyperbola has two real points at infinity, a parabola two coincident points, and an ellipse two imaginary points.

3. *A pair of straight lines is a conic.*

For let the cutting plane be taken through the vertex, so as to cut the cone in two lines. Then these lines are a section of the cone, i.e. a conic.

But properties of a pair of lines cannot be directly obtained by projection from a circle. For let the cutting plane meet the circle in the points $P$ and $Q$. Then the projection of every point on the circle except $P$ and $Q$ is at the vertex, whilst the projection of $P$ is any point on the line $VP$ and the projection of $Q$ is any point on the line $VQ$. Now if we take any point $R'$ on one of the lines $VP$ and $VQ$, its projection is $P$ or $Q$ unless $R'$ is at the vertex and then its projection is some point on the rest of the circle.

To get over this difficulty we take a section of the cone parallel to the section through the vertex. Then however near the vertex this plane is, the theorem is true for the hyperbolic section; hence the theorem is true in the limit when the section passes through the vertex and the hyperbola becomes a pair of lines.

4. *A pair of points is a conic.*

This follows by Reciprocation. (See VIII.) For the reciprocal of two lines is two points and the reciprocal of a conic is a conic. Hence two points is a conic.

Clearly however we cannot obtain two points by the section of a circular cone.

**5.** As in the case of the circle we define *the polar of a point for a conic* as the locus of the fourth harmonics of the point for the conic.

*The polar of a point for a conic is a line.*

Through the given point $U$ draw a chord $PP'$ of the conic and on this chord take the point $R$, such that $(PP', UR)$ is harmonic. We have to show that the locus of $R$ is a line. Now by hypothesis the conic is the projection of a circle. Suppose the range $(PP', UR)$ is the projection of $(pp', ur)$ in the figure of the circle. Then since $(PP', UR)$ is harmonic, so is $(pp', ur)$. Hence $r$ is on the locus of the fourth harmonics of $u$ for the circle; hence the locus of $r$ is a line. Hence by projection the locus of $R$ is a line.

As in the case of the circle, if the line $u$ is the polar of $U$ for a conic, then $U$ is defined to be the *pole* of $u$ for the conic; and $U$ and $u$ are said to be *pole and polar* for the conic.

We have proved above implicitly that *The projection of a pole and polar for a circle is a pole and polar for the conic which is the projection of the circle.*

The following theorems now follow at once by projection.

*If $P$ be outside the conic, the polar of $P$ is the chord of contact of tangents from $P$.*

*If $P$ be on the conic, the polar of $P$ is the tangent at $P$, and the pole of a tangent is the point of contact.*

Note that a point is said to be inside or outside a conic according as the tangents from the point are imaginary or real, i.e. according as the polar of the point meets the curve in imaginary or real points. When the point is on the conic, its polar, viz. the tangent, meets the curve in coincident points and the tangents from the point coincide with the tangent at the point.

**Ex. 1.** *PQ is a chord of a conic through the fixed point U, and u is the polar of U; show that* $(P, u)^{-1} + (Q, u)^{-1}$ *is constant.*
viz. $= 2 \cdot (U, u)^{-1}$ by similar triangles.

**Ex. 2.** *If further a be any line, show that*
$$\frac{(P, a)}{(P, u)} + \frac{(Q, a)}{(Q, u)} = 2 \cdot \frac{(U, a)}{(U, u)}.$$
Take the meet of *PQ* and *a* as origin.

**Ex. 3.** *From any point on the line u, tangents p and q are drawn to a conic, and U is the pole of u, and A is any point; show that*
$$\frac{(A, p)}{(U, p)} + \frac{(A, q)}{(U, q)} = 2 \cdot \frac{(A, u)}{(U, u)}.$$
Take *U* on the range *UA* as origin.

**6.** Since a pole and polar project into a pole and polar, the whole theory of conjugate points and conjugate lines for a conic follows at once by projection from the theory of conjugate points and conjugate lines for a circle. Hence all the theorems enunciated in III. 10-12 for a circle follow for a conic by projection.

**Ex. 1.** *If a series of conics be drawn touching two given lines at given points, the polar of every point on the chord of contact is the same for all.*
Let the conics touch *TL* and *TM* at *L* and *M*. The polar of *P* on *LM* passes through *T* the pole of *LM* and passes through the fourth harmonic of *P* for *LM*.

**Ex. 2.** *The pole of any line through T is the same for all.*

**Ex. 3.** *TP, TQ touch a conic at P and Q, and on PQ is taken the point U such that TU bisects the angle PTQ, and through U is drawn any chord RUR' of the conic; show that TU also bisects the angle RUR'.*
Draw *TU'* perpendicular to *TU*; then *TU'* is the polar of *U*. Hence (*ZU, RR'*) is harmonic, *Z* being on *TU'*.

**Ex. 4.** *A is a fixed point, P is a point on the polar of A for a given conic. The tangents from P meet a fixed line in Q, R. AR, PQ meet in X; and AQ, PR in Y. Show that XY is a fixed line.*
Viz. the fourth harmonic of *BR* for *BP* and *BA*; *B* being the meet of *QR* with the polar of *A*.

**Ex. 5.** *The polar of any point taken on either of two conjugate lines for a conic meets the lines and the conic in pairs of harmonic points.*
For if *P* be the point, its polar meets the other line in the pole of the line on which *P* is.

**Ex. 6.** *A, B, C are three points on a conic and CT is the tangent at C; if C (TD, AB) be harmonic, show that CD passes through the pole of AB.*

**Ex. 7.** *TP, TQ touch a conic at P, Q; the tangent at R meets PQ in N, PT in L, QT in M; show that (LM, RN) is harmonic.*

**Ex. 8.** *A and B are two fixed points; a line through A cuts a fixed conic,*

E

in $C$ and $D$, $BD$ cuts the polar of $A$ in $F$, and $BC$ cuts the polar in $E$; show that $DE$ and $CF$ meet in a fixed point.

Viz. the fourth harmonic of $B$ for $A$ and the meet of $AB$ with the polar of $A$.

**Ex. 9.** *Through $U$, the mid-point of a chord $AB$ of a conic is drawn any chord $PQ$. The tangents at $P$ and $Q$ cut $AB$ in $L$ and $M$. Prove that $AL = BM$.*

If $R$ be the pole of $PQ$, then $R\Omega$ is the polar of $U$, $\Omega$ being the point at infinity upon $AB$. Hence $UL = UM$.

**Ex. 10.** *The tangents $TP$, $TP'$ to a conic are cut by the tangent at $Q$ (which is parallel to the chord of contact $PP'$) in $L$, $L'$; show that $LQ = QL'$.*

**Ex. 11.** *Through the point $U$ is drawn the chord $PQ$ of a conic and $UY$ is drawn perpendicular to the polar of $U$; show that $UY$ bisects the angle $PYQ$ or its supplement.*

**7.** The theory of *self-conjugate triangles for a conic* follows at once by projection from a circle, since the theory involves only the theory of poles and polars.

*Of the three vertices of a self-conjugate triangle two are outside and one inside the conic.*

Let $UVW$ be the vertices of the given triangle. Then if $U$ is outside, $VW$, being the polar of $U$, cuts the conic. Also $V$, $W$ form a harmonic pair with the meets of $VW$ with the conic; hence $V$ or $W$ is outside the conic.

If $U$ is inside, $VW$ does not cut the conic, and hence $V$ and $W$ are both outside the conic.

**Ex. 1.** *Of the three sides of a self-conjugate triangle two meet the conic and one does not.*

**Ex. 2.** *The joins of $n$ points on a conic meet again in three times as many points as there are combinations of $n$ things taken four together, and of these meets one-third lie within and two-thirds without the conic.*

**Ex. 3.** *Show that one vertex of a triangle self-conjugate for a given conic is arbitrary, that the second vertex may be taken anywhere on the polar of the first, and that the third vertex is then known.*

**Ex. 4.** *Show that one side may be taken arbitrarily and complete the construction.*

**8.** *The harmonic points of a quadrangle inscribed in a conic form a triangle which is self-conjugate for the conic.*

*The harmonic lines of a quadrilateral circumscribed to a conic form a triangle which is self-conjugate for the conic.*

*If a quadrilateral be circumscribed to a conic, the harmonic triangle of this quadrilateral coincides with the harmonic triangle of the inscribed quadrangle formed by the points of contact.*

## Harmonic Properties of a Conic.

For these propositions are true for the circle, and they follow for the conic by projection. So also—*the quadrangle construction for the polar of a point applies to a conic.*

*Through a given point $P$ draw a pair of tangents to a conic.*

By the quadrangle construction obtain the polar of $P$ for the conic, and join $P$ to the points where this polar cuts the conic. The joining lines are the tangents from $P$ to the conic.

**Ex. 1.** *$A, B, C, D$ are four points on a conic; $AB$, $CD$ meet in $E$, and $AC$ $BD$ meet in $H$, and the tangents at $A$ and $D$ meet in $G$; show that $E, G, H$ are collinear.*

**Ex. 2.** *A system of conics touch $AB$ and $AC$ at $B$ and $C$. $D$ is a fixed point and $BD$, $CD$ meet one of the conics in $P, Q$. Show that $PQ$ meets $BC$ in a fixed point.*

Viz. the pole of $AD$.

**Ex. 3.** *Through the fixed point $A$ is drawn the variable chord $PQ$ of a conic, and the chords $PU$, $QV$ pass through the fixed point $B$. Show that $UV$ passes through a fixed point.*

Viz. the fourth harmonic of $A$, for $B$ and the polar of $B$.

**Ex. 4.** *$PP'$, $QQ'$ are chords of a conic through $C$, and $A$ and $B$ are the points of contact of tangents from $C$. Show that a conic which touches the four lines $PQ$, $P'Q'$, $P'Q$, $PQ'$ and passes through $B$, touches $BC$ at $B$.*

For $AB$ is the polar of $C$ for the new conic.

**Ex. 5.** *The lines $AB$, $BC$, $CD$, $DA$ touch a conic at $a$, $b$, $c$, $d$, and $AB$ and $CD$ are parallel. If $ac$, $bd$ meet at $E$, and $AD$, $BC$ meet at $F$, show that $FE$ bisects $AB$ and $CD$.*

For if $AB$ and $CD$ meet at $\Omega$, then $FE$ and $F\Omega$ are conjugate lines.

**Ex. 6.** *Through one of the vertices $U$ of a triangle $UVW$ self-conjugate for a conic are drawn a pair of chords of the conic harmonic with $UV$ and $UW$. Show that the lines joining the ends of these chords all pass through $V$ or $W$.*

Through $U$ draw the chord $PQ$, and join $Q$ to $V$.

**9.** *If one point on a conic be given and also a triangle self-conjugate for the conic, then three other points are known.*

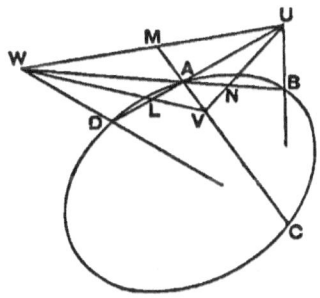

Let $A$ be the given point and $UVW$ the given self-conjugate triangle. Let $UA$ cut $WV$ in $L$. Then the other point $D$ in which $UA$ cuts the conic is known since $(UALD)$ is harmonic. Similarly the points $C$ and $B$ where $VA$ and $WA$ cut the conic are known.

E 2

The four points $A, B, C, D$ form an inscribed quadrangle of which $UVW$ is the harmonic triangle.

By construction $(UALD)$ is harmonic; hence $W(UAVD)$ is harmonic. Similarly $W(UAVC)$ is harmonic. Hence $WD$ and $WC$ coincide, i.e. $WD$ passes through $C$. Similarly $UB$ passes through $C$. Hence the pole of $UW$ is the meet of $AC$ and $BD$. But the pole of $UW$ is $V$. Hence $BD$ passes through $V$.

**Ex. 1.** *Show that if one tangent of a conic be given and also a self-conjugate triangle, then three other tangents are known; and that the four tangents together form a circumscribed quadrilateral of which the given triangle is the harmonic triangle.*

**Ex. 2.** *If two sides of a triangle inscribed in a conic pass through two vertices of a triangle self-conjugate for the conic, then the third side will pass through the third vertex.*

**10.** Properties peculiar to the parabola follow from the fact that the line at infinity touches the parabola.

The lines $TQ$, $TQ'$ touch a parabola at $Q$, $Q'$, and $TV$ bisects $QQ'$ in $V$ and meets the curve in $P$; show that $TP = PV$.

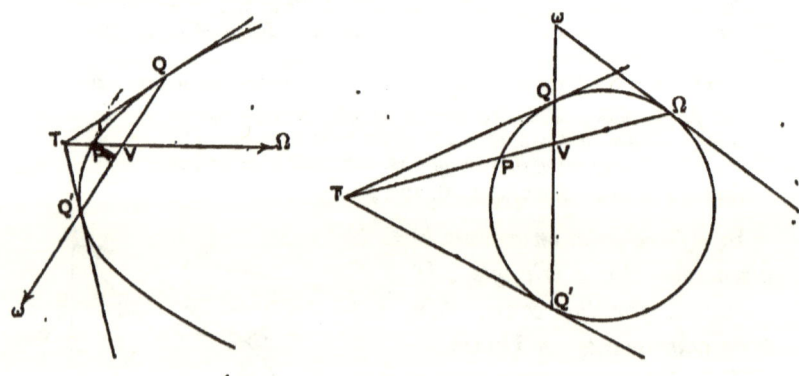

Take the point at infinity $\omega$ on $QQ'$. Then since $\omega$ lies on the polar of $T$, hence the polar of $\omega$ passes through $T$. Since $(\omega V, QQ')$ is harmonic, hence the polar of $\omega$ passes through $V$. Hence $TV$ is the polar of $\omega$. Now suppose the line at infinity to touch the parabola in $\Omega$. Then $\omega$ is on the polar of $\Omega$, viz. the line at infinity; hence $TV$ passes through $\Omega$. Also $P$ and $\Omega$ being points on the curve, therefore $(TV, P\Omega)$ is harmonic; hence $TP = PV$.

For clearness the figure is drawn of which the above figure is the projection. In this case, as in other cases, the theorem might have been proved directly by projection.

**Ex. 1.** *The line half-way between a point and its polar for a parabola touches the parabola.*

**Ex. 2.** *The lines joining the middle points of the sides of a triangle self-conjugate for a parabola touch the parabola.*

**Ex. 3.** *The nine-point circle of a triangle self-conjugate for a parabola passes through the focus.*

**Ex. 4.** *Through the vertices of a triangle circumscribing a parabola are drawn lines parallel to the opposite sides; show that these lines form a triangle self-conjugate for the parabola.*

Being the harmonic triangle of the circumscribing quadrilateral formed by the sides of the triangle and the line at infinity.

**Ex. 5.** *No two tangents of a parabola can be parallel.*

For if possible let them meet at $\omega$ on the line at infinity; then three tangents are drawn from $\omega$ to the conic, viz. the two tangents and the line at infinity.

11. We define the pole of the line at infinity for a conic as the *centre* of the conic. Hence *the centre of a parabola is at infinity*. For since the line at infinity touches the parabola, the centre is the point of contact and therefore is on the line at infinity, i.e. is at infinity. The centre of a hyperbola is outside the curve since the polar of the centre cuts the hyperbola in real points; and the centre of an ellipse is inside the curve since the polar of the centre cuts the ellipse in imaginary points. The hyperbola and ellipse are called *central conics*.

*The centre of a central conic bisects every chord through it.*

Let the chord $PP'$ pass through the centre $C$ of a conic; then $PC = CP'$. For let $PP'$ meet the line at infinity in $\omega$. Then since $\omega$ is on the polar of $C$, hence $(C\omega, PP')$ is harmonic. Hence $PC = CP'$.

*A conic is its own reflexion in its centre.*

For if we join any point $P$ on the conic to the centre $C$ and produce $PC$ backwards to $P'$, so that $CP' = PC$; then $P'$ is another point on the conic.

**Ex. 1.** *All conics circumscribing a parallelogram have their centres at the centre of the parallelogram.*

For by the quadrangle construction for a polar, the polar of the intersection of diagonals is the line at infinity.

**Ex. 2.** *ABC is a triangle circumscribed to a conic, and the point P of contact of BC bisects BC; show that the centre of the conic is on AP.*

For $AP$ is the polar of the point at infinity upon $BC$.

**Ex. 3.** *$QQ'$ is the chord of contact of tangents from $T$ to a conic, and $CT$ cuts $QQ'$ in $V$ and the conic in $P$; show that $CV \cdot CT = CP^2$.*

For $(PP', TV)$ is harmonic.

**Ex. 4.** *Given the centre $O$ of a conic and a self-conjugate triangle $ABC$, construct six points on the conic.*

**12.** *The locus of the middle points of parallel chords of a conic is a line* (called a *diameter*).

Let $QQ'$ be one of the parallel chords bisected in $V$. The system of chords parallel to $QQ'$ passes through a point $\omega$ at infinity. Also since $(\omega V, QQ')$ is harmonic, $V$ is on the polar of $\omega$. Hence the locus required is the polar of $\omega$.

*All diameters of a central conic pass through the centre.*

*All diameters of a parabola are parallel.*

For since a diameter is the polar of a point on the line at infinity, it passes through the pole of the line at infinity. Hence in a central conic it passes through the centre, and in a parabola it passes through a fixed point at infinity, viz. the point of contact of the line at infinity.

**Ex. 1.** *The tangents at the ends of a diameter are parallel to the chords which the diameter bisects.*

Being the tangents from $\omega$.

**Ex. 2.** *A diameter contains the poles of all the chords it bisects.*

Viz. the poles of lines through $\omega$.

**Ex. 3.** *If the tangents at the ends of a chord are parallel, the chord is a diameter.*

**Ex. 4.** *Two chords of a conic which bisect one another are diameters.*

**13.** Conjugate lines at the centre of a conic are called *conjugate diameters.*

*Each of two conjugate diameters bisects chords parallel to the other.*

Let $PCP'$ and $DCD'$ be conjugate diameters. Then by definition the pole of $CP$ is on $CD$. But $CP$ passes through

## Harmonic Properties of a Conic.

the centre; hence the pole of $CP$ is at infinity. Hence the pole of $CP$ is the point $\omega$ at infinity on $CD$. Through $\omega$, i.e. parallel to $DD'$, draw the chord $QQ'$ meeting $CP$ in $V$. Then since $PP'$ is the polar of $\omega$, hence $(QQ', V\omega)$ is harmonic, i.e. $QV = VQ'$. Hence $PP'$ bisects every chord parallel to $DD'$. So $DD'$ bisects every chord parallel to $PP'$.

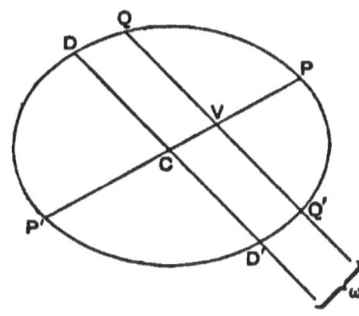

**Ex. 1.** *A pair of conjugate diameters form with the line at infinity a self-conjugate triangle.*

**Ex. 2.** *In the hyperbola one and only one of a pair of conjugate diameters cuts the curve in real points.*

**Ex. 3.** *The polar of a point is parallel to the diameter conjugate to the diameter containing the point.*

**Ex. 4.** *The tangents at the end of a diameter are parallel to the conjugate diameter.*

**Ex. 5.** *The line joining any point to the middle point of its chord of contact passes through the centre.*

**Ex. 6.** *The sides of a parallelogram inscribed in a conic are parallel to a pair of conjugate diameters; and the diagonals meet at the centre.*

**Ex. 7.** *The diagonals of a parallelogram circumscribing a conic are conjugate diameters; and the points of contact are the vertices of a parallelogram whose sides are parallel to the above diagonals.*

**Ex. 8.** *A tangent cuts two parallel tangents in $P$ and $Q$, show that $CP$ and $CQ$ are conjugate diameters.*

For, reflecting the figure in the centre $C$, this reduces to Ex. 7.

**14.** *If each diameter of a conic be perpendicular to its conjugate diameter, the conic is a circle.*

Take any two points $P$, $Q$ on the conic. Bisect $PQ$ in $V$ and join $CV$. Then $CV$ is the diameter bisecting chords parallel to $PQ$, i.e. $CV$ and $PQ$ are parallel to conjugate diameters. Hence $CV$ and $PQ$ are perpendicular. Also $PV = VQ$. Hence $CP = CQ$. Hence all radii of the conic are equal, i.e. the conic is a circle.

**15.** The *asymptotes* of a conic are the tangents from the centre. They are clearly the joins of the centre to the points at infinity on the conic. In the hyperbola they are real and distinct, in the parabola they coincide with the line at infinity, and in the ellipse they are imaginary. *The asymptotes are harmonic with every pair of conjugate diameters.* For the tangents from any point are harmonic with any pair of conjugate lines through the point.

*Any line cuts off equal lengths between a hyperbola and its asymptotes.*

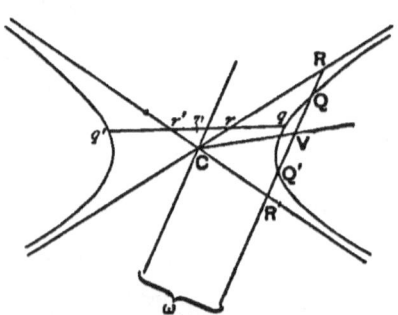

Let a line cut the hyperbola in $Q$, $Q'$ and its asymptotes in $R$, $R'$; then $RQ = Q'R'$.

On $RR'$ take the point at infinity $\omega$ and bisect $QQ'$ in $V$. Then since $(QQ', V\omega)$ is harmonic, the polar of $\omega$ passes through $V$. Since $\omega$ is at infinity, its polar passes through $C$. Hence $CV$ is the polar of $\omega$. Hence $CV$ and $C\omega$ are conjugate lines. And $CR$, $CR'$ are the tangents from $C$. Hence $C(RR', V\omega)$ is harmonic. Hence $(RR', V\omega)$ is harmonic. Hence $RV = VR'$. But $QV = VQ'$. Hence $RQ = Q'R'$. The proof applies whether we take $QQ'$ to cut the same branch in two points or (as in the case of $qq'$) to cut different branches of the hyperbola.

*The intercept made by any tangent between the asymptotes is bisected at the point of contact.*

For let $Q$ and $Q'$ coincide; then $RQ = QR'$.

**Ex. 1.** *Given the asymptotes and one point on a hyperbola, construct any number of points on the curve.*

**Ex. 2.** *Given the asymptotes and one tangent of a hyperbola, construct any number of points and tangents of the curve.*

**Ex. 3.** *Two of the diagonals of a quadrilateral formed by two tangents of a hyperbola and the asymptotes are parallel to the chord joining the points of contact of the tangents.*

Consider the harmonic triangle of the quadrangle formed by the points of contact and the points at infinity on the hyperbola.

**Ex. 4.** *If a hyperbola be drawn through two opposite vertices of a parallelogram*

*Harmonic Properties of a Conic.*

with its asymptotes parallel to the sides, show that the centre lies on the join of the other vertices.

**16.** *A rectangular hyperbola* is defined to be a hyperbola whose asymptotes are perpendicular.

*Conjugate diameters of a rectangular hyperbola are equally inclined to the asymptotes.*

For they form a harmonic pencil with the asymptotes, which are perpendicular.

> **Ex.** *The lines joining the ends of any diameter of a rectangular hyperbola to any point on the curve are equally inclined to the asymptotes.*

**17.** A *principal axis* of a conic is a diameter which bisects chords perpendicular to itself.

*All conics have a pair of principal axes; but one of the principal axes of a parabola is at infinity.*

Consider first the *hyperbola*. Then the asymptotes are real and distinct. Now the bisectors of the angles between the asymptotes are harmonic with the asymptotes and are therefore conjugate diameters. But the bisectors are also perpendicular. Hence they are a pair of conjugate diameters at right angles. Each of the bisectors is therefore a principal axis; for each bisects chords parallel to the other, i.e. perpendicular to itself.

Consider next the *parabola*. We might say that here the asymptotes are coincident with the line at infinity; and the bisectors of the angles between a pair of coincident lines are the line with which they coincide and a perpendicular to it. Hence the principal axes of a parabola are the line at infinity and another line called *the axis of the parabola*.

Or thus—All the diameters of a parabola are parallel. Draw chords perpendicular to a diameter, then the diameter bisecting these chords is perpendicular to them and is called *the axis of the parabola*. The other principal axis (like the diameter conjugate to any of the other parallel diameters) is the line at infinity.

Consider last *the ellipse*. Here the asymptotes are imaginary and this method fails. But it will be proved under

## Harmonic Properties of a Conic.

Involution that there is always a pair of conjugate diameters of any conic at right angles. Hence the ellipse also has a pair of principal axes. (See XIX. 4.)

*An axis cuts the conic at right angles.*

For the tangent at the end of an axis is the limit of a bisected chord.

*A central conic is symmetrical for each axis.*

For the principal axis $AL$ bisects chords perpendicular to itself.

Let $PMP'$ be such a chord. Then $P'$ is clearly the reflexion of $P$ in $AL$, i.e. the conic is symmetrical for $AL$.

The same proof shows that

*A parabola is symmetrical for its axis.*

**Ex. 1.** *The tangent at P meets the axis CA in T and PN is the perpendicular on CA; show that* $CN \cdot CT = CA^2$.

For $PN$ is the polar of $T$.

**Ex. 2.** *PQ, PR touch a conic at Q, R. PM is drawn perpendicular to either axis. Show that PM bisects the angle QMR.*

# CHAPTER VI.

### CARNOT'S THEOREM.

**1.** *The sides $BC, CA, AB$ of a triangle cut a conic in the points $A_1 A_2$, $B_1 B_2$, $C_1 C_2$, show that*

$$AC_1 . AC_2 . BA_1 . BA_2 . CB_1 . CB_2$$
$$= AB_1 . AB_2 . BC_1 . BC_2 . CA_1 . CA_2.$$

By definition a conic is the projection of a circle. Let the points $ABC A_1 A_2 \ldots$ be the projections of $A'B'C'A_1'A_2'\ldots$ in the figure of the circle.

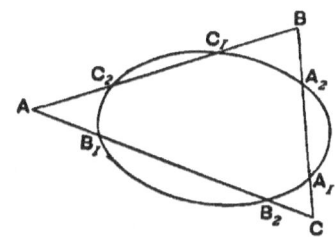

Now in the circle we have
$$A'C_1'. A'C_2'. B'A_1'. B'A_2'. C'B_1'. C'B_2'$$
$$= A'B_1'. A'B_2'. B'C_1'. B'C_2'. C'A_1'. C'A_2'$$
for $A'C_1'. A'C_2' = A'B_1'. A'B_2'$, and so on.

Let $V$ be the vertex of projection.

Then
$$\frac{AC_2}{BC_2} = \frac{\triangle AVC_2}{\triangle BVC_2}$$
$$= \frac{AV. C_2V. \sin AVC_2}{BV. C_2V. \sin BVC_2}$$
$$= \frac{AV}{BV} \cdot \frac{\sin AVC_2}{\sin BVC_2}$$

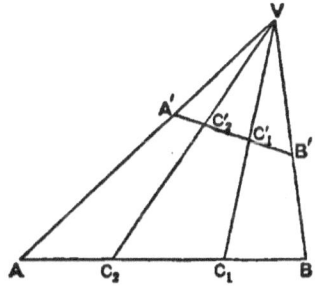

and so for each ratio.

Hence $\dfrac{AC_1 \cdot AC_2 \ldots}{AB_1 \cdot AB_2 \ldots} = \dfrac{\sin AVC_1 \cdot \sin AVC_2 \ldots}{\sin AVB_1 \cdot \sin AVB_2 \ldots}$

where each segment is replaced by the sine of the corresponding angle. Also the last expression

$= \dfrac{\sin A'VC_1' \cdot \sin A'VC_2' \ldots}{\sin A'VB_1' \cdot \sin A'VB_2' \ldots}$, and this equals $\dfrac{A'C_1' \cdot A'C_2' \ldots}{A'B_1' \cdot A'B_2' \ldots}$

by exactly the same reasoning as before, and this has been proved equal to unity. Hence

$AC_1 \cdot AC_2 \cdot BA_1 \cdot BA_2 \cdot CB_1 \cdot CB_2$
$\qquad = AB_1 \cdot AB_2 \cdot BC_1 \cdot BC_2 \cdot CA_1 \cdot CA_2.$

**Ex. 1.** *The sides $AB$, $BC$, $CD$, ... of a polygon meet a conic in $A_1 A_2$; $B_1 B_2$, $C_1 C_2$, ...; show that*
$$AA_1 \cdot AA_2 \cdot BB_1 \cdot BB_2 \cdot CC_1 \cdot CC_2 \ldots = BA_1 \cdot BA_2 \cdot CB_1 \cdot CB_2 \cdot DC_1 \cdot DC_2 \ldots$$

**Ex. 2.** *By taking the conic to be a line and the line at infinity, deduce Menelaus's theorem from Carnot's theorem.*

**Ex. 3.** *If a conic touch the sides of the triangle $ABC$ in $A_1$, $B_1$, $C_1$; then $AA_1$, $BB_1$, $CC_1$ are concurrent.*

For $AB_1^2 \cdot CA_1^2 \cdot BC_1^2 = AC_1^2 \cdot BA_1^2 \cdot CB_1^2$;
and we cannot have $AB_1 \cdot CA_1 \cdot BC_1 = + AC_1 \cdot BA_1 \cdot CB_1$, for then $A_1 B_1 C_1$ would cut the conic in three points.

**Ex. 4.** *If the vertex $A$ in Carnot's theorem be on the conic, show that the ratio $AC_2 : AB_1$ must be replaced by $\sin TAC : \sin TAB$, $AT$ being the tangent at $A$.*

For $B_1 C_2$ is ultimately the tangent at $A$.

**Ex. 5.** *What does Carnot's theorem reduce to when $A$, $B$, and $C$ are on the curve?*

**Ex. 6.** *If through fixed points $A$, $B$ we draw the chords $AB_1 B_2$, $BA_2 A_1$ of a conic meeting in the variable point $C$, then the ratio*
$$BA_1 \cdot BA_2 \cdot CB_1 \cdot CB_2 \div AB_1 \cdot AB_2 \cdot CA_1 \cdot CA_2 \text{ is constant.}$$

**Ex. 7.** *Deduce the corresponding theorem when $B$ is at infinity.*

**Ex. 8.** *$A$, $B$, $C$ are three points on a conic; the tangents at $ABC$ meet in $GHK$; points $DEF$ are taken on $BC$, $CA$, $AB$ such that $AD$, $BE$, $CF$ are concurrent: show that $GD$, $HE$, $KF$ are concurrent.*

For $\sin DGB/\sin DGC = DB/DC \div BG/CG$. Now use two forms of Ceva's theorem.

**Ex. 9.** *$AC$ touches a conic at $A$, $AB$ meets it again in $C_1$, and $BC$ meets it in $A_2$, $A_1$; if the circle of curvature at $A$ meet $AB$ in $C'$, show that*
$$AC' \cdot CA_1 \cdot CA_2 \cdot BC_1 \cdot BA = AC_1 \cdot BA_1 \cdot BA_2 \cdot CA^2.$$

Consider the circle of curvature as the limit of the circle through $B_1 B_2 C_2$.

**Ex. 10.** *If $A_1 A_2$ be parallel to the tangent at $A$, this reduces to*
$$AC' \cdot BC_1 \cdot BA = AC_1 \cdot BA_1 \cdot BA_2.$$

**Ex. 11.** *Deduce the expression $2 CD^2 \div CP$ for the central chord of curvature.*

**Ex. 12.** *A conic cuts the sides BC, CA, AB of a triangle in $P_1P_2$, $Q_1Q_2$, $R_1R_2$; $BQ_2$ and $CR_2$ meet in $X$, $AP_1$ and $CR_1$ in $Y$, and $AP_2$ and $BQ_1$ in $Z$; show that $AX$, $BY$, $CZ$ are concurrent.*

**2.** *If, on the sides BC, CA, AB of a triangle, the pairs of points $A_1A_2$, $B_1B_2$, $C_1C_2$ be taken, such that*

$$AC_1 . AC_2 . BA_1 . BA_2 . CB_1 . CB_2$$
$$= AB_1 . AB_2 . BC_1 . BC_2 . CA_1 . CA_2,$$

*then the six points $A_1$, $A_2$, $B_1$, $B_2$, $C_1$, $C_2$ lie on a conic.*

Through the five points (XXIV. 2) $A_1$, $A_2$, $B_1$, $B_2$, $C_1$ draw a conic. If this conic does not pass through $C_2$, let $AB$ cut the conic again in $\gamma_2$. Then we have

$$AC_1 . A\gamma_2 . BA_1 . BA_2 . CB_1 . CB_2$$
$$= AB_1 . AB_2 . BC_1 . B\gamma_2 . CA_1 . CA_2.$$

Dividing the given relation by this relation we have

$$AC_2 / A\gamma_2 = BC_2 / B\gamma_2.$$

Hence $C_2$ and $\gamma_2$ coincide. Hence the six points $A_1$, $A_2$, $B_1$, $B_2$, $C_1$, $C_2$ lie on a conic.

**Ex. 1.** *If from any two points the vertices of a triangle be projected upon the opposite sides, the six projections lie on a conic.*

**Ex. 2.** *The parallels through any point to the sides of a triangle meet the sides in six points on a conic.*

**Ex. 3.** *If a conic which has two sides of a triangle as asymptotes touch the third side, the point of contact bisects the side.*

**Ex. 4.** *A conic can be drawn to touch the three sides of a triangle at their middle points.*

**3.** Newton's theorem—*If two chords of a conic $UPQ$, $ULM$ be drawn in given directions through a variable point $U$, show that the ratio of $UP . UQ$ to $UL . UM$ is independent of the position of $U$.*

Let $U'P'Q'$, $U'L'M'$ be another position of the chords $UPQ$, $ULM$. Then $PQ$ is parallel to $P'Q'$ and $LM$ to $L'M'$. Let $PQ$, $P'Q'$ meet at infinity in $\omega'$, and $LM$, $L'M'$ at infinity in $\omega$. Apply Carnot's theorem to the triangle $\omega'U'V$. Then

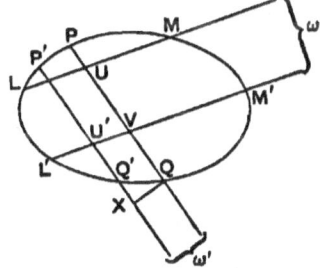

$$\omega'Q' . \omega'P' . U'L . U'M' . VQ . VP$$
$$= \omega'Q . \omega'P . VL' . VM' . U'Q' . U'P'.$$

From $Q$ drop the perpendicular $QX$ on $Q'\omega'$.
Then $\omega'Q'/\omega'Q = (\omega'X+XQ')/\omega'X = 1+XQ'/\omega'X = 1$.
So $\omega'P' = \omega'P$. Hence
$$U'L'.\ U'M'.\ VQ.\ VP = VL'.\ VM'.\ U'Q'.\ U'P'$$
i. e. $U'P'.\ U'Q' \div U'L'.\ U'M' = VP.\ VQ \div VL'.\ VM'$

In exactly the same way the triangle $\omega UV$ gives us
$$VP.\ VQ \div VL'.\ VM' = UP.\ UQ \div UL.\ UM.$$
Hence $UP.\ UQ \div UL.\ UM = U'P'.\ U'Q' \div U'L'.\ U'M'$,
i. e. $UP.\ UQ \div UL.\ UM$ is independent of the position of $U$.

**Ex.** *If the tangents from $T$ to the conic touch at $P$ and $Q$, show that*
$$TP:TQ :: CP' : CQ',$$
*where $CP'$, $CQ'$ are the semi-diameters parallel to $TP$, $TQ$.*
Take $U$ at $T$ and $C$ successively.

**4.** *In a parabola $QV^2 = 4.\ SP.\ PV$.*

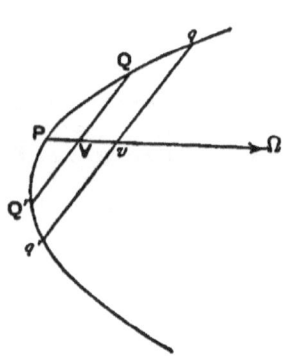

Besides $QVQ'$ draw a second double ordinate $qvq'$ of the diameter $PV$. Now $PV$ meets the parabola again at $\Omega$, a point at infinity. Also by Newton's theorem we have
$$\frac{VQ.\ VQ'}{VP.\ V\Omega} = \frac{vq.\ vq'}{vP.\ v\Omega}.$$
But $V\Omega = v\Omega$. Hence
$$VQ.\ VQ' \div VP = vq.\ vq' \div vP,$$
i. e. $QV^2 \div PV$ is constant. To obtain the value of this constant take $qq'$ through the focus $S$. Then by Geometrical Conics $qq' = 4.\ SP$ and $Pv = SP$. Hence $QV^2 \div PV = 4.\ SP$.

Note that the theorem also follows directly from Carnot's theorem by using the triangle contained by $QV$, $Vv$, $vq$.

**5.** *In an ellipse $QV^2 : PV.\ VP' :: CD^2 : CP^2$.*

In the figure of V. 13, we have by Newton's theorem,
$$VQ.\ VQ' : VP.\ VP' :: CD.\ CD' : CP.\ CP',$$
i. e. $QV^2 : PV.\ VP' :: CD^2 : CP^2$.

**6.** *In a hyperbola* $QV^2 : PV \cdot VP' :: CD^2 : CP^2$.

Besides $QVQ'$ draw a second double ordinate $qvq'$ of the diameter $PCP'$. Then by Newton's theorem $VQ \cdot VQ' : VP \cdot VP'$
$:: vq \cdot vq' : vP \cdot vP'$,
i.e. $QV^2 : PV \cdot VP'$ is constant.

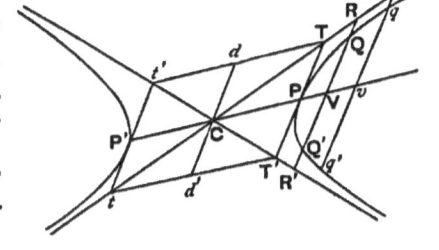

To obtain the value of this constant, take $V$ at $C$, and let $D$ be the position of $Q$.

Then $QV^2 = CD^2$ and $PV \cdot VP' = PC \cdot CP' = CP^2$.

Hence $QV^2 : PV \cdot VP' :: CD^2 : CP^2$,

the formula required.

But this is not the formula given in books on Geometrical Conics; for in the above formula either $P$ or $D$ is imaginary, since, of two conjugate diameters of a hyperbola, one only meets the curve in real points. Take $P$ real and $D$ imaginary. Then $CD^2$ is negative, otherwise $D$ would be real. On $CD$ take the point $d$, such that $Cd^2 = -CD^2$. Then $d$ is real, for $Cd^2$ is positive.

Then $QV^2 : PV \cdot VP' :: -Cd^2 : CP^2$,

i.e. $QV^2 : PV \cdot P'V :: Cd^2 : CP^2$,

which is the formula given in books on Geometrical Conics, the $d$ here replacing the $D$ of the books.

We may call $CD$ the true and $Cd$ the conventional semi-diameter conjugate to $CP$.

It is sometimes convenient to employ the symbol $D$ for the conventional point $d$ when the meaning is clear from the context.

Note that the locus of $d$ is the so-called conjugate hyperbola.

The theorems of § 5 and § 6 may also be obtained directly from Carnot's theorem by using the triangle contained by $DC$, $VC$, $VQ$.

**7.** *If the diameter conjugate to $PCP'$ meet the curve in* the imaginary points $D$ and $D'$, and if the tangent at $P$ meet an asymptote in $T$, then $CD^2 = -PT^2$, i.e. $PT$ is equal to the conventional $CD$ and parallel to it.

In the figure of § 6 let $RQQ'$ be parallel to the tangent at $P$, and let $\Omega$ be the point at infinity on the asymptote $CR$. Then by Newton's theorem $RQ \cdot RQ' \div R\Omega^2 = rq \cdot rq' \div r\Omega^2$, $rqq'$ being parallel to $RQQ'$. But $R\Omega = r\Omega$. Hence

$$RQ \cdot RQ' = rq \cdot rq'.$$

Now take $R$ at $T$, then $RQ \cdot RQ' = TP^2$. Again, take $r$ at $C$, then $rq \cdot rq' = -rq^2 = -CD^2$. Hence $TP^2 = -CD^2 = Cd^2$. Hence $TP = Cd$, i.e. $TP$ represents $Cd$ in magnitude and direction.

Notice that we have incidentally proved the theorem—*If a chord $QQ'$ of a hyperbola drawn in a fixed direction cut one of the asymptotes in $R$, then $RQ \cdot RQ'$ is constant and the same whichever asymptote is taken.*

For $RQ \cdot RQ' = TP^2 = T'P^2 = R'Q \cdot R'Q'$.

It follows that $RQ \cdot QR'$ and $RQ' \cdot Q'R'$ are constant and equal. For $RQ' = QR'$.

Also $RQ \cdot RQ' = R'Q \cdot R'Q' = Cd^2$, $Cd$ being parallel to $RQQ'R'$.

**Ex. 1.** *Pd is parallel to an asymptote.*

For by reflexion in $C$ we get the complete parallelogram $TT'tt'$, and clearly $Pd$ is parallel to $t'T'$.

**Ex. 2.** *Given in magnitude and position a pair of conjugate diameters of a hyperbola, construct the asymptotes.*

**Ex. 3.** *Through any point $R$ on an asymptote of a hyperbola is drawn a line parallel to the real diameter $P'CP$ cutting the curve in $QQ'$, show that*

$$RQ \cdot RQ' = -CP^2.$$

**Ex. 4.** *If the same line cut the other asymptote in $R'$, show that*

$$QR \cdot QR' = Q'R \cdot Q'R' = CP^2.$$

**Ex. 5.** *Given a pair of conjugate diameters of a hyperbola in magnitude and position, construct the axes in magnitude and position.*

Use Ex. 2 and Ex. 4.

**Ex. 6.** *The tangent at $Q$ to a hyperbola meets a diameter $CD$ or $Cd$ (which meets the curve in imaginary points) in $T$, and the parallel through $Q$ to the conjugate diameter $CP$ meets $CD$ in $V$, show that $CV \cdot CT = CD^2 = -Cd^2$.*

For $(DD', TV)$ is harmonic, and $C$ bisects $DD'$. Since $CV \cdot CT$ is negative, $V$ and $T$ are on opposite sides of $C$. Of the above harmonic range notice that $DD'$ are imaginary points and $TV$ real points.

**Ex. 7.** *Given a pair of conjugate diameters of a hyperbola in position and a tangent and its point of contact, construct the axes in magnitude and position.*
$CV \cdot CT = CP^2$ gives the lengths of the diameters.

**Ex. 8.** *The polar of $d$ is $d'T'$.*
Consider the chords intercepted on $dT$ and $dP$.

**Ex. 9.** *If through a variable point $U$ a chord $PUQ$ be drawn in a fixed direction, and also a chord $UR$ parallel to one of the asymptotes in the case of the hyperbola, or parallel to the axis in the case of the parabola, then $UP \cdot UQ \div UR$ is constant.*

**8.** *In a rectangular hyperbola, conjugate diameters are equal and equally inclined to the asymptotes. Also diameters which are perpendicular are equal.*

Since conjugate diameters are harmonic with the asymptotes which are perpendicular, they are equally inclined to the asymptotes. Again $CD = PT$. But in the r. h. $TCT'$ is a right angle and

$$TP = PT'.$$

Hence $CP = PT$, hence $CP = CD$.
Draw $CQ$ perpendicular to $CP$. Then

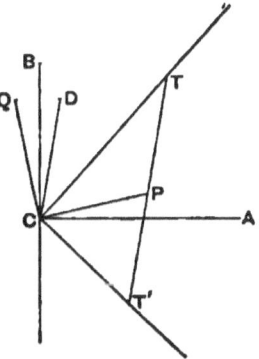

$\angle BCQ = 90° - \angle BCP = \angle ACP = \angle BCD$, since $\angle PCT = \angle DCT$; hence $CQ$ is the reflexion of $CD$ in $CB$. Hence $CQ = CD$. Hence $CQ = CP$. Similarly $CD$ is equal to the semidiameter perpendicular to it.

Notice that the true formulae are $CP^2 = -CD^2 = -CQ^2$; so that if a diameter meet a r. h. in real points, the perpendicular diameter meets the curve in imaginary points.

**Ex.** *The perpendicular chords $LM$, $L'M'$ of a r. h. meet in $U$, show that*
$$UL \cdot UM = -UL' \cdot UM'.$$
For $UL \cdot UM : UL' \cdot UM'$ as the squares of the parallel diameters, i. e. as $CP^2 : -CQ^2$.

**9.** *Every rectangular hyperbola which circumscribes a triangle passes through the orthocentre.*

Let $ABC$ be the triangle and $P$ its orthocentre. Suppose a r. h. through $ABC$ cuts the perpendicular $AD$ in $Q$. Then from the r. h. we have $DQ \cdot DA = -DB \cdot DC$. And from

Elementary Geometry we have $DP \cdot DA = -DB \cdot DC$. Hence $DQ = DP$, i.e. $Q$ coincides with $P$, i. e. the r. h. passes through the orthocentre.

For the converse see XXI. 9.

**Ex. 1.** *If a triangle PQR which is right angled at Q be inscribed in a r. h., the tangent at Q is the perpendicular from Q on PR.*

**Ex. 2.** *If a r. h. circumscribe a triangle, the triangle formed by the feet of the perpendiculars from the vertices on the opposite sides is self-conjugate for the r. h.*

Being the harmonic triangle of $ABCP$.

# CHAPTER VII.

### FOCI OF A CONIC.

1. A *focus* of a conic is a point at which every two conjugate lines are perpendicular.

A *directrix* of a conic is the polar of one of the foci. The polar of a focus is called the corresponding directrix.

From the definition of a focus it at once follows that *every two perpendicular lines through a focus are conjugate.*

The theory of the foci of a conic is given in Chapter XXVIII. It is there shown that—

*Every conic has four foci.*

*All the foci are inside the curve.*

*The foci lie, two by two, on the principal axes; the pair $SS'$ on one axis* (called the focal axis) *are real, and those $FF'$ on the other axis are imaginary; also $SS'$ are equidistant from the centre on opposite sides, and so are $FF'$.*

*One real focus of a parabola is at infinity on the axis of the parabola.*

*All the foci of a circle coincide with the centre.*

Note that the *focal axis* is the major axis in an ellipse, the transverse axis in a hyperbola, and the axis in a parabola.

**Ex. 1.** *Tangents at the ends of a focal chord meet on the directrix.*

**Ex. 2.** *If $CT$ meet the directrix in $Z$, then $SZ$ is perpendicular to the polar of $T$.*

· Being perpendicular to the polar of $Z$.

**Ex. 3.** *$PSQ$ is a focal chord of a conic. $UOV$ is any chord of the conic through the middle point $O$ of $PQ$. Parallels through $U$, $V$ to $PQ$ meet the directrix corresponding to $S$ in $M$, $N$. Show that $PQ$ bisects the angle $MSN$.*

Let the polar of $O$ (which is parallel to $PQ$) meet the directrix in $R$; then $SR$ and $SP$ are conjugate lines.

**2.** *If from any point P on a conic, a perpendicular PM be drawn to the directrix which corresponds to a focus S, then SP ÷ PM is constant.*

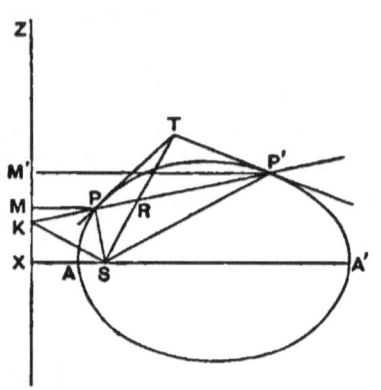

Take any two points P and P' on the conic. Let the tangents at P and P' meet in T. Let PP' meet the corresponding directrix in K, and ST in R. From P and P' drop the perpendiculars PM and P'M' on the directrix.

Now SK and ST are conjugate lines at the focus; for the polar of K, which lies on PP' and on the directrix, is TS. Hence SK is perpendicular to ST. Also (KPRP') is harmonic, since K is the pole of ST. Hence S(KPRP') is harmonic. Hence SK and ST, being perpendicular, are the bisectors of the angle PSP'. Now since SK bisects the angle PSP' (externally in the figure), we have $SP : SP' :: PK : P'K :: PM : P'M'$. Hence $SP : PM :: SP' : P'M'$; in other words, $SP : PM$ is constant.

*In the parabola, SP = PM.*

For let SA be the axis. Then SA meets the parabola again at infinity, at $\Omega$, say. Hence $(XAS\Omega)$ is harmonic, since XZ is the polar of S. Hence $SA = AX$.

But $SP : PM :: SA : AX$, for A is on the parabola.

Hence $SP = PM$.

*In the ellipse, SP < PM.*

Since a focus is an internal point, S must lie between A and A'.

Let A be the vertex between S and X. Then since A' is a point on the ellipse, we have $SP : PM :: SA' : A'X$.

But $SA' < A'X$, hence $SP < PM$.

*In the hyperbola, SP > PM.*

Since the focus is an internal point, S must lie outside the segment AA'.

As before $SP : PM :: SA' : A'X > 1$.

*The corresponding property in the circle is that the radius is constant.*

For the focus is the centre. Hence the directrix is the line at infinity. Hence $PM = P'M'$. Hence $SP = SP'$, i.e. $CP = CP'$.

**Ex. 1.** *Any two tangents to a conic subtend at a focus angles which are either equal or supplementary.*

**Ex. 2.** *Show that it is not true conversely that 'if any two tangents to a conic subtend at a point on an axis angles which are equal or supplementary, then this point is a focus.'*

The foot of the perpendicular from $T$ on the axis is such a point.

**3.** Assuming from Chapter XXVIII that a conic has a real focus, we have just shown that this focus possesses the $SP : PM$ property by which a focus is defined in books on Geometrical Conics. This opens up to us all the proofs given in such books. It will be assumed that these proofs are known to the reader; and the results will be quoted when convenient. Properties of Conics which can be best treated by the methods of Geometrical Conics will be usually omitted from this treatise.

**4.** *In any conic, the semi-latus rectum is equal to the harmonic mean between the segments of any focal chord.*

Let the focal chord $P'SP$ cut the directrix in $K$.

Then $(KPSP')$ is harmonic since $S$ is the pole of $XK$. Hence

$$2(KS)^{-1} = (KP)^{-1} + (KP')^{-1}.$$

But
$$KP : KS : KP' :: PM : SX : P'M'$$
$$:: SP : SL : SP',$$

for $SP : PM :: SL : LU :: SL : SX$.

Hence
$$2(SL)^{-1} = (SP)^{-1} + (SP')^{-1}.$$

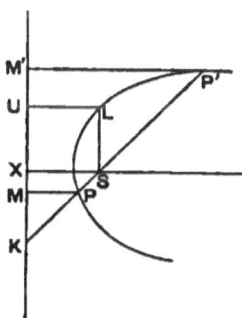

**Ex. 1.** *If $T$ be the pole of the focal chord $PQ$ of a parabola, show that*
$$PQ \propto ST^2.$$

**Ex. 2.** *A focal chord of a central conic is proportional to the square of the parallel diameter.*

**5.** *If the tangent at P meet the tangents at the vertices AA′ of the focal axis in UU′, then UU′ subtends a right angle at S and S′. Also if US, U′S′ cut in E, and US′, U′S cut in F, then EF is the normal at P.*

For since $AU$ and $PU$ subtend equal angles at $S$ and since $A'U'$ and $PU'$ subtend equal angles at $S$, it follows that $USU'$ is a right angle. Similarly $UU'$ subtends a right angle at $S'$.

Again, $F$ is the orthocentre of the triangle $UEU'$. Hence $EF$ is at right angles to $UU'$. Let $PU$ cut the axis in $T$ and draw the ordinate $PN$. Then $(TUPU') = (TANA')$ is harmonic. Also if $EF$ cut $UU'$ in $P'$, then since $UU'$ is a harmonic side of the quadrilateral $SF, FS', S'E, ES$, we have $(TUP'U')$ harmonic. Hence $P'$ and $P$ coincide, i.e. $EF$ passes through $P$. Hence $EF$ is the normal at $P$.

**Ex. 1.** *If a circle through the foci cut the tangent at the vertex A in U, V and the tangent at the vertex A′ in U′, V′, show that the diagonals of the rectangle UU′V′V touch the conic.*

**Ex. 2.** *Given the focal axis AA′ in magnitude and position and one tangent, construct the foci.*

**6.** *If the tangent at a point P of a central conic cut the focal axis in T, and if the normal at P cut the same axis in G, then $CG \cdot CT = CS^2$.*

For since the tangent and normal bisect the angle $SPS'$, it follows that $P(SS', TG)$ is harmonic; hence
$$CG \cdot CT = CS^2.$$

**Ex. 1.** *Given the axes in position and one tangent and its point of contact, construct the foci.*

**Ex. 2.** *In the parabola, S bisects GT.*

For $S'$ is at infinity.

**Ex. 3.** *Given the axis of a parabola in position and one tangent and its point of contact, construct the focus.*

### Confocal Conics.

**7.** Confocal conics (or briefly confocals) are conics which have the same foci. If one of the given foci is at infinity, we have confocal parabolas, which may also be defined as parabolas having the same focus and the same axis.

*Two confocals can be drawn through any point, one an ellipse and one a hyperbola, and these cut at right angles.*

Join the given point $P$ to the foci $S$, $S'$, and draw the bisectors $PL$ and $PL'$ of the angle $SPS'$. Since both foci are finite, the conic must be an ellipse or a hyperbola. If it be an ellipse, then $Q$ being any point on the ellipse,

$$SQ + S'Q = SP + S'P;$$

so that one and only one ellipse can be drawn through $P$ with $S$ and $S'$ as foci. Similarly one and only one hyperbola can be drawn. And the two conics cut at right angles, for $PL$ and $PL'$ are their tangents at $P$.

If one focus is at infinity, the ellipses and hyperbolas become parabolas, and we get the theorem—

*Of the system of parabolas which have the same focus and the same axis, two pass through any point and these are orthogonal.*

This can be easily proved directly.

**8.** *One confocal and one only can be drawn to touch a given line.*

Take $\sigma$, the reflexion of $S$, in the given tangent. Then $\sigma S'$ cuts the given line in the point of contact $P$ of the given line. If the given line cuts $SS'$ internally, the required conic is a hyperbola, viz. the locus of $Q$ where $S'Q - SQ = S'P - SP$. If the given line cuts $SS'$ externally, the required conic is an ellipse, viz. the locus of $Q$ where $S'Q + SQ = S'P + SP$.

If one focus is at infinity we get the theorem—

*Of a system of confocal parabolas, one and one only touches a given line.*

This can be easily proved directly.

**9.** *The locus of the poles of a given line for a system of confocals is a line.*

Let the given line be $LM$, and let $V$ be the point of contact of the confocal which touches $LM$. Draw $VL'$ perpendicular to $VL$. Then $VL'$ contains the pole of $LM$ for any confocal.

Since $V$ is the pole of $LM$ for the confocal which touches $LM$, the pole of $LM$ for this confocal is on $VL'$. From $V$

draw the tangents $VT$ and $VT'$ to any other confocal. Now $VL$ and $VL'$ bisect $SVS'$, for they are the tangent and normal to the confocal touching $LM$. Also $\angle TVS = \angle T'VS'$ by Geometrical Conics. Hence $VL$ and $VL'$ are the bisectors of $TVT'$, i.e. $VL$ and $VL'$ are harmonic with $VT$ and $VT'$. Hence $VL$, $VL'$ are conjugate for this confocal, i.e. for any confocal of the system. Hence the pole of $VL$ for any confocal lies on $VL'$.

The theorem follows for the confocals to which real tangents cannot be drawn from $V$ by the principle of continuity.

We have incidentally proved the proposition—

*If $V$ be any point in the plane of a conic whose foci are $S$ and $S'$, then the bisectors of the angle $SVS'$ are conjugate for the conic.*

If one focus is at infinity, we get the theorem—

*The locus of the poles of a given line for a system of confocal parabolas is a line.*

*If $V$ be any point in the plane of a parabola whose focus is $S$, and if $VM$ be parallel to the axis, the bisectors of the angle $SVM$ are conjugate for the parabola.*

**Ex. 1.** *If a triangle be inscribed in one conic and circumscribed to a confocal, the points of contact are the points of contact of the escribed circles.*

Let $ABC$ be the triangle. Let the tangents at $A$ and $B$ meet in $R$. Then the locus of the poles of $AB$ is the normal at the point of contact $N$ of $AB$, i.e. $RN$ is perpendicular to $AB$. And $R$ is the centre of the escribed circle because the external angles at $A$ and $B$ are bisected.

**Ex. 2.** *From $T$ are drawn the tangents $TP$, $TP'$ to a conic and the tangents $TQ$, $TQ'$ to a confocal; show that the angle $QPQ'$ is bisected by the normal at $P$.*

For the normal at $P$ meets $QQ'$ in the pole of $TP$ for the other conic.

### Focal Projection.

**10.** *To project a given conic into a circle so that a focus of the conic may be projected into the centre of the circle; and to show that angles at the focus are projected into equal angles at the centre.*

Let $S$ be the focus to be projected into the centre of the circle; and let $XZ$ be the corresponding directrix. Since $S$ is to be projected into the centre, its polar $XZ$ must be projected to infinity. Rotate $S$ about $XZ$ into any position out of the plane of the conic, and take this position as the position of the vertex of projection $V$. With $V$ as vertex project the conic on to a plane parallel to $VXZ$. Now the projection of a conic is a conic. Also $C$, the projection of $S$, is the centre of the new conic; for the polar of $S$ is projected to infinity, hence $C$ is the pole of the line at infinity. Again, the angle $LSM$ at $S$ is superposable to the angle $LVM$; and the projection of $SL$ is parallel to $VL$, and the projection of $SM$ to $VM$. Hence $LSM$ is projected into an equal angle at $C$; so every angle at $S$ is projected into an equal angle at $C$. Also conjugate lines at $S$ are projected into conjugate lines at $C$. Hence the perpendicular conjugate lines at $S$ are projected into perpendicular conjugate lines at $C$, i.e. every two conjugate lines through the centre $C$ are perpendicular. Hence the new conic is a circle.

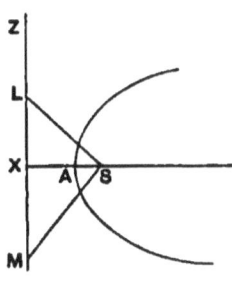

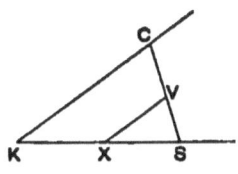

**Ex. 1.** *Project a conic into a conic so that one focus of the one shall project into one focus of the other.*

**Ex. 2.** *Project a circle into a conic so that the centre of the circle shall project into a focus of the conic.*

Take any line as vanishing line, and to get $V$ rotate $C$ about the vanishing line.

11. *Find the envelope of a chord of a conic which subtends a constant angle at a focus of the conic.*

Project the conic $a$ into a circle $\beta$ so that the focus $S$ may project into the centre $C$ of the circle. Then if the chord $PQ$ of the conic subtend a constant angle at $S$, its projection $P'Q'$ will subtend the same angle and therefore a constant angle at $C$. Hence the envelope of $P'Q'$ is a concentric circle $\beta'$. The required envelope is therefore the conic $a'$ of which $\beta'$ is the projection.

Now $S$ is the focus of $a'$; for the perpendicular conjugate lines of $\beta'$ at $C$ are the projections of perpendicular conjugate lines of $a'$ at $S$, since angles at $S$ project into equal angles at $C$. Also the line at infinity is the polar of $C$ for $\beta'$; hence the vanishing line, i.e. the directrix corresponding to $S$ in the given conic $a$, is the polar of $S$ for $a'$. Hence the envelope $a'$ of $PQ$ is a conic having the given focus as focus and having as corresponding directrix the directrix corresponding to the focus in the given conic.

**Ex.** *In the above, find the locus of the pole of PQ.*

Note that these and all other examples of this method can be more easily dealt with by Reciprocation.

# CHAPTER VIII.

## RECIPROCATION.

1. IF we have any figure determined by points $A, B, C, \ldots$ and lines $l, m, n, \ldots$, we can form another figure called a *reciprocal figure* in the following way. Choose any conic $\Gamma$ called the *base conic*. Take the polar $a$ of $A$ for this conic, the polar $b$ of $B$, the polar $c$ of $C, \ldots$; also take the pole $L$ of $l$ for this conic, the pole $M$ of $m$, the pole $N$ of $n, \ldots$; then the figure determined by the lines $a, b, c, \ldots$ and the points $L, M, N, \ldots$ is said to be reciprocal to the figure determined by the points $A, B, C, \ldots$ and the lines $l, m, n, \ldots$; also the point $A$ and the line $a$ are said to be *reciprocal*, so also $B$ and $b$, $C$ and $c$, ..., $l$ and $L$, $m$ and $M$, $n$ and $N$, ....

The name reciprocal arises from the following property—

*If the reciprocal of the figure a be the figure a', then the reciprocal of a' is a.*

For let $A$ be a point of the figure $a$. The reciprocal of $A$ is the polar $a$ of $A$ for the base conic $\Gamma$. Hence $a$ is one of the lines of $a'$ the reciprocal of $a$. Again, in obtaining $a''$, the reciprocal of $a'$, we should obtain the pole of $a$ (a line of $a'$) for $\Gamma$; but the pole of $a$ is $A$. Hence $A$ is a point in $a''$. Hence every point belonging to $a$ belongs also to $a''$. So every line belonging to $a$ belongs also to $a''$. Hence $a$ and $a''$ coincide.

*The reciprocal of the join of two points $A$, $B$ is the meet of the reciprocal lines $a$, $b$; and the reciprocal of the meet of two lines $l$, $m$ is the join of the reciprocal points $L$, $M$.*

By definition the reciprocal of $AB$ is the pole of $AB$ for the base conic $\Gamma$. But the pole of $AB$ is the meet of the polars of $A$ and $B$ for $\Gamma$, i.e. is the meet of the reciprocal lines $a$ and $b$. Similarly the second part follows.

**2.** A curve may be considered either as the locus of points on it or as the envelope of tangents to it. Hence the *reciprocal of a curve* may be defined either as the envelope of the polars for the base conic $\Gamma$ of points on the given curve or as the locus of the poles for $\Gamma$ of the tangents to the given curve. These definitions determine the same curve.

For take two points $P$ and $Q$ on the given curve $a$ and the polars $p$ and $q$ of $P$ and $Q$ for the base conic $\Gamma$. Then by the

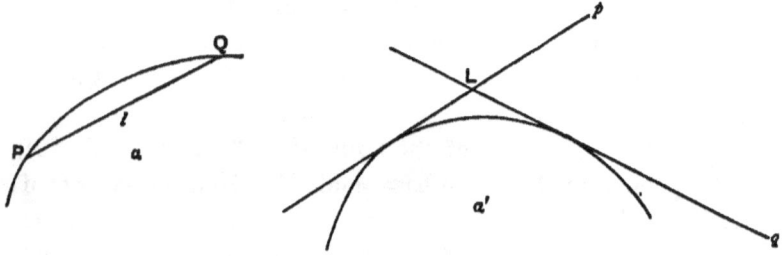

first definition $p$ and $q$ touch the reciprocal curve $a'$ of $a$. Now the reciprocal of $l$, the join of $P$ and $Q$ in $a$, is the meet $L$ of $p$ and $q$ in $a'$. Also when $P$ and $Q$ coincide, $PQ$ becomes a tangent to $a$. At the same time $p$ and $q$ coincide and $L$ becomes a point on $a'$. Hence the reciprocal of a tangent to $a$ is a point on $a'$. Which agrees with the second definition.

From the above we see that—*the reciprocals of a point $P$ on a curve and the tangent $l$ to the curve at $P$ are a tangent $p$ to the reciprocal curve and the point of contact $L$ of $p$.*

*The reciprocal of a point of intersection of two curves is a common tangent to the reciprocal curves.*

For let $l$ and $m$ be the tangents to the curves $a$ and $\beta$ at their meet $P$. In the reciprocal figure we shall have two curves $a'$ and $\beta'$ which have one tangent $p$ with different points of contact $L$ and $M$.

*The reciprocal of two curves touching is two curves touching.*

For the reciprocal of $l$ touching both $a$ and $\beta$ at $P$ is $L$, the point of contact of $p$ with both $a'$ and $\beta'$.

**Ex. 1.** *The reciprocal of a conic, taking the conic itself as base conic, is the conic itself.*

**Ex. 2.** *The reciprocal of a circle, taking a concentric circle as base conic, is a circle concentric with both.*

**3.** *Whatever base conic is taken, the reciprocal of a conic is a conic.*

From any point can be drawn two tangents real or imaginary to the given conic. Hence every line meets the reciprocal curve in two points real or imaginary; hence the reciprocal curve is a conic. (For another proof see XIII. 2.)

More generally. *If the degree of a curve is $m$ and its class $n$, then the class of the reciprocal curve is $m$ and its degree is $n$.*

For a line cuts the given curve in $m$ points; hence from any point can be drawn $m$ tangents to the reciprocal curve. Also from any point can be drawn $n$ tangents to the given curve; hence any line cuts the reciprocal curve in $n$ points.

**Ex. 1.** *The reciprocal of two conics having double contact is two conics having double contact.*

**Ex. 2.** *The reciprocal of a common chord of two conics is a meet of common tangents of the reciprocal conics.*

**4.** *If the point $P$ be the pole of the line $l$ for the conic $a$ and if $p$, $L$, $a'$ be the reciprocals of $P$, $l$, $a$ for any base conic, then the line $p$ is the polar of the point $L$ for the conic $a'$;* or briefly— *the reciprocal of a pole and polar for any conic is a polar and pole for the reciprocal conic.*

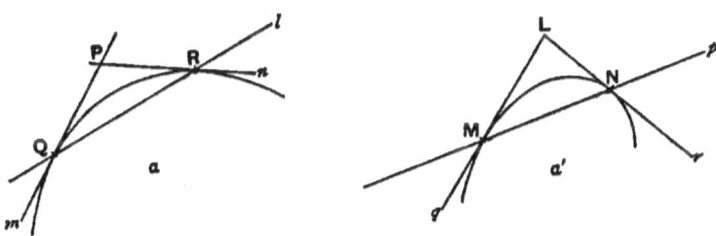

From $P$ draw the real or imaginary tangents $m$, $n$ to $a$ touching in $Q$, $R$. Then $QR$ is $l$, the polar of $P$ for $a$. The reciprocals of $Q$ and $m$ in $a$ are a tangent $q$ to $a'$ and its point of contact $M$; so for $r$ and $N$. The reciprocal of the meet $P$ of the tangents $m$ and $n$ at $Q$ and $R$ is the join $p$ of the points of contact $M$ and $N$ of the tangents $q$ and $r$. Again, the reciprocal of $l$, the join of $Q$ and $R$, is the meet of $q$ and $r$, i.e. is $L$. Hence the reciprocals of $P$ and $l$ which are pole and polar for $a$ are $p$ and $L$ which are polar and pole for $a'$. (For another proof see XIII. 3.)

*The reciprocals of conjugate points are conjugate lines.*

For if the point $P$ is conjugate to the point $Q$, then the polar $l$ of $Q$ passes through $P$. Hence in the reciprocal figure the pole $L$ of $q$ lies on $p$, i.e. the reciprocals $p$ and $q$ of $P$ and $Q$ are conjugate lines. Similarly—

*The reciprocals of conjugate lines are conjugate points.*

**Ex.** *The reciprocal of a triangle self-conjugate for a conic is a triangle self-conjugate for the reciprocal conic.*

**5.** It will be found that all geometrical theorems occur in pairs called *reciprocal theorems*. Thus the theorems (i) '*The harmonic points of a quadrangle inscribed in a circle are the vertices of a triangle self-conjugate for the circle*,' and (ii) '*The harmonic lines of a quadrilateral circumscribed to a circle are the sides of a triangle self-conjugate for the circle*,' are reciprocal theorems. The reason of the name is that each can be derived from the other by reciprocation. Hence we need only have proved half the theorems in the former part of the book; the other half might have been deduced by reciprocation. This method will be often used in future to duplicate a theorem.

For example, to deduce the second of the above theorems from the first, reciprocate, taking the given circle as base conic. The reciprocals of four points on the circle are the polars of these points for the circle, i.e. are the tangents at these points, and so on step by step; and the triangle obtained is self-conjugate because the reciprocal of a self-conjugate triangle is a self-conjugate triangle.

**6.** If one conic only is involved it is best to reciprocate for this conic itself, as then a theorem about a circle gives a

## Reciprocation.

theorem about a circle, a theorem about a parabola gives a theorem about a parabola, and so on. In this way we get a theorem as general as the given one.

**7.** *Write down the Reciprocals of the following propositions— in other words—obtain the corresponding new propositions by Reciprocation.*

1. If two vertices of a triangle move along fixed lines while the sides pass each through a fixed point, the locus of the third vertex is a conic section.
If however the points lie on a line, the locus is a line.
In what other case will the locus be a line?

2. If a triangle be inscribed in a conic, two of whose sides pass through fixed points, the envelope of the third side is a conic, having double contact with the given conic.

3. Given two points on a conic and two tangents, the line joining the points of contact of these tangents passes through one or other of two fixed points.

4. Given four tangents to a conic, the locus of the poles of a fixed line is a line.

5. Given four points on a conic, the locus of the poles of a given line is a conic.

6. Inscribe in a conic a triangle whose sides shall pass through three given points.

7. If three conics have two points common or if they have each double contact with a fourth, the six meets of common tangents lie three by three on the same lines.

8. The meets of each side of a triangle with the corresponding side of the triangle formed by the polars of the vertices for any conic lie on a line.

9. If through the point of contact of two conics which touch, any chord be drawn, the tangents at its ends will meet on the common chord of the two conics.

10. If on a common chord of two conics, any two points be taken, and from these, tangents be drawn to the conics,

the diagonals of the quadrilateral so formed will pass through one or other of the meets of the common tangents of the conics.

11. If $a$ and $\beta$ be two conics having each double contact with the conic $\gamma$, the chords of contact of $a$ and $\beta$ with $\gamma$ and their common chords with each other meet in a point.

12. If $a$, $\beta$, $\gamma$ be three conics, having each double contact with the conic $\sigma$, and if $a$ and $\beta$ both touch $\gamma$, the line joining the points of contact will pass through a meet of the common tangents of $a$ and $\beta$.

### Point Reciprocation.

**8.** If the base conic is a circle (the most common case), the reciprocation is generally called *point reciprocation*, the centre $O$ of the base circle is called the *origin of reciprocation*, and the radius $k$ of the base circle is called the *radius of reciprocation*. The reason of the name point reciprocation is that the value of $k$ is usually of no importance. By reciprocation is meant point reciprocation unless the contrary is stated or implied in the context.

*In point reciprocation, the angle between two lines is equal to the angle subtended by the reciprocal points at the origin of reciprocation.*

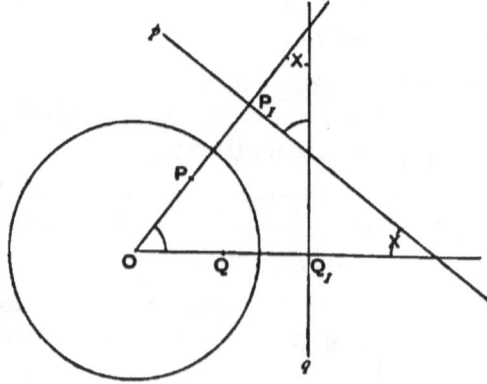

Let $p$ and $q$ be the lines, and $P$ and $Q$ the reciprocals of $p$ and $q$. Let $O$ be the origin of reciprocation. Then $P$ being the pole of $p$ for a circle whose centre is $O$, $OP$ is perpendicular to $p$. So $OQ$ is perpendicular to $q$. Hence $POQ$ is equal to the angle between $p$ and $q$.

*In point reciprocation, the angle between a line $p$ and the line joining the origin $O$ of reciprocation to a point $Q$, is equal to the angle between the line $q$ and $OP$, $P$ and $q$ being the reciprocals of $p$ and $Q$.*

This follows at once, as before, from the above figure.

*In point reciprocation, if $P$ be the reciprocal of $p$ and if $O$ be the origin of reciprocation, then $OP$ is inversely proportional to the perpendicular from $O$ on $p$.*

For $OP \cdot OP_1 = OP \cdot (O, p) = k^2$.

**9.** The reciprocal of a figure for a given point $O$ and a given radius $k$ may be obtained without considering a circle at all. To obtain the reciprocal of $P$—on $OP$ take a point $P_1$, such that $OP \cdot OP_1 = k^2$, and through $P_1$ draw a perpendicular $p$ to $OP$. To obtain the reciprocal of $p$—drop the perpendicular $OP_1$ from $O$ to $p$, and on $OP_1$ take the point $P$, such that

$$OP \cdot OP_1 = k^2.$$

Instead of taking $OP \cdot OP_1 = k^2$, we may take

$$OP \cdot OP_1 = -k^2,$$

i.e. we may take $P$ and $P_1$ on opposite sides of $O$. This is called *negative reciprocation*, and is equivalent to reciprocating for an imaginary circle whose radius is $k\sqrt{-1}$.

**Ex. 1.** *The reciprocal of the origin of reciprocation is the line at infinity; and conversely, the reciprocal of the line at infinity is the origin.*

For the polar of the centre of the base circle is the line at infinity; and conversely.

**Ex. 2.** *The reciprocal of a line through the origin is a point at infinity; and conversely.*

**Ex. 3.** *Reciprocate a quadrangle into a parallelogram.*

Take $O$ at one of the harmonic points.

**Ex. 4.** *The reciprocal of the meet of $OP$ and $m$ is the line through $M$ parallel to $p$.*

**Ex. 5.** *If $P$ and $Q$ be points on a curve such that $PQ$ passes through $O$, then in the reciprocal for $O$, $p$ and $q$ are parallel tangents.*

**Ex. 6.** *The reciprocal for $O$ of the foot of the perpendicular from $O$ on $p$ is the line through $P$ perpendicular to $OP$.*

82               *Reciprocation.*                [CH.

**Ex. 7.** *The reciprocal of a triangle for its orthocentre is a triangle having the same orthocentre.*

**Ex. 8.** *On the sides, BC, CA, AB of a triangle are taken points P, Q, R such that the angles POA, QOB, ROC are right, O being a fixed point; show that PQR are collinear.*

Reciprocating for $O$, we have to prove that the three perpendiculars from the vertices on the opposite sides meet in a point.

**Ex. 9.** *The reciprocal of the curve $p = f(r)$ for the origin is $k^2/r = f(k^2/p)$.*

Let $b$ be the tangent at $A$ to the given curve. Then $B$ is on the reciprocal curve and $a$ touches it. Hence.

$p = (O, b) = k^2/OB = k^2/r'$, and $r = OA = k^2(O, a) = k^2/p'$.

### Reciprocation of a conic into a circle.

**10.** *The reciprocal of a circle, taking a circle with centre $O$ as base conic, is a conic having a focus at $O$.*

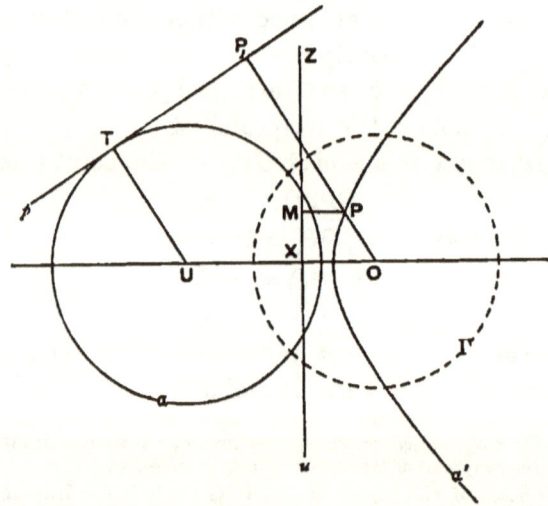

Let $U$ be the centre of the given circle $a$. Take $u$ the reciprocal of $U$, i.e. the polar of $U$ for the base circle $\Gamma$ whose centre is $O$. Let $p$ be any tangent to $a$ touching at $T$. Take $P$ the reciprocal of $p$. Draw the perpendicular $PM$ from $P$ to $u$.

Then since $p$ is the polar of $P$ and $u$ the polar of $U$ for $\Gamma$, we have by Salmon's theorem (III. 9)

$OP/(P, u) = OU/(U, p)$,   i.e.   $OP/PM = OU/UT.$

Hence $OP/PM$ is constant, i.e. the locus of $P$ is a conic with $O$ as focus. But the reciprocal of $a$ for $\Gamma$ is the locus of the poles for $\Gamma$ of the tangents to $a$, i.e. is the locus of $P$. Hence the reciprocal of a circle $a$ for the circle $\Gamma$ whose centre is $O$ is a conic $a'$ having a focus at $O$.

Briefly, *the reciprocal of a circle for a point $O$ is a conic having a focus at $O$.*

Since $e = OP/PM = OU/UT$, we see that the reciprocal of a circle for a circle whose centre is $O$, is an ellipse, parabola or hyperbola according as $OU < = > UT$, i.e. according as $O$ is inside, on or outside the given circle. This is a particular case of a general theorem. (See § 21.)

Let $OU = \delta$, $UT = R$, and let $k$ be the radius of the base circle. Then $e = \delta/R$. Also $OX \cdot OU = k^2$.

Hence $k^2/\delta = OX = a/e - ae$. Hence $a = k^2 R/(R^2 - \delta^2)$.

**Ex.** *Show that the semi-latus rectum $l = k^2/R$.*

This follows from $l = a(1-e^2)$; or directly by noticing that an end of the latus rectum through $O$ reciprocates into a tangent of $a$ parallel to $OU$.

Notice that $l$ is independent of $\delta$, i.e. of the relative positions of the circles.

11. Conversely, *the reciprocal of a conic, taking any circle whose centre is at a focus as base conic, is a circle.*

Let $O$ be the given focus, and $XZ$ or $u$ the corresponding directrix. Take any point $P$ on the conic $a'$, and let $p$ be its reciprocal, i.e. the polar of $P$ for the base circle $\Gamma$ whose centre is at $O$. Draw the perpendicular $PM$ from $P$ to $u$. Take the reciprocal $U$ of $u$. Draw the perpendicular $UT$ from $U$ to $p$.

Then since $p$ is the polar of $P$ and $u$ the polar of $U$ for the conic $\Gamma$, we have by Salmon's theorem

$$OU/UT = OP/PM = e.$$

Hence $OU/UT$ is constant. Also $U$ is a fixed point; hence $UT$ is of constant length. Hence the perpendicular from $U$ on $p$ is constant, i.e. $p$ envelopes a fixed circle $a$. But the reciprocal of $a'$ for $\Gamma$ is the envelope of the polars for $\Gamma$

of the points on $a'$. Hence the reciprocal of the conic $a'$ for a circle $\Gamma$ whose centre is at one of the foci $O$ of the conic is a circle $a$.

Briefly, *the reciprocal of a conic for one of its foci is a circle.*

**Ex. 1.** *The envelope of the polar for a of the centre of a circle which touches two given circles a and β is a circle.*

**Ex. 2.** *Deduce a construction for the centre of a circle touching three given circles.*

**Ex. 3.** *Given four points $A$, $B$, $C$, $D$, show that, with $D$ as focus, one conic can be drawn touching $BC$, $CA$, $AB$, and four conics through $ABC$. Show also that, if $ADB$ be a right angle, a conic, with focus at $D$, can be found to touch the five conics.*

In a right-angled triangle the nine-point circle touches the circumcircle.

**Ex. 4.** *Of the above four conics, the sum of the latera recta of three is equal to the latus rectum of the fourth.*

**Ex. 5.** *The reciprocals of equal circles are conics having equal parameters.*

**Ex. 6.** *Reciprocate for the orthocentre of $ABC$ the theorem—'If $DEF$ be the feet of the perpendiculars from $A$, $B$, $C$ on $BC$, $CA$, $AB$, then the radius of the circle about $ABC$ is double the radius of the circle about $DEF$.'*

**Ex. 7.** *Four conics $a$, $β$, $γ$, $σ$ have one focus and one tangent $t$ in common. A second common tangent to $a$ and $σ$ meets the corresponding directrix of $a$ at a point on $t$; similarly for $βσ$ and $γσ$. Show that the other common tangents of $aβ$, $βγ$, $γa$ are concurrent.*

**Ex. 8.** *Three conics $a$, $β$, $γ$ which have a focus in common are such that $a$ touches $β$ in $R$, $β$ touches $γ$ in $P$, and $γ$ touches $a$ in $Q$. Show that the tangents at $P$, $Q$, $R$ meet the corresponding directrices of $a$, $β$, $γ$ in three collinear points.*

**Ex. 9.** *Reciprocate the centres of similitude of two circles.*

The two circles reciprocate into conics having a common focus $S$. Let $u$, $u'$ be the directrices corresponding to $S$. Then two common chords pass through the meet of $u$ and $u'$; and these chords are the reciprocals of the centres of similitude.

**Ex. 10.** *The reciprocal of two circles for either centre of similitude is two similar and similarly situated conics with a common focus as centre of similitude.*

Reciprocate a pair of parallel tangents.

**12.** The figures of the reciprocals of an ellipse, a parabola and a hyperbola are given below. In the first figure in each case the curves are in their proper relative positions; the second figure represents the circle separately and the third figure represents the conic separately, so that if one figure

VIII.]    *Reciprocation.*    85

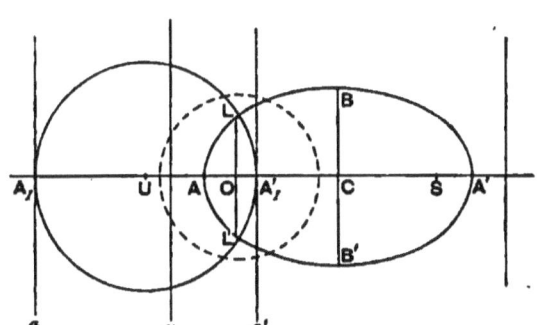

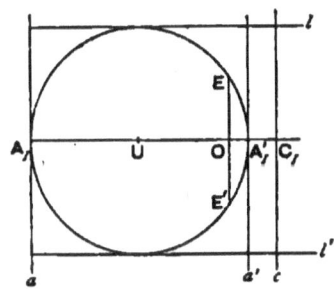
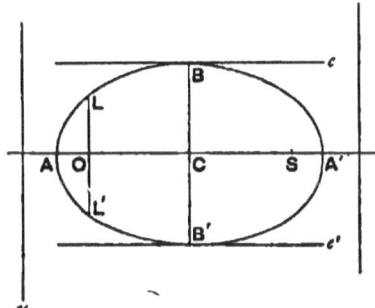

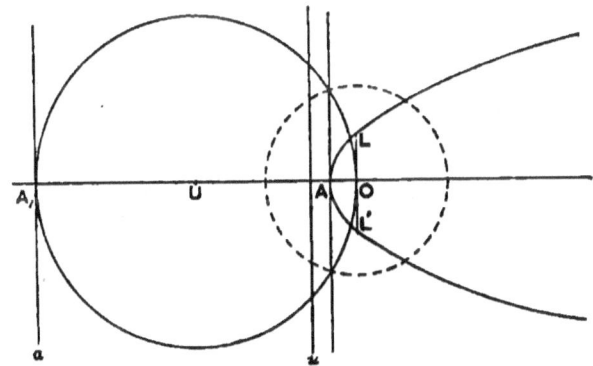

86                    *Reciprocation.*                    [CH.

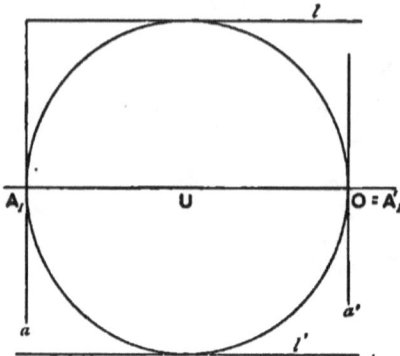
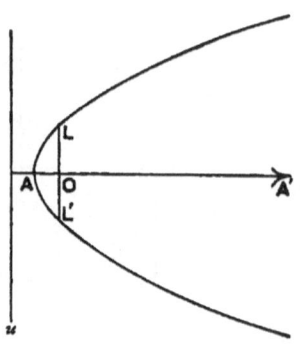

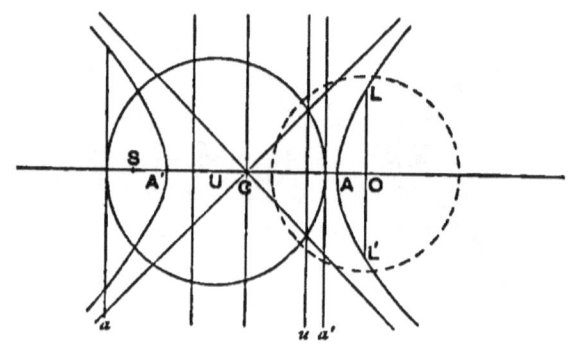

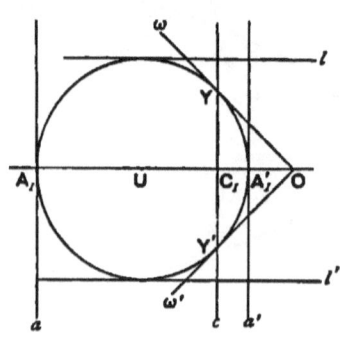
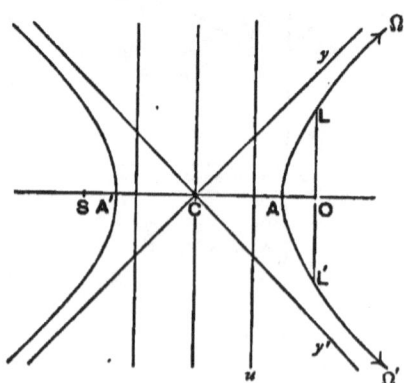

be slid on to the other, so that $O$ in one comes on $O$ in the other, we get the proper figure as in the first figure. To avoid complication the figures will generally be separated as in the second and third figures.

**13.** We already know that the reciprocal of $O$ is the line at infinity and the reciprocal of the *line at infinity* is $O$. Also that the reciprocal of the directrix $u$ corresponding to $O$ is the centre $U$ of the circle.

The *centre* $C$ of the conic is the pole of the line at infinity for the conic. Hence the reciprocal of the centre is the polar $c$ of $O$ for the circle.

The *asymptotes* $y$, $y'$ are the tangents from $C$ to the conic. Hence the reciprocals of the asymptotes are the points in which $c$ meets the circle; i.e. the points in which the polar of $O$ for the circle meets the circle.

The reciprocals of the *vertices* $A$, $A'$ are clearly the tangents at the points where $OU$ meets the circle. In the parabola $A'$ is at infinity; hence its reciprocal is the tangent at $O$.

The reciprocals of the *vertices* $B$, $B'$ are clearly the tangents to the circle at $E$, $E'$, the points where the perpendicular through $O$ to $OU$ meets the circle.

The reciprocals of $L$, $L'$, *the ends of the latus rectum* $LOL'$, are clearly the tangents $l$, $l'$ of the circle parallel to $OU$.

**Ex. 1.** *The reciprocal of the second focus $S$ is the line half-way between $O$ and its polar for the circle.*

For $OS = 2.OC$; hence $OC_1 = 2.OS_1$, where $C_1$ and $S_1$ are the points where the reciprocals of $C$ and $S$ meet $OU$.

**Ex. 2.** *$ACB$ is the diameter of a circle whose centre is $C$. Two equal parabolas are drawn with foci at $C$ and vertices at $A$ and $B$. A hyperbola is drawn having a focus at $C$, and a vertex at $D$ one of the ends of the diameter perpendicular to $AB$, and touching the parabolas. The corresponding directrix of this hyperbola meets $DC$ in $E$, and the hyperbola meets $DC$ again in $F$. Show that*

$$CF = 2.CE = 3.CD.$$

Reciprocate for the circle $ABD$, and notice that $CF_1 = \frac{1}{2}.CE_1 = \frac{1}{3}.CD$.

**Ex. 3.** *If $EE'$ be the chord of the given circle which passes through $O$ and is perpendicular to $OU$, then the minor axis of the reciprocal conic is $2k^2 \div OE$.*

**Ex. 4.** *The reciprocals of coaxal circles for any point on the radical axis are conics having equal minor axes.*

**14.** *If the polar of a point $T$ for a conic meet the conic in $P$, $Q$*

88        *Reciprocation.*        [CH.

and a directrix in $K$, then, $O$ being the corresponding focus, the bisectors of the angle $POQ$ are $OT$ and $OK$.

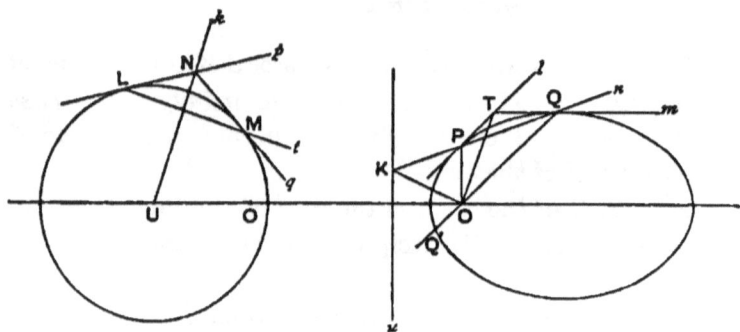

Let the two tangents $l$ and $m$ of the conic touch in $P$ and $Q$ and meet in $T$, and let $n$ be the chord of contact. Let $O$ be a focus of the conic and $u$ the corresponding directrix, and let $PQ$ meet $u$ in $K$. Then we have to prove that $OT$ and $OK$ are the internal and external bisectors of $POQ$.

Reciprocate the conic for a circle with centre at $O$. Then in the reciprocal figure $p$ and $q$ touch the circle at $L$ and $M$ and meet in $N$, and $t$ is the chord of contact. Also the reciprocal of $K$, the meet of $n$ and $u$, is $NU$.

Now $\angle POT = \angle tp$: so $\angle TOQ = \angle tq$. But $\angle tp = \angle tq$. Hence $\angle POT = \angle TOQ$. Again

$\angle POK = \angle pk = 180° - \angle qk = 180° - \angle QOK = \angle KOQ'$,

if we produce $QO$ to $Q'$. Hence $OT$ bisects $\angle POQ$, and $OK$ bisects the supplement $\angle POQ'$.

Note that if $TP$ and $TQ$ had been drawn to touch different branches of a hyperbola, $OT$ would have been the external bisector and $OK$ the internal, instead of as above.

**Ex. 1.** *Reciprocate for any point the theorem—'The tangent to a circle is perpendicular to the radius through the point of contact.'*

If the tangent at $P$ meet $u$ in $K$, then $\angle POK = 90°$.

**Ex. 2.** *Reciprocate for any point the theorem—'The angle between the tangent to any circle and a chord through the point of contact is equal to the angle in the alternate segment.'*

**Ex. 3.** *Two conics which have a common focus S touch at P. From any point Q on one of the conics, tangents are drawn to the other, meeting the tangent at P in UV. The tangent at Q meets the tangent at P in T. Show that TU and TV subtend equal angles at S.*

**Ex. 4.** *The common tangent of an ellipse and its circle of curvature at P meets the tangent at P in a point T, such that SP and ST are equally inclined to the join of the focus S to the centre of curvature.*

Reciprocating for $S$ we get a circle and an ellipse having three-point contact.

**Ex. 5.** *The polar of T for a conic meets in Q a conic which has the same focus S and corresponding directrix. The perpendicular to SQ through S meets the directrix in Z, and SQ and TZ meet in P. Show that the locus of P is a conic having the same focus and directrix. Show also that the eccentricity of the locus is a third proportional to those of the two given conics.*

Reciprocate for $S$ and notice that the envelope reduces to a locus.

**Ex. 6.** *If the chord PQ of a conic subtend at the focus O a constant angle, the envelope of PQ is a conic having O as a focus; and the directrices corresponding to O in the two conics coincide.*

For if $\angle POQ$ is constant, then $\angle pq$ is constant; hence the locus of $N$ is a circle having $U$ as centre. Hence the envelope of $n$ is a conic having $O$ as focus and $u$ as corresponding directrix.

**Ex. 7.** *Find the locus of T when $\angle POQ$ is constant.*

**Ex. 8.** *From two conjugate points on the directrix of a conic are drawn four tangents to the conic. Show that the locus of each of the other meets of the tangents is a single conic; and that the given directrix is a directrix of this conic, and that the corresponding foci of the two conics coincide.*

**Ex. 9.** *The parameter of any conic is a harmonic mean between the segments of any focal chord of the conic.*

For if perpendiculars $OP_1$ and $OQ_1$ be drawn from any point $O$ to two parallel tangents of a circle, then the radius $= \frac{1}{2}(OP_1 + OQ_1)$. If $O$ is outside, $OQ_1$ must be considered negative.

**Ex. 10.** *A pair of parallel tangents to a conic meet a perpendicular to them through a focus in Y and Z and the corresponding directrix in M and N. Show that MZ and NY touch the conic.*

For the angle in a semicircle is a right angle.

**Ex. 11.** *On the tangent at P to a conic is taken a point Q, such that PQ subtends at a focus S a given angle; show that the locus of Q is a conic having a focus at S. Show also that its eccentricity is to the eccentricity of the given conic as its parameter is to the parameter of the given conic.*

For $e : e' :: R' : R :: l : l' :: \sec\theta : 1$.

**Ex. 12.** *Reciprocate for any point the theorem—'If $PP'$, $QQ'$ be two pairs of inverse points for a circle, then $PP'QQ'$ are concyclic.*

Notice that inverse points are conjugate points whose join passes through the centre.

**Ex. 13.** *'If two circles touch one another at C and be touched by a common tangent in A and B, then ACB is a right angle.' Reciprocate this theorem (i) for any point, (ii) for A, (iii) for C, and (iv) for the centre of one of the circles.*

**Ex. 14.** *Reciprocate for any point the theorem—'The locus of the points of contact of tangents from a fixed point to a system of concentric circles is a circle through the fixed point and through the common centre.'*

90  *Reciprocation.*  [CH.

**Ex. 15.** *Reciprocate for the centre of the given circle*—'*The joins of two fixed points on a given circle with the ends of a variable diameter meet at P on a fixed circle through the fixed points and orthogonal to the given circle. Also the tangent at P to the locus is parallel to the diameter.*'

**Ex. 16.** *Reciprocate for any point*—'*The bisectors of the angles of a triangle meet, three by three, in the centres of the four circles touching the sides.*'

**Ex. 17.** *Also*—'*The chord of a circle which subtends a right angle at a fixed point on the circle passes through the centre.*'

**Ex. 18.** *If a circle be reciprocated into a hyperbola, taking a circle with centre O as base conic, then $BC = k^2/OT$, OT being the tangent from O to the circle.*

**15.** *The triangles subtended at the focus of a parabola by any two tangents are similar.*

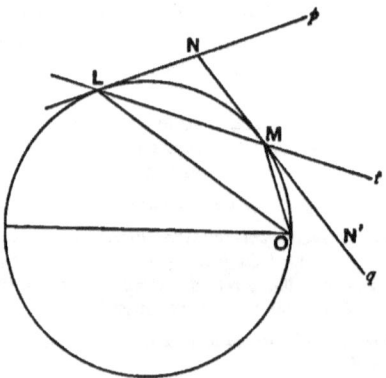

The reciprocal of the parabola for its focus $O$ is a circle through $O$.

We have to prove that
$$\angle PTO = \angle TQO \text{ and } \angle POT = \angle TOQ.$$

Now $\angle PTO$, the angle between the line $l$ and the radius $OT$, is equal to the angle between the radius $OL$ and the line $t$, i.e. equals $\angle OLM$. So $\angle TQO$ is equal to the angle between $OM$ and $q$, i.e. equals $\angle OMN'$. But $\angle OLM = \angle OMN'$. Hence $\angle PTO = \angle TQO$. As before, $\angle POT = \angle TOQ$ follows from $\angle NLM = \angle NML$.

**Ex. 1.** *Obtain a property of a circle from the theorem*—'*The orthocentre of a triangle circumscribing a parabola is on the directrix.*'

**Ex. 2.** *Reciprocate the property of a circle obtained in Ex. 1 (i) for the circle itself, (ii) for any circle.*

**Ex. 3.** *Reciprocate for O the theorem*—'*If from any point O on a circle perpendiculars be drawn to the sides of an inscribed triangle, the feet lie on a line.*'

We get—'*If O be the focus of a parabola and PQR the vertices of a circumscribed triangle, then the perpendiculars through $P$, $Q$, $R$ to $OP$, $OQ$, $OR$ meet in a point.' Calling this point $D$, we have proved that the points $A$, $B$, $C$, $O$ lie on the circle on $OD$ as diameter. Hence 'The circle about a triangle circumscribing a parabola passes through the focus.'

**Ex. 4.** *Reciprocate the same theorem for any point.*

**Ex. 5.** *Find by reciprocation the locus of the meet of tangents to a parabola which meet* (i) *at a given angle,* (ii) *at right angles.*

**16.** *Find the envelope of a chord of a circle which is bisected by a given line.*

Let the chord $p$ of the circle be bisected by the fixed line $l$ in the point $Q$. Take $O$ the centre of the circle; then $OQ$ is perpendicular to $p$. Reciprocate for the circle itself. Then $P$ is the foot of the perpendicular from $O$ on the variable line $q$ through the fixed point $L$. Hence the locus of $P$ is a circle on $OL$ as diameter, i.e. a circle through $O$ and having the opposite point at $L$. Hence the required envelope is a parabola with focus at $O$ and having its vertex at $L_1$ the foot of the perpendicular from $O$ on $l$. Hence the envelope is completely determined.

**Ex. 1.** $A, B, C, D$ *are four points on a circle, and* $AC$, $BD$ *are perpendicular; show that* $AB$, $BC$, $CD$, $DA$ *envelope one and the same conic.*

Let $AC$, $BD$ meet in $O$. Reciprocate for $O$ and we obtain the property of the director circle.

**Ex. 2.** *The envelope of the base $BC$ of a triangle $ABC$ whose vertex $A$ and vertical angle $BAC$ are given and whose base angles move on fixed lines is a conic one of whose foci is $A$.*

Reciprocate for $A$.

**Ex. 3.** *Find the envelope of the asymptotes of a system of hyperbolas having the same focus and corresponding directrix.*

**17.** $O$ *is a fixed point, and* $Q$ *is a variable point on a fixed circle.* $QR$ *is drawn such that the angle $OQR$ is constant. Find the envelope of $QR$.*

Let $QR$ be called $p$. Reciprocate for $O$. Then we have to find the locus of a point $P$ taken on a tangent $q$ to a conic one of whose foci is $O$, given that the angle between $OP$ and $q$ is constant. Draw $OY$ the perpendicular from $O$ on $q$. Then since the locus of $Y$ is a circle and since $OY:OP$ is constant and $\angle YOP$ is constant, hence the locus of $P$ is a circle. Hence the envelope of $p$ is a conic with $O$ as one focus.

**Ex.** *If the locus of Q be a line instead of a circle, find the envelope of QR.*

**18.** To investigate bifocal properties of a conic by reciprocation we reflect the figure in the centre of the conic. For example—

*In any central conic the pair of tangents from a point make equal angles with the focal radii to the point.*

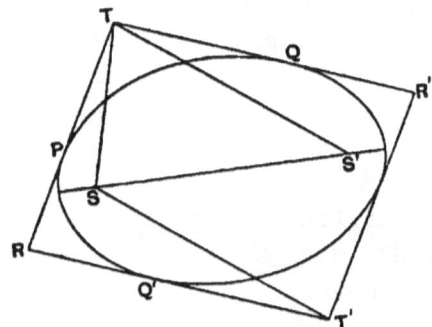

Let the tangents from $T$ to a conic touch in $P$ and $Q$. We have to prove that $PTS = QTS'$. Reflect the whole figure in the centre $C$. The tangents at $P$ and $Q$ with their reflexions form a parallelogram $RTR'T'$. Then $T'$ is the reflexion of $T$, $Q'$ of $Q$, $T'Q'$ of $TQ$, $T'S$ of $TS'$. Hence the angle $QTS'$ is equal to its reflexion, the angle $Q'T'S$. Hence we have to prove that $\angle STP$ and $\angle ST'Q'$ are equal. Reciprocating for $S$ this reduces to 'angles in the same segment of a circle are equal.'

Prove by reciprocation that—

**Ex. 1.** *The focal radii to a point on a conic make equal angles with the tangent at the point.*

**Ex. 2.** *The product of the perpendiculars from the foci of a conic on any tangent is equal to the square of the semi-axis minor.*

**Ex. 3.** *If two opposite vertices of a parallelogram circumscribed to a conic move on the directrices, the other two vertices move on the auxiliary circle.*

That is, if tangents $a$, $b$ be drawn from any point on a directrix of a conic and $a'$, $b'$ be the parallel tangents; then, $S$ being the corresponding focus, $S(a'b)$ is perpendicular to $a'$ and $S(ab')$ to $b'$. Now reciprocate for $S$.

**Ex. 4.** *The sum of the reciprocals of the perpendiculars from any point $O$ within a circle to the tangents from any point on the polar of $O$ is constant.*

**19.** *To reciprocate a system of coaxal circles into a system of confocal conics.*

If we reciprocate the system of coaxal circles for any point $O$, we get a system of conics having one focus $O$ in common. In order that the other focus may be common to all, the conics must have the same centre, i.e. the line at infinity must have the same pole for each conic. Hence in the figure of the circles, $O$ must have the same polar for each circle, i.e. $O$ must be one of the limiting points of the coaxal system. Now reciprocate the coaxal system for the limiting point $L$. Then the reciprocal conics have a focus and centre in common, and hence are confocal.

**20.** *To reciprocate a system of confocal conics into a system of coaxal circles.*

Since each conic is to be reciprocated into a circle, we must reciprocate for one of the common foci. Reciprocate for the focus $O$. Then since the conics have the same centre, the reciprocal circles have the same polar of $O$. We have to show that a system of circles each of which has the same polar of $O$ is coaxal. Drop the perpendicular $OO'$ on the polar of $O$. Bisect $OO'$ in $X$. Let $OO'$ cut one of the circles in $A$, $A'$. Then since $(OO', AA')$ is harmonic, and $X$ bisects $OO'$, hence $XA \cdot XA' = XO^2$, a constant. Hence $X$ has the same power for all the circles. And the centres all lie on the line $OO'$. Hence the circles are coaxal, $X$ being the foot of the radical axis.

Note that $O$, $O'$ are the limiting points of the coaxal system.

*The reciprocal of the other focus $S$ is the radical axis.*

For $OS = 2 \cdot OC$; hence $OS_1 = \frac{1}{2} \cdot OC_1$. But $C_1$ is the $O'$ of the above proof. Hence $S_1$ is the $X$ of the above proof.

**Ex. 1.** *The reciprocal of the minor axis is the other limiting point.*

**Ex. 2.** *$S$ and $H$ are the foci of a system of confocal conics. A parabola with $S$ as focus touches the minor axis. Show that its directrix passes through $H$; and that if $P$, $Q$ be the points of contact of a tangent to one of the confocals and the parabola, then $PSQ$ is a right angle.*

**Ex. 3.** *Prove by reciprocation that the circle of similitude of two circles is coaxal with them.*

The circle of similitude is symmetrical for the line of centres and passes through the meets on this line of the common tangents. Now reciprocate for a limiting point $O$. The circles become an ellipse and hyperbola with the same foci $O$ and $S$, which have a pair of common chords $l$ and $l'$ perpendicular to $OS$. We have to show that a conic which is symmetrical for $OS$, which has $O$ as a focus and which touches $l$ and $l'$, has $S$ as its other focus. This is obvious.

**Ex. 4.** *Deduce properties of coaxal circles from—(i) 'Confocal conics meet at right angles,' (ii) 'Tangents from any point to two confocals are equally inclined to each other.'*

**Ex. 5.** *Deduce a property of confocal conics from—'The polars of a fixed point for a system of coaxal circles meet at another fixed point; and the two points subtend a right angle at either limiting point.'*

**Ex. 6.** *If the sides of a polygon touch a conic, and all but one of the vertices lie on confocal conics, the last vertex also lies on a confocal conic.*

Reciprocate Poncelet's theorem respecting coaxal circles.

## Reciprocation for any conic.

**21.** Having discussed the particular case of two reciprocal conics, one of which is a circle, we return to the general case of the reciprocal of a conic, taking any base conic.

*The reciprocal of a conic, taking a conic with centre $O$ as base conic, is a hyperbola, parabola, or ellipse, according as $O$ is outside, on or inside the given conic.*

Let $a$ be the given conic and $\Gamma$ the base conic, and $a'$ the reciprocal conic. Then $a'$ is a hyperbola, parabola, or ellipse, according as the line at infinity cuts $a'$ in real, coincident or imaginary points. Now the reciprocal of the line at infinity is the pole of the line at infinity for $\Gamma$, i.e. is $O$. Hence the reciprocals of the points in which $a'$ meets the line at infinity are the tangents to $a$ from $O$. And the tangents from $O$ are real if $O$ be outside, coincident if $O$ be on, and imaginary if $O$ be inside $a$.

The reciprocal of *the centre* of the given conic, i.e. of the pole of the line at infinity for $a$, is the polar of $O$ for $a'$. The reciprocal of the *asymptotes* of the given conic, i.e. of the tangents to $a$ from the pole of the line at infinity for $a$, are the points of meet with $a'$ of the polar of $O$ for $a'$, i. e. are the points of contact of tangents from $O$ to $a'$.

**Ex. 1.** *The axes of the reciprocal of a conic for a point O are parallel to the bisectors of the angles between the tangents from O to the conic.*

**Ex. 2.** *If $2\theta$ be the angle between these tangents, show that $\operatorname{cosec} \theta$ is the eccentricity of the reciprocal conic, and deduce the formula $e = OU \div UT$ of § 10.*

**Ex. 3.** *The reciprocal of a parabola for any point on the directrix is a rectangular hyperbola.*

For since the points of contact of tangents from $O$ to $a$ subtend a right angle at $O$, hence the asymptotes of $a'$ are perpendicular.

**Ex. 4.** *From 'The orthocentre of a triangle circumscribed to a parabola lies on the directrix,' deduce by reciprocation · The orthocentre of a triangle inscribed in a rectangular hyperbola is on the curve.'*

Reciprocate for the orthocentre.

**Ex. 5.** *The reciprocal of a rectangular hyperbola for any point O is a conic whose director passes through O.*

**Ex. 6.** *Reciprocate for any point—'A diameter of a rectangular hyperbola and the tangent at either end are equally inclined to either asymptote.'*

Let $CP = r$ be the diameter, $q$ the tangent at $P$, and $y$ the asymptote. Then we have to reciprocate that $\angle ry = \angle qy$. We get—'If $c$ be the polar of any point $O$ on the director of a conic, and if from the point $R$ on $c$ a tangent be drawn touching in $Q$; then $Y$ being either of the points in which $c$ cuts the conic, $RY$ and $QY$ subtend equal angles at $O$.'

**Ex. 7.** *Reciprocate for any point O—a focus of a conic.*

A line such that every pair of conjugate points upon it subtend a right angle at a given point $O$. Hence given a conic and a point $O$, there are four such lines.

**Ex. 8.** *Reciprocate for any point—a directrix of a conic.*

The pole of such a line.

**Ex. 9.** *If the chord PQ of a conic subtend a right angle at a fixed point O on the conic, then PQ passes through a fixed point* (called the Frégier point of $O$ for the conic).

Reciprocate for the fixed point; and we have to prove that the locus of the meet of perpendicular tangents of a parabola is a line (the directrix).

**Ex. 10.** *Obtain by reciprocating Ex. 9 a property of a circle.*

**Ex. 11.** *The reciprocal for O of the focus of a parabola is the polar of the Frégier point of O for the reciprocal conic.*

**Ex. 12.** *O, D, E are fixed points on a conic, and P a variable point. PD, PE meet the polar of the point in which chords which subtend a right angle at O meet, in B and C; show that $\angle BOC = \angle DOE$.*

**Ex. 13.** *The envelope of a chord of a conic which subtends a right angle at a fixed point O, not on the conic, is a conic having a focus at O.*

**Ex. 14.** *A system of four-point conics or four-tangent conics can be reciprocated into concentric conics.*

Take as origin one of the vertices of the common self-conjugate triangle.

**Ex. 15.** *The reciprocal of a central conic, taking a concentric circle as base conic, is a similar conic.*
For $OA.OA_1 = OB.OB_1 = k^2$; hence $OA_1 : OB_1 :: OB : OA$.

**Ex. 16.** *Reciprocate for any point—a system of coaxal circles.*
That is, a system of circles passing through the same two points, real or imaginary.

**Ex. 17.** *Reciprocate for any point O—'The directors of a system of conics touching the same four lines are coaxal.'*

**Ex. 18.** *Also—'The locus of the centres of a system of rectangular hyperbolas passing through the same three points is a circle.'*

**22.** *Reciprocate Carnot's theorem, taking any circle as base conic.*

Let $O$ be the origin of reciprocation. Then, as in VI. 1, Carnot's theorem gives

$$\sin AOC_1 . \sin AOC_2 \ldots = \sin AOB_1 . \sin AOB_2 \ldots$$

Now $\angle AOC_1 = \angle ac_1$, and so on. Hence the reciprocal theorem is—'The sides $a, b, c$ of a triangle meet in the points $P, Q, R$; and from $P, Q, R$ are drawn the pairs of tangents $a_1 a_2, b_1 b_2, c_1 c_2$ to any conic; then

$$\sin ac_1 . \sin ac_2 . \sin ba_1 . \sin ba_2 . \sin cb_1 . \sin cb_2$$
$$= \sin ab_1 . \sin ab_2 . \sin bc_1 . \sin bc_2 . \sin ca_1 . \sin ca_2,$$

where $ac_1$ denotes the angle between the lines $a$ and $c_1$, and so on. And conversely if this relation hold, then the six lines $a_1 a_2 \, b_1 b_2 \, c_1 c_2$ touch the same conic.'

**Ex. 1.** *If the sides of a triangle ABC meet a conic in $A_1 A_2, B_1 B_2, C_1 C_2$, then the six lines $AA_1, AA_2, BB_1, BB_2, CC_1, CC_2$ touch a conic; and conversely, if the latter touch a conic, the former are on a conic.*

**Ex. 2.** *Reciprocate the extension of Carnot's theorem given in Ex. 1 of VI. 1.*

**Ex. 3.** *Reciprocate the theorem—'The lines joining the vertices of a triangle to any two points meet the opposite sides in six points which lie on a conic.'*

NOTE.

**23.** The following theory would have been preferable in some ways to that employed in the text.

Prove by § 3 or XIII. 2 that the reciprocal of a conic for a point (i.e. for a circle with centre at this point) is a conic.

*The reciprocal of a circle for any point O is a conic one of whose foci is O.*

For in the circle, every pair of conjugate points on the line at infinity subtends a right angle at the centre of the circle and therefore at $O$. Hence in the reciprocal conic every pair of conjugate lines at $O$ is orthogonal, i.e. $O$ is a focus of the reciprocal conic.

Also since the centre of the circle is the pole of the line at infinity, the reciprocal of the centre of the circle is the polar of the origin, i.e. is the corresponding directrix.

*The reciprocal of a conic for one of its foci is a circle.*

Every pair of conjugate points on the corresponding directrix subtends a right angle at the focus. Hence in the reciprocal conic, every pair of conjugate lines at the pole of the line at infinity, i.e. at the centre, is orthogonal. Hence every pair of conjugate diameters of the reciprocal conic is orthogonal; hence the reciprocal conic is a circle.

*In any conic $SP:PM$ is constant.*

For as in § 10, $OP:PM::OU:UT$. Hence $OP:PM$ is constant. Hence the eccentricity of the reciprocal conic is $\delta \div R$, for

$$e = SP:PM.$$

Notice that we have here given by Reciprocation an independent proof of the $SP:PM$ property of a conic.

# CHAPTER IX.

## ANHARMONIC OR CROSS RATIO.

**1.** ONE of the *anharmonic or cross ratios* of the four collinear points $A, B, C, D$ is $\dfrac{AB}{BC} \div \dfrac{AD}{DC}$. This is denoted by $(AC, BD)$. So every other order of writing the letters gives us a cross ratio of the points, e.g. another cross ratio is

$$(BA, CD) = \frac{BC}{CA} \div \frac{BD}{DA}.$$

**Ex. 1.** *If* $(AB, CD) = (AB, C'D')$, *then* $(AB, CC') = (AB, DD')$.

**Ex. 2.** *If* $(AC, A'B) = (A'C', AB')$, *then* $(AC, C'B) = (A'C', CB')$.

**Ex. 3.** *If* $(AB, CD) = (A'B', C'D')$, *and* $(AB, CE) = (A'B', C'E')$, *show that* $(AB, DE) = (A'B', D'E')$.

**Ex. 4.** *If $OA$, $OB$, $OC$ cut $BC$, $CA$, $AB$ in $P$, $Q$, $R$, and if any line cut $BC$, $CA$, $AB$ in $P'$, $Q'$, $R'$, then*
$$(BC, PP') \times (CA, QQ') \times (AB, RR') = -1;$$
*and conversely, if this relation hold, and if $PA, QB, RC$ be concurrent, then $P', Q', R'$ are collinear, and if $P', Q', R'$ be collinear, then $PA, QB, RC$ are concurrent.*

**Ex. 5.** *If $OA$, $OB$, $OC$ cut the sides of the triangle $ABC$ in $P$, $Q$, $R$, and $O'A$, $O'B$, $O'C$ cut the sides in $P'$, $Q'$, $R'$, or if two transversals cut the sides in $P, Q, R$ and $P', Q', R'$, then*
$$(BC, PP') \times (CA, QQ') \times (AB, RR') = 1;$$
*and conversely, if this relation hold, and if $PA, QB, RC$ be concurrent, then $P'A, Q'B, R'C$ are concurrent, and if $P, Q, R$ be collinear, then $P', Q', R'$ are collinear.*

**Ex. 6.** *A cross ratio is not altered by inversion for a point on the line.*
For given $OA.OA' = OB.OB' = \ldots = k^2$,
we have $AB = OB - OA = k^2/OB' - k^2/OA'$
$= -k^2 . A'B'/OA' . OB'$.

**Ex. 7.** *The tangent at $O$ to a conic meets the sides of a circumscribed triangle in $A, B, C$ and the sides of the triangle formed by the points of contact in $A', B', C'$; show that $(OA, BC) = (OA', B'C')$.*

## Anharmonic or Cross Ratio.

Since $(OA', BC)$ is harmonic, we have $a' = 2bc \div (b+c)$. So for $b'$ and $c'$. Now substitute in $(OA', B'C')$, viz. in

$$\frac{OB'}{B'A'} \times \frac{C'A'}{OC'} = \frac{b'}{a'-b'} \times \frac{a'-c'}{c'},$$

and we get $(OA, BC)$.

**2.** *A cross ratio is equal to any other, in which any two points being interchanged, the other two are also interchanged.*

Let $(AC, BD)$ be the cross ratio. We may interchange $A$ with $B$, $C$ or $D$. Hence we have to prove that

$$(AC, BD) = (BD, AC) = (CA, DB) = (DB, CA),$$

or that

$$\frac{AB}{BC} \cdot \frac{DC}{AD} = \frac{BA}{AD} \cdot \frac{CD}{BC} = \frac{CD}{DA} \cdot \frac{BA}{CB} = \frac{DC}{CB} \cdot \frac{AB}{DA}.$$

**3.** *There are 24 cross ratios of four points; and these can be divided into 3 groups of 8, such that every cross ratio in a group is equal to or the reciprocal of every other in the group.*

Let the points be $ABCD$. Take the three cross ratios $(AB, CD)$, $(AC, DB)$ and $(AD, BC)$. Now

$$(AB, CD) = (BA, DC) = (CD, AB) = (DC, BA)$$

by IX. 2. Also it is easy to prove that $(AB, CD)$ is the reciprocal of $(AB, DC)$, $(BA, CD)$, $(CD, BA)$, $(DC, AB)$. Hence we get a group of 8 connected with $(AB, CD)$. Similarly there is a group of 8 connected with $(AC, DB)$ and with $(AD, BC)$. And no ratio can belong to two groups; for in the first group $AB$ are together and $CD$, so in the second group $AC$ are together and $DB$, and in the third group $AD$ and $BC$.

**4.** *If* $\lambda = (AB, CD)$, $\mu = (AC, DB)$, $\nu = (AD, BC)$,

*then*
$$\lambda + \frac{1}{\mu} = \mu + \frac{1}{\nu} = \nu + \frac{1}{\lambda} = -\lambda\mu\nu = 1.$$

For $\lambda + \dfrac{1}{\mu} - 1 = \dfrac{AC}{CB} \cdot \dfrac{DB}{AD} + \dfrac{DC}{AD} \cdot \dfrac{AB}{BC} - 1$

$$= \frac{AC \cdot DB - DC \cdot AB - CB \cdot AD}{CB \cdot AD}$$

$$= \frac{(c-a)(b-d) - (c-d)(b-a) - (b-c)(d-a)}{CB \cdot AD}$$

$$= \frac{cb - cd - ab + ad - cb + ca + db - da - bd + ba + cd - ac}{CB \cdot AD}$$
$$= 0.$$

Also $\lambda \cdot \mu \cdot \nu = \dfrac{AC}{CB} \cdot \dfrac{DB}{AD} \cdot \dfrac{AD}{DC} \cdot \dfrac{BC}{AB} \cdot \dfrac{AB}{BD} \cdot \dfrac{CD}{AC} = -1.$

We have now shown that the three fundamental cross ratios $\lambda$, $\mu$, $\nu$ are connected by the above four relations. Two of these are independent and give $\mu$, $\nu$ in terms of $\lambda$. The other two can be derived from these. Hence given any one cross ratio of four points, the other 23 can be calculated.

**Ex. 1.** *Given* $\lambda + \dfrac{1}{\mu} = 1$, $\mu + \dfrac{1}{\nu} = 1$, *show that*

$$\nu + \dfrac{1}{\lambda} = 1 \quad \text{and} \quad \lambda\mu\nu = -1.$$

**Ex. 2.** *Given* $\lambda + \dfrac{1}{\mu} = 1$, $\lambda\mu\nu = -1$, *show that*

$$\nu + \dfrac{1}{\lambda} = 1 \quad \text{and} \quad \mu + \dfrac{1}{\nu} = 1.$$

**Ex. 3.** *If $(AB, CD) = 1$, show that either $A$ and $B$ coincide, or $C$ and $D$; and conversely, if $A$ and $B$ coincide, or $C$ and $D$, then $(AB, CD) = 1$.*

**Ex. 4.** *If two points of a range of four points coincide, each of the cross ratios is equal to 0, 1, or $\infty$; and no cross ratio can equal 0 or 1 unless two points coincide.*

**Ex. 5.** *Show that no real range can be found of which all the cross ratios are equal.*

**Ex. 6.** *Of the three $\lambda$, $\mu$, $\nu$, two are positive and one negative.*

**Ex. 7.** *If any cross ratio of the range $ABCD$ is equal to the corresponding cross ratio of the range $A'B'C'D'$, then every two corresponding cross ratios of the ranges are equal.*

For if $\lambda = \lambda'$, then $\mu = \mu'$ and $\nu = \nu'$.

Two such ranges are said to be *homographic*, and we denote the fact by the equation $(ABCD) = (A'B'C'D')$.

**Ex. 8.** *If $(ABB'C) = (A'B'BC')$ and $(ABB'D) = (A'B'BD')$, show that*
$$(BB'CD) = (B'BC'D').$$

Divide $(BB', AC) = (B'B, A'C')$ by $(BB', AD) = (B'B, A'D')$.

**5.** *If $(AC, BD)$ be harmonic, then $(AC, BD) = -1$.*

For $\dfrac{AB}{BC} = -\dfrac{AD}{DC}$, hence $\dfrac{AB}{BC} \div \dfrac{AD}{DC} = -1.$

*If $(AC, BD)$ be harmonic, then $(AC, BD) = (AC, DB)$; and conversely, if $(AC, BD) = (AC, DB)$, then either $(AC, BD)$ is harmonic or two points coincide.*

# Anharmonic or Cross Ratio.

For if $(AC, BD) = (AC, DB)$,

then $\dfrac{AB}{BC} \cdot \dfrac{DC}{AD} = \dfrac{AD}{DC} \cdot \dfrac{BC}{AB}$, hence $\left(\dfrac{AB}{BC} \cdot \dfrac{DC}{AD}\right)^2 = 1$,

hence $\dfrac{AB}{BC} \cdot \dfrac{DC}{AD} = \pm 1$, i.e. $(AC, BD) = \pm 1$.

If $(AC, BD) = +1$, then $A$ and $C$, or $B$ and $D$ coincide; and if $(AC, BD) = -1$, then $(AC, BD)$ is harmonic.

**Ex.** *If a range of four points be harmonic, each of its 24 cross ratios is equal to $-1$, $\frac{1}{2}$, or 2; and if any one of the cross ratios of four points be equal to $-1$ or $\frac{1}{2}$ or 2, then the four points form a harmonic range.*

**6.** *If $A, B, C, D, D'$ be collinear points, such that*
$$(AC, BD) = (AC, BD'),$$
*then $D$ and $D'$ coincide.*

For $\dfrac{AB}{BC} \cdot \dfrac{DC}{AD} = \dfrac{AB}{BC} \cdot \dfrac{D'C}{AD'}$, hence $\dfrac{DC}{AD} = \dfrac{D'C}{AD'}$;

i.e. $AC$ is divided in the same ratio at $D$ and $D'$; hence $D$ and $D'$ coincide.

**7.** *If four lines $a, b, c, d$ passing through the same point $V$ be cut by two transversals in $ABCD$ and $A'B'C'D'$, then*
$$(ABCD) = (A'B'C'D').$$

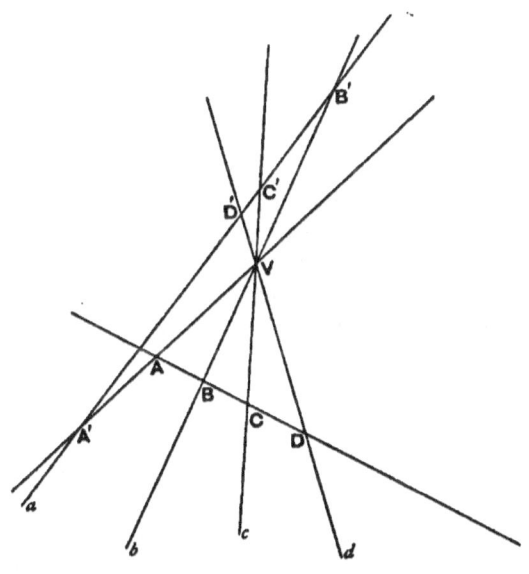

It is sufficient to prove that
$$(AC, BD) = (A'C', B'D').$$
Now $(AC, BD) = \dfrac{AB}{BC} \cdot \dfrac{DC}{AD} = \dfrac{\triangle AVB}{\triangle BVC} \cdot \dfrac{\triangle DVC}{\triangle AVD}$

$$= \dfrac{VA \cdot VB \cdot \sin AVB}{VB \cdot VC \cdot \sin BVC} \cdot \dfrac{VD \cdot VC \cdot \sin DVC}{VA \cdot VD \cdot \sin AVD}$$

$$= \dfrac{\sin AVB}{\sin BVC} \cdot \dfrac{\sin DVC}{\sin AVD}.$$

Similarly,
$$(A'C', B'D') = \dfrac{\sin A'VB'}{\sin B'VC'} \cdot \dfrac{\sin D'VC'}{\sin A'VD'}.$$

Now $AVB$ is equal to either $A'VB'$ or its supplement. In either case, $\sin AVB = \sin A'VB'$. And so on. Hence
$$(AC, BD) = (A'C', B'D'), \text{ i.e. } (ABCD) = (A'B'C'D').$$

We may enunciate the above theorem in the form—*Every transversal cuts a pencil of four lines in the same cross ratio.*

The cross ratio $(AC, BD)$ of the pencil is written
$$V(AC, BD) \text{ or } (ac, bd).$$

Also, by the above,
$$(ac, bd) = \dfrac{\sin ab}{\sin bc} \div \dfrac{\sin ad}{\sin dc}.$$

**Ex. 1.** *Show that the fundamental cross ratios* $\lambda, \mu, \nu$ *of the range* $(ABCD)$ *are equal to* $\operatorname{cosec}^2 \phi$, $-\tan^2 \phi$ *and* $\cos^2 \phi$, *where* $2\phi$ *is the angle at which the circles on* $AC$ *and* $BD$ *as diameters intersect.*

For $\lambda = \dfrac{AC}{CB} \cdot \dfrac{DB}{AD} = \dfrac{\sin APC}{\sin CPB} \cdot \dfrac{\sin DPB}{\sin APD}$,

and $\angle APC = \dfrac{\pi}{2}$, $\angle DPB = -\dfrac{\pi}{2}$, $\angle CPB = -\phi$, $\angle APD = \pi - \phi$.

**Ex. 2.** *Express* $(ac, bd)$ *as a ratio of two segments of a line.*

Draw a transversal parallel to $d$. Then $(ac, bd) = AB : CB$, for $AD = CD$, $D$ being at infinity.

**Ex. 3.** *Given the three points* $A, B, C$; *find* $D$ *so that* $(AB, CD)$ *may have a given value* $\lambda$.

Take any line $AB'$, and divide it in $C'$ so that $-AC' \div C'B' = \lambda$. Let $BB'$, $CC'$ meet in $V$. Through $V$ draw $VD$ parallel to $AB'$. Then $(AB, CD) = (AB', C'\Omega')$, [where $\Omega'$ is the point at infinity upon $AB'$,] $= -AC' \div C'B' = \lambda$.

**Ex. 4.** *Through a given point* $O$ *draw a transversal to cut the sides of a given triangle* $ABC$ *in points* $A', B', C'$, *such that* $(OA', B'C')$ *may have a given value.*

IX.] *Anharmonic or Cross Ratio.* 103

Let $OA$ cut $BC$ in $O'$. Then $(OA', B'C') = A(OA', B'C') = (O'A', CB)$. Hence $A'$ is known. .

**Ex. 5.** *If $AA'$, $BB'$, $CC'$ meet in a point $O$ and if $(AC, BD) = (A'C', B'D')$, then $DD'$ passes through $O$.*

**8.** *A cross ratio of a range of four points is unaltered by projection.*

Let the range $ABCD$ be joined to the vertex $V$, and let the joining plane cut the plane of projection in $A'B'C'D'$. Then since $A'B'C'D'$ is a section of the pencil $V(ABCD)$, it follows that $(ABCD) = A'B'C'D')$.

**Ex.** *If the points $a$, $b$, $c$,... be taken on the sides $AB$, $BC$, $CD$,... of a polygon; show that the continued product of such ratios as $Aa/aB$ is unaltered by projection.*

Let any transversal cut the sides $AB$, $BC$, $CD$, ... in $a$, $\beta$, $\gamma$, ...; then the continued product of $A\alpha/\alpha B$ is numerically unity. Hence, dividing, we have to prove that the continued product of $Aa/aB \div A\alpha/\alpha B$ is unaltered by projection, i.e. the continued product of certain cross ratios.

**9.** *A cross ratio of a pencil of four lines is unaltered by projection.*

Join the pencil $O(ABCD)$ to the vertex $V$, and let the joining planes cut the plane of projection in the pencil $O'(A'B'C'D')$. Through $V$ draw any plane cutting the pencils in $abcd$ and $a'b'c'd'$. Then

$$O(ABCD) = (abcd) = V(abcd) = V(a'b'c'd')$$
$$= (a'b'c'd') = O'(A'B'C'D').$$

Hence the pencils $O(ABCD)$ and $O'(A'B'C'D')$ have the same cross ratios.

**Ex. 1.** *If through the vertices $A$, $B$, $C$,... of a polygon there be drawn any lines $Aa$, $Bb$, $Cc$, ..., then the continued product of the ratios $\sin ABb/\sin bBC$ is unaltered by projection.*

Take any point $O$ and consider the cross ratio
$$\sin ABb/\sin bBC \div \sin ABO/\sin OBC.$$

**Ex. 2.** *The figure $ABCD$ consisting of four points joined by four lines can be projected into any figure $A'B'C'D'$ of the same kind.*

Let $AC$, $BD$ meet in $U$, and $A'C'$, $B'D'$ meet in $U'$. Take $X$ on $AC$ so that $(XAUC) = (\Omega'A'U'C')$, and $Y$ on $BD$ so that $(YBUD) = (\Omega B'U'D')$, where $\Omega$ and $\Omega'$ are at infinity. Now project $XY$ to infinity, and the angles $AUB$, $BAU$ into angles of magnitude $A'U'B'$, $B'A'U'$. Let the projections of $ABCDUXY$ be $a'b'c'd'u'\omega'\omega$, where $\omega$ and $\omega'$ are at infinity. Then $(\omega'a'u'c') = (XAUC) = (\Omega'A'U'C')$. Hence $a'u':u'c'::A'U':U'C$; so $b'u':u'd'::B'U':U'D'$; also $\angle a'u'b' = \angle A'U'B'$ and $\angle b'a'u' = \angle B'A'U'$.

Hence the figures $a'b'c'd'u'$ and $A'B'C'D'U'$ are similar. If they are not equal, we proceed as in IV. 7.

Note that this construction fails if $XY$ as constructed be at infinity; in other cases, by IV. 6, the construction is real.

*Cross ratio of four planes meeting in a line.*

**10.** *Any transversal cuts four planes which pass through the same line in four points whose cross ratio is constant.*

Let two transversals cut the planes in $ABCD$ and $A'B'C'D'$. Join $ABCD$ to any point $O$ on the meet of the planes, and $A'B'C'D'$ to any other point $O'$ on this meet. Then the meet of the planes $OABCD$ and $O'A'B'C'D'$ is a line which cuts the four given planes in the points $a$, $\beta$, $\gamma$, $\delta$, say.

Then $(ABCD) = O(ABCD) = O(a\beta\gamma\delta) = (a\beta\gamma\delta)$
$= O'(a\beta\gamma\delta) = O'(A'B'C'D') = (A'B'C'D')$.

Hence $(ABCD)$ is constant.

**Ex.** *Any plane cuts four planes which meet in a line in four lines whose cross ratio is constant.*

*Homographic ranges and pencils.*

**11.** Two *ranges* of points $ABCD\ldots$ and $A'B'C'D'\ldots$ on the same or different lines, in which to each point ($A$ say) of one range corresponds a point ($A'$) of the other, are said to be *homographic* if the range formed by every four points ($ABCD$) of one range is homographic with the range formed by the corresponding four points ($A'B'C'D'$) of the other. (See Ex. 7 of § 4.)

Two *pencils* of rays at the same or different vertices are said to be *homographic* when any two sections of them are homographic.

It is convenient to use the notation

$$(ABCD\ldots) = (A'B'C'D'\ldots),$$

to denote that the ranges $(ABCD\ldots)$ and $(A'B'C'D'\ldots)$ are homographic; and the notation $V(ABCD\ldots) = V'(A'B'C'D'\ldots)$ to denote that the pencils $V(ABCD\ldots)$ and $V'(A'B'C'D'\ldots)$ are homographic.

A *range* is said to be *homographic* with a *pencil* when the range is homographic with a section of the pencil. This is denoted by $(ABCD\ldots) = V'(A'B'C'D'\ldots)$.

**Ex. 1.** *Two ranges (or pencils) which are homographic with the same range (or pencil) are homographic.*

**Ex. 2.** *If $UX : V'X'$ be given, $U$ being a fixed point and $X$ a variable point on one line, and $V'$, $X'$ on another line; then $X$ and $X'$ generate homographic ranges on these lines.*

Let $A, B, C, D$ be four positions of $X$, and $A', B', C', D'$ the corresponding four positions of $X'$.
Then $AC = UC - UA = \lambda (V'C' - V'A') = \lambda . A'C'$.
Hence $(AB, CD) = (A'B', C'D')$.

**Ex. 3.** *The same is true if $UX . V'X'$ be given.*
For $AC = UC - UA = \lambda/V'C' - \lambda/V'A'$
$= -\lambda . A'C' \div V'C' . V'A'$.

**Ex. 4.** *A variable circle passes through a fixed point and cuts a given line at a given angle; show that it determines on the line two homographic ranges.*

For the pencils at the point are superposable.

**12.** *To form two homographic ranges on different lines.*

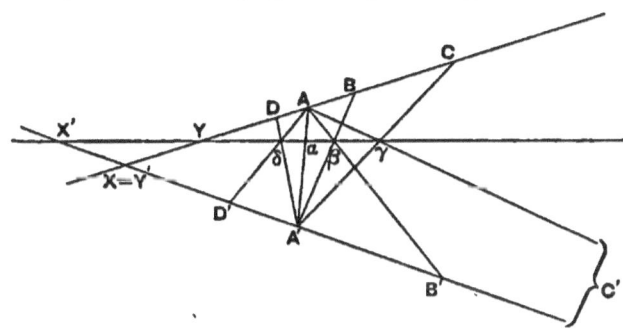

Take any range $ABCDE \ldots$ on one of the lines, and take any three points $A', B', C'$ arbitrarily on the other line to correspond to $ABC$.

Let $AB'$ and $A'B$ meet in $\beta$, $AC'$ and $A'C$ in $\gamma$; let $\beta\gamma$ meet $AA'$ in $\alpha$ and $A'D$ in $\delta$; let $A\delta$ meet $A'B'$ in $D'$. Similarly construct the points $\epsilon$, $E'$, &c. Then the range $A'B'C'D'E' \ldots$ is homographic with the range $ABCDE \ldots$

For take any four points of the first range, viz. $LMNR$, and the corresponding four points of the other, viz. $L'M'N'R'$. Then

$(L'M'N'R') = A (L'M'N'R') = (\lambda\mu\nu\rho) = A'(\lambda\mu\nu\rho) = (LMNR)$.

Hence every range of four points of one range is homographic with the range of the corresponding four points of the other range, i.e. the ranges are homographic.

**13.** *To form two homographic ranges on the same line.*

Take the range $ABCDE\ldots$ on one line. Take any section $A''B''C''D''E''\ldots$ of the pencil joining any point $V$ to $ABCDE\ldots$. Then with any three points $A'B'C'$ on the given line to correspond to $A''B''C''$, construct a range $A'B'C'D'E'\ldots$ homographic with $A''B''C''D''E''\ldots$. Then

$$(A'B'C'D'E'\ldots) = (A''B''C''D''E''\ldots),\text{ by construction}$$

$$= (ABCDE\ldots)\text{ by projection.}$$

Hence the range $A'B'C'D'E'\ldots$ is homographic with the range $ABCDE\ldots$ on the same line. Also the three points $A'B'C'$ which correspond to $ABC$ are taken arbitrarily.

*To form two homographic pencils at the same or different vertices.*

Join the vertices to any two homographic ranges.

Notice that in this case also, if one pencil be given, the rays in the other pencil corresponding to three rays in the given pencil may be taken arbitrarily.

**14.** *Two ranges $ABC\ldots$ and $A'B'C'\ldots$ on different lines are said to be in perspective* when the lines $AA'$, $BB'$, $CC'$,... joining corresponding points meet in a point (called the *centre of perspective*).

*Two pencils $V(ABC\ldots)$ and $V'(A'B'C'\ldots)$ at different vertices are said to be in perspective* when the meets of corresponding rays lie on a line (called the *axis of perspective*.)

*Two ranges in perspective are homographic.*

For let the centre of perspective be $O$. Then

$$(LMNR) = O(LMNR) = O(L'M'N'R') = (L'M'N'R').$$

*Two pencils in perspective are homographic.*

For let $VA$, $V'A'$ meet in $a$, and so on. Then

$$V(LMNR) = (\lambda\mu\nu\rho) = V'(L'M'N'R').$$

**15.** *If two homographic ranges on different lines have the meet of the lines as a point corresponding to itself in the two ranges, then the ranges are in perspective.*

Let the ranges be $(ABCD\ldots) = (AB'C'D'\ldots)$.

IX.]   *Anharmonic or Cross Ratio.*   107

Let $BB'$, $CC'$ meet in $O$, and let $OD$ meet $AB'$ in $D''$. Then $(AB'C'D') = (ABCD)$ by hypothesis $= (AB'C'D'')$ by

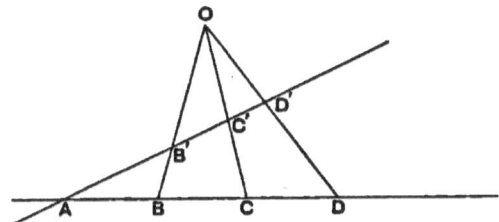

projection. Hence $(AB'C'D') = (AB'C'D'')$, i.e. $D'$ and $D''$ coincide, i.e. the join $DD'$ of any pair of corresponding points passes through $O$.

**Ex. 1.** *If $A$ be the meet of two corresponding rays of two homographic pencils, then any two transversals through $A$ will cut the pencils in ranges in perspective.*

**Ex. 2.** *If a cross ratio of the range $ABCD$ be equal to the corresponding cross ratio of the range $A'B'C'D'$, show that every two corresponding cross ratios are equal.* (See also § 4, Ex. 7.)

Place the two ranges so that $A$ and $A'$ coincide and that the lines $AB$ and $A'B'$ do not coincide. Then, as above, the ranges are in perspective; and hence every cross ratio is equal to the corresponding cross ratio.

**Ex. 3.** *If $(ABCD) = (A'B'C'D')$ and $(ABCE) = (A'B'C'E')$ and so on, then $(ABCDE...)$ and $(A'B'C'D'E'...)$ are homographic ranges.*

**Ex. 4.** *If $(UV, AA') = (UV, BB') = (UV, CC') = \cdots$, show that $(ABC...) = (A'B'C'...)$.*

For $(UV, AB) = (UV, A'B')$.

**Ex. 5.** *If $P$ be a variable point on the line joining two fixed points $A$, $B$, and $P'$ a variable point on the line joining the fixed points $A'$, $B'$, such that*
$$AP/BP \div A'P'/B'P'$$
*is constant, then $P$ and $P'$ generate homographic ranges.*

For if $C$ be a position of $P$ and $C'$ of $P'$, we have
$$AC/BC \div AP/BP = A'C'/B'C' \div A'P'/B'P',$$
i.e. $(ABCP) = (A'B'C'P')$.

**Ex. 6.** *If $VA$, $VB$, $VP$ and $V'A'$, $V'B'$, $V'P'$ be such that*
$$\sin AVP/\sin BVP \div \sin A'V'P'/\sin B'V'P'$$
*is constant, then $VP$ and $V'P'$ generate homographic pencils.*

**Ex. 7.** *Also if $\tan AVP/\tan A'V'P'$ be constant.*
Take $AVB$ and $A'V'B'$ right angles.

**Ex. 8.** *If $AP.B'P' \div BP$ be constant, then $P$ and $P'$ generate homographic ranges.*

For $AP.B'P' \div BP.\Omega'P' = AC.B'C' \div BC.\Omega'C'$,
hence $(AB, CP) = (\Omega'B', C'P')$.

**Ex. 9.** *If the triangle ABC be circumscribed to the triangle LMN; show that an infinite number of triangles can be drawn which are inscribed in the triangle LMN and at the same time circumscribed to the triangle ABC.*

Take any point $R$ on $LM$; let $AR$ cut $NL$ in $Q$, and let $BR$ cut $NM$ in $P$. It will be sufficient to prove that $PQ$ passes through $C$. Let $BC$ cut $NM$ in $X$, let $CA$ cut $LN$ in $Y$, and let $AB$ cut $ML$ in $Z$.

Then $(NMPX) = B(NMPX) = (ZMRL) = A(ZMRL) = (NYQL)$.

Hence the ranges $(NMPX)$ and $(NYQL)$ are in perspective. Hence $MY$, $PQ$, $XL$ meet in a point, i.e. $PQ$ passes through $C$. Hence $PQR$ is inscribed in $LMN$ and circumscribed to $ABC$.

**Ex. 10.** *Six points $A$, $B$, $C$, $D$, $E$, $F$ are taken, such that $AB$, $FC$, $ED$ meet in a point $G$, and also $FA$, $EB$, $DC$ in $H$; show that $BC$, $AD$, $FE$ also meet in a point.*

Let $BE$ and $CF$ meet in $P$, $CF$ and $AD$ in $R$, and $AD$ and $BE$ in $Q$. Then $(BPQE) = G(BPQE) = (ARQD) = H(ARQD) = (FRPC) = (CPRF)$.

**Ex. 11.** *$AO$ meets $BC$ in $D$, $BO$ meets $AC$ in $E$, $CO$ meets $AB$ in $F$. $X$, $Y$, $Z$ are taken such that $(AD, OX) = (BE, OY) = (CF, OZ) = -1$; show that the triangle $XYZ$ circumscribes the triangle $ABC$.*

For $(AD, OX) = (EB, OY)$.

**Ex. 12.** *The points $A$ and $B$ move on fixed lines through $O$, and $U$ and $V$ are fixed points collinear with $O$; if $UA$ and $VB$ meet on a fixed line, show that $AB$ passes through a fixed point.*

Take several positions of the point $A$, viz. $A_1 A_2 A_3 \ldots$. Join $A_1 U$ cutting the given line in $C_1$, and join $C_1 V$ cutting $OB$ in $B_1$. Similarly construct $C_2 C_3 \ldots$ and $B_2 B_3 \ldots$. Then
$(A_1 A_2 A_3 \ldots) = U(A_1 A_2 A_3 \ldots) = (C_1 C_2 C_3 \ldots) = V(C_1 C_2 C_3 \ldots) = (B_1 B_2 B_3 \ldots)$.
Hence the ranges $(A_1 A_2 A_3 \ldots)$ and $(B_1 B_2 B_3 \ldots)$ are homographic. Also when $A$ is at $O$, $B$ is also at $O$. Hence the ranges are in perspective. Hence $A_1 B_1$, $A_2 B_2$, $A_3 B_3$, $\ldots$ meet in a point, i.e. $AB$ passes through a fixed point.

**Ex. 13.** *If the points $A, B, C$ move on fixed lines through $O$, and $AB$ turn about a fixed point $P$, and $BC$ turn about a fixed point $Q$, show that $CA$ turns about a fixed point.*

**Ex. 14.** *If the vertices of a polygon move on fixed concurrent lines, and all but one of the sides pass through fixed points, this side and every diagonal will pass through a fixed point.*

**16.** *If two homographic pencils at different vertices have the ray joining the vertices as a ray corresponding to itself in the two pencils, then the pencils are in perspective.*

Let the two homographic pencils be $V(V'ABC\ldots)$ and $V'(VA'B'C'\ldots)$. Let $VA$ cut $V'A'$ in $a$. Let $VB$ cut $V'B'$ in $\beta$. Let $a\beta$ cut $VV'$ in $v$. If $a\beta$ does not cut $VC$ and $V'C'$ in the same point, let $a\beta$ cut $VC$ in $\gamma$ and $V'C'$ in $\gamma'$.

Now $V(V'ABC\ldots) = V'(VA'B'C'\ldots)$. Hence

$$(va\beta\gamma) = (va\beta\gamma'),$$

## Anharmonic or Cross Ratio.

by considering the sections of these pencils by $\alpha\beta$. Hence $\gamma$ and $\gamma'$ coincide. Hence $VC$, $V'C'$ meet on $\alpha\beta$. So every

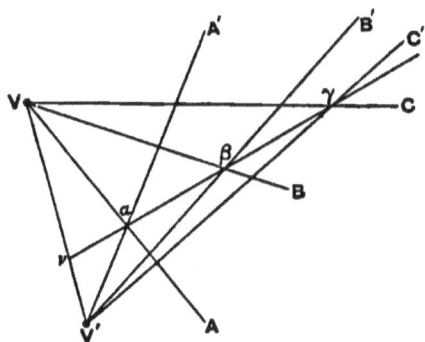

pair of corresponding rays meet on $\alpha\beta$. Hence the pencils are in perspective.

**Ex. 1.** *If* $(ABCD...)$ *and* $(A'B'C'D'...)$ *be two homographic ranges, and any two points $V$, $V'$ be taken on $AA'$, show that the meets of $VB$ and $V'B'$, of $VC$ and $V'C'$, of $VD$ and $V'D'$, &c., all lie on a line.*

**Ex. 2.** *If $AB$ pass through a fixed point $U$, and $A$ and $B$ move on fixed lines meeting in $O$, and if $V$, $W$ be fixed points collinear with $O$, show that the locus of the meet of $AV$ and $BW$ is a line.*

Let $AV$ and $BW$ cut in $P$. Take several positions $A_1 A_2 ...$ of $A$, $B_1 B_2 ...$ of $B$, $P_1 P_2 ...$ of $P$. Then
$$V(OP_1 P_2 ...) = (OA_1 A_2 ...) = U(OA_1 A_2 ...) = (OB_1 B_2 ...) = W(OP_1 P_2 ...).$$
Now the pencils $V(OP_1 P_2 ...)$ and $W(OP_1 P_2 ...)$ have a common ray; hence they are in perspective. Hence all the meets $(VP_1; WP_1)$, $(VP_2; WP_2)$, ... lie on a line.

**Ex. 3.** *Show that the meet of $UV$ and $OB$, and the meet of $UW$ and $OA$ lie on the locus.*

**Ex. 4.** *If $A$ and $B$ move on fixed lines through $O$, and $AB$, $BP$, and $AP$ pass through fixed collinear points $U$, $V$, $W$, show that the locus of $P$ is a line through $O$.*

**Ex. 5.** *If each side of a polygon pass through one of a set of collinear points whilst all but one of its vertices slide on fixed lines, then will the remaining vertex and every meet of two sides describe a line.*

**17.** *If* $(ABC...)$ *and* $(A'B'C'...)$ *be two homographic ranges on different lines, then the meet of $AB'$ and $A'B$, of $BC'$ and $B'C$, and generally of $PQ'$ and $P'Q$, where $PP'$, $QQ'$ are any two pairs of corresponding points, all lie on a line* (called the homographic axis).

Let the two lines meet in a point which we shall call

$X$ or $Y'$, according as we consider it to belong to the range $(ABC...)$ or to $(A'B'C'...)$. Take the points $X'$ and $Y$ corresponding to the point $X (= Y')$ in the two ranges. Then every cross meet such as $(PQ'; P'Q)$ lies on $X'Y$. (See figure of § 12.) For by hypothesis $(XYABC...) = (X'Y'A'B'C'...)$. Hence $A'(XYABC...) = A(X'Y'A'B'C'...)$; and these two pencils have the common ray $AA'$; hence they are in perspective; hence $(A'X; AX')$, $(A'Y; AY')$, $(A'B; AB'), ...$ all lie on a line. But $(A'X; AX')$ is $X'$, and $(A'Y; AY')$ is $Y$. Hence $(A'B; AB')$ lies on the fixed line $X'Y$; i.e. every cross meet lies on a fixed line, for $AA'$, $BB'$ are any two pairs of corresponding points.

**18.** By Reciprocation, or by a similar proof, we show that *if $V(ABCD...)$ and $V'(A'B'C'D'...)$ be homographic pencils, then all the cross joins such as the join of $(VB; V'C')$ with $(V'B'; VC)$ pass through a fixed point* (called *the homographic pole*).

**Ex. 1.** *If $A$, $B$, $C$ be any three points on a line, and $A'$, $B'$, $C'$ be any three points on another line, show that the meets of $AB'$ and $A'B$, of $AC'$ and $A'C$, and of $BC'$ and $B'C$, are collinear.*

Consider $X (= Y')$ as above.

**Ex. 2.** *When two ranges are in perspective, the axis of homography is the polar of the centre of perspective for the lines of the ranges.*

### Projective ranges and pencils.

**19.** If *range* $a$ is in perspective with range $\beta$, and range $\beta$ with range $\gamma$, and range $\gamma$ with range $\delta$, and so on; then each of the ranges $a, \beta, \gamma, \delta...$ is said to be *projective* with every other.

If *pencil* $a$ is in perspective with pencil $\beta$, and pencil $\beta$ with pencil $\gamma$, and pencil $\gamma$ with pencil $\delta$, and so on; then each of the pencils $a, \beta, \gamma, \delta...$ is said to be *projective* with every other.

*Projective ranges are homographic.*

For the range $a$ is homographic with the range $\beta$, being in perspective with it; so $\beta$ with $\gamma$, $\gamma$ with $\delta$, and so on; hence each is homographic with every other.

*Projective pencils are homographic.*

For the pencil $a$ is homographic with the pencil $\beta$, being in perspective with it; and so on.

*Homographic ranges are projective.*

For they can be put in perspective with the same range on the homographic axis.

*Homographic pencils are projective.*

For they can be put in perspective with the same pencil at the homographic pole.

*A range and a pencil* are said to be *projective*, when the range is projective with a section of the pencil.

Hence *a range and a pencil which are projective are homographic;* and *a range and a pencil which are homographic are projective.*

# CHAPTER X.

## VANISHING POINTS OF TWO HOMOGRAPHIC RANGES.

1. The points corresponding to the two points at infinity in two homographic ranges are called the *vanishing points*.

*To construct the vanishing points.*

Let the ranges be $(\Omega IABC...) = (J'\Omega'A'B'C'...)$, where $\Omega$ and $\Omega'$ are the points at infinity, and $I$ and $J'$ are the vanishing points.

First, suppose the ranges to be on different lines.

Through $A'$ draw $A'\omega$ parallel to $AB$ (and therefore passing through $\Omega$) cutting the homographic axis in $\omega$. Then $A\omega$ will cut $A'B'$ in the vanishing point $J'$. Similarly $I$ can be constructed.

Second, suppose the ranges to be on the same line.

Join $\Omega ABC...$ to any point $V$, not on the line; and let the joining lines cut any other line in $oabc....$ Then $Vo$ is parallel to $AA'$. By using the homographic axis of the two homographic ranges $abc...$ and $A'B'C'...$, find the point $J'$ in $A'B'C'...$ corresponding to $o$ in $abc....$ Then $J'$ is the vanishing point belonging to the range $A'B'C'....$ For

$$(\Omega ABC...) = (oabc...) = (J'A'B'C'...).$$

Similarly $I$ can be constructed.

2. *In two homographic ranges* $(ABCP...)$ *and* $(A'B'C'P'...)$, *on the same or different lines, if $I$ correspond to the point $\Omega'$ at infinity in the range* $(A'B'...)$, *and $J'$ correspond to the point $\Omega$ at infinity in the range* $(AB...)$, *then* $IP.J'P'$ *is the same whatever corresponding points $P$ and $P'$ are taken.*

## Vanishing Points of two Homographic Ranges. 113

For we have $(I \Omega ABCP...) = (\Omega'J'A'B'C'P'...)$;
hence $(AP, I\Omega) = (A'P', \Omega'J')$,
i.e. $AI/IP \div A\Omega/\Omega P = A'\Omega'/\Omega'P' \div A'J'/J'P'$.
But $A\Omega/\Omega P = -1$ and $A'\Omega'/\Omega'P' = -1$.
$\therefore AI/IP = J'P'/A'J'$,
$\therefore IP . J'P' = IA . J'A'$, which is constant.

Conversely, *if $IP . J'P'$ be constant, then $P$ and $P'$ generate ranges which are homographic, and $I$ and $J'$ are the points corresponding to the points at infinity in the ranges.*

For let $A$ and $A'$ be any two positions of $P$ and $P'$, then $IP . J'P' = IA . J'A'$. Hence retracing the above steps, we get $(AP, I\Omega) = (A'P', \Omega'J')$. Hence $P$ and $P'$ are corresponding points in the ranges determined by $AI\Omega$ and $A'\Omega'J'$, and $I$ and $J'$ correspond to $\Omega'$ and $\Omega$ in these ranges.

**Ex. 1.** *If through the centre of perspective $O$ of the two ranges $(ABC...)$ and $(A'B'C'...)$, there be drawn a parallel to $AB'$ meeting $AB$ in $I$ and a parallel to $AB$ meeting $AB'$ in $J'$, prove geometrically that*
$$IA . J'A = IB . J'B' = \cdots = IO . J'O.$$
*Deduce the formula $IP . J'P'$ for any two homographic ranges.*

**Ex. 2.** *If $OP . OP'$ be constant, $O$ being the meet of the lines on which $P$ and $P'$ lie, show that $P$ and $P'$ generate homographic ranges.*

**Ex. 3.** *If $I$ and $J'$ be the vanishing points of the homographic ranges*
$$(ABCP...) = (A'B'C'P'...),$$
show that   (a)   $AP : AI :: A'P' : J'P'$;
(b)   $AP/BP \div A'P'/B'P' = AI/BI.$

**Ex. 4.** *Show also that $AP . J'P' \div A'P'$ is independent of the position of $P$.*
For $(A\Omega, PQ) = (A'J', P'Q')$.

**Ex. 5.** *If $O$, $A$, $B$ be fixed points on the fixed line $OAB$, and $O$, $A'$, $B'$ be fixed points on the line $OA'B'$ which may have any direction in space, show that the meet of $AA'$ and $BB'$ describes a sphere.*

Through the meet $V$ of $AA'$ and $BB'$ draw $VI$ parallel to $A'B'$. Then $I$ is a fixed point, for $(IOAB) = V(IOAB) = (\Omega'OA'B')$. Again, through $V$ draw $VJ'$ parallel to $AB$. Then $J'$ is a fixed point on $OA'$, i.e. $OJ'$ is known, i.e. $IV$ is known.

**Ex. 6.** *If one of two copolar triangles be rotated about the axis of homology, show that the centre of homology describes a circle, whose centre is on the axis.*

Viz. the meet of the spheres determined by $AB$, $A'B'$ and by $AC_1$, $A'C'$, whose centres are on the axis.

3. Take any two origins $U$ and $V'$ on the lines of the ranges. Then $IP = UP - UI = x - a$, say;
and   $J'P' = V'P' - V'J' = x' - a'$, say.

Hence we get $(x-a)(x'-a') = $ constant,

or $xx' - a'x - ax' + aa' = $ constant,

a relation of the form $kxx' + lx + mx' + n = 0$.

*Hence the distances $x$ and $x'$ of corresponding points in two homographic ranges from any fixed points on the lines of the ranges are connected by a relation of the form*

$$kxx' + lx + mx' + n = 0,$$

*where $k$, $l$, $m$, $n$ are constants.*

Conversely, *if the distances be connected by this relation, the points generate homographic ranges.*

For if $kxx' + lx + mx' + n = 0$,

$$\text{then } k\left(x + \frac{m}{k}\right)\left(x' + \frac{l}{k}\right) = \frac{lm}{k} - n,$$

or $IP \cdot J'P' = $ constant, where $m/k = IU$ and $l/k = J'V'$.

The above relation assumes a neat form if we take $V'$ to coincide with $U'$. For then

$$IP \cdot J'P' = IU \cdot J'U', \therefore xx' - a'x - ax = 0,$$

or $a/x + a'/x' = 1$, or $UI/UP + U'J'/U'P' = 1$.

**4.** *If $P$ and $P'$ be connected by the relation $lx + mx' + n = 0$, $P$ and $P'$ generate homographic ranges in which the vanishing points are at infinity; and conversely, the corresponding points of two homographic ranges whose vanishing points are at infinity, are connected by a relation of the form $lx + mx' + n = 0$.*

(The reasoning employed in the general case does not apply here because $I$ and $J'$ are at infinity, and hence we cannot start with the equation $IP \cdot J'P' = $ constant.)

If $lx + mx' + n = 0$, then $x' = \beta x + \gamma$ (say).

Hence $P'Q' = V'Q' - V'P' = y' - x'$ (say) $= \beta(y - x) = \beta \cdot PQ$.

Hence the two lines are divided proportionally by the two sets of points, which therefore form homographic ranges.

Also putting $x = \infty$, we get $x' = \infty$; hence $\Omega$ and $\Omega'$ are corresponding points, i.e. $I$ and $J'$ are at infinity.

Conversely, if $I$ and $J'$ are at infinity, then

$$(AB, P\Omega) = (A'B', P'\Omega').$$

Hence $\quad \dfrac{AP}{PB} = \dfrac{A'P'}{P'B'}\quad$ or $\quad\dfrac{x-a}{b-x} = \dfrac{x'-a'}{b'-x'}$

$$\therefore\; x(b'-a') + x'(a-b) + a'b - ab' = 0,$$

which is of the form $\quad lx + mx' + n = 0$.

Or, we may consider the equation $lx + mx' + n = 0$ as the limit of the relation $kxx' + lx + mx' + n = 0$ when $k$ decreases indefinitely. Since the latter equation determines two homographic ranges however small $k$ is, we may assume this to be true in the limit when $k = 0$.

Two homographic ranges in which the vanishing points are at infinity may be called *similar homographic ranges*.

**Ex. 1.** *If* $(\Omega\, IAB...) = (J'\Omega'A'B'...)$, *and* $AB = A'B'$; *show that*
$AB = AI + A'J'$, *and* $AI = -B'J'$.

**Ex. 2.** *Through the vertex $V$ of the parallelogram $VIOJ'$ is drawn a line cutting $OI$ in $A$ and $OJ'$ in $A'$, show that $OI/OA + OJ'/OA' = 1$.*

**Ex. 3.** *Find the values of the constants in the relation*
$$xx' + lx + mx' + n = 0.$$
The relation is $\quad 1 + l/x' + m/x + n/xx' = 0$.
Put $x = \infty$; $\therefore\; l = -x' = -V'J'$; so $m = -UI$.
Again, put $x = 0$ and $x' = V'U'$; $\therefore\; n = V'U'.UI$.
Hence $\quad UP.V'P' - V'J'.UP - UI.V'P' + V'U'.UI = 0$.
Another value of $n$ is $UV.V'J'$.
These values come also at once from
$$IP.J'P' = IU.J'U' \text{ or } IV.J'V'.$$

**Ex. 4.** *Deduce the formula when the vanishing points are at infinity.*
Dividing by $UI$ and putting $UI = \infty$, we get $\; c.UP + V'P' = V'U'$ (where $c$ is the limit of $V'J'/UI$), or $lx + mx' + n = 0$.

**Ex. 5.** *Show that the formula $lx + mx' + n = 0$ can be written*
$$UP/UV + V'P'/V'U' = 1.$$
Put $P = U$ and $P' = V'$ successively.

**Ex. 6.** *Show that by properly choosing $V'$, the general relation can be thrown into the form* $xx' + l(x-x') + n = 0$.

**Ex. 7.** *Show that corresponding points $PP'$ of two homographic ranges on the same line are connected by a relation of the form*
$$UP.V'P' + \gamma.PP' + \delta = 0,$$
*provided* $UV = U'V'$. *Show also that* $\gamma = IU = V'J'$ *and*
$$\delta = UI.UU' = V'J'.V'V.$$

**5. The following are geometrical applications.**

**Ex. 1.** *If $O$, $O'$, $A$, $A'$, $U$, $V'$ be fixed points of which $OAA'O'$ are collinear, and if points $P$ and $P'$ be taken on $AU$ and $A'V'$ such that*
$$a.UP/AP + \beta.V'P'/A'P' = \gamma,$$

where $a, \beta, \gamma$ are constants, show that the locus of the meet of $OP$ and $O'P'$ is a line.

Reducing the given relation to any origins on $AU$ and $A'V'$, it is clearly of the form $xx' + lx + mx' + n = 0$. Hence $P$ and $P'$ generate homographic ranges. Also putting $P = A$, we get $P' = A'$. Hence in the two homographic pencils $O(P...)$ and $O'(P'...)$, $OO'$ is a common ray. Hence the locus is a line.

**Ex. 2.** *The same is true if any one of the following relations hold—*
 (i) $a \cdot UP/AP + \beta/A'P' = \gamma$, $V'$ being at infinity;
 (ii) $a/AP + \beta/A'P' = \gamma$, $U$ and $V'$ being at infinity;
 (iii) $a \cdot UP/AP + \beta \cdot V'P' = \gamma$, $A'$ being at infinity;
 (iv) $a \cdot UP + \beta \cdot V'P' = \gamma$, $A$ and $A'$ being at infinity.

**Ex. 3.** *If $\gamma = 0$ in any of these relations, the locus passes through the meet of $OU$ and $O'V'$.*

**Ex. 4.** *Obtain the Cartesian equation of a line, viz. $Ax + By + C = 0$.*
Consider the pencils at the points at infinity on the axes.

**Ex. 5.** *If $PM$, $PM'$ drawn in given directions from $P$ meet given lines $OM$ and $OM'$ in $M$ and $M'$ so that $a \cdot PM + \beta \cdot PM' = \gamma$, show that $P$ moves on a line.*
For $PM'$ and $PM$ are proportional to the $x$ and $y$ of Ex. 4.

**Ex. 6.** *If $O, U, V'$ be fixed points, and if points $P$ and $P'$ be taken on $OU$ and $OV'$ such that* $a \cdot UP/OP + \beta \cdot V'P'/OP' = \gamma$, *then $PP'$ passes through a fixed point.*

**Ex. 7.** *The same is true if any one of the following relations hold—*
 $a/OP + \beta \cdot V'P'/OP' = \gamma$, $U$ being at infinity;
 $a/OP + \beta/OP' = \gamma$, $U$ and $V'$ being at infinity;
 $a \cdot UP + \beta \cdot V'P' = \gamma$, $O$ being at infinity.

**Ex. 8.** *If $\gamma = 0$, the point is on $UV'$.*

**Ex. 9.** *If $p, q, r$, the perpendiculars from $A, B, C$ on a line, be connected by the relation $\lambda \cdot p + \mu \cdot q + \nu \cdot r = 0$, then the line passes through a fixed point.*
Divide by $p$ and use Ex. 6.

**6.** If $P$ and $P'$ be connected by a relation which can be reduced to the form $kxx' + lx + mx' + n = 0$, we have proved that $P$ and $P'$ generate homographic ranges. The following converse is very important, viz.

*Any relation which can be reduced to the form*

$$kxx' + lx' + mx' + n = 0$$

*is true of every pair of corresponding points of two homographic ranges, provided it is true of three pairs of corresponding points.*

Let the two homographic ranges be $(ABCD...)$ and $(A'B'C'D'...)$. Suppose the above relation (in which $x = UP$ and $x' = V'P'$) is satisfied when $P$ is at $A$ and $P'$ at $A'$, and when $P$ is at $B$ and $P'$ at $B'$, and when $P$ is at $C$ and $P'$ at $C'$. Then it will be satisfied when $P$ is at $D$ and $P'$ at $D'$, $D$ and $D'$ being any other two corresponding points of the ranges.

For if not, suppose that when $P$ is at $D$, the above relation gives $E'$ as the position of $P'$. Then since the given relation determines two homographic ranges, we have

$$(ABCD) = (A'B'C'E');$$

but $(ABCD) = (A'B'C'D')$ by hypothesis. Hence $D'$ and $E'$ coincide, i.e. the given relation is true for every pair of corresponding points of the two ranges.

**Ex. 1.** *If the point $P$ on the line $AB$ and the point $P'$ on the line $B'C'$ be connected by the relation*

$$\lambda \cdot AP/BP + \mu \cdot C'P'/B'P' = 1,$$

*show that $P$ and $P'$ generate homographic ranges, and that $B$ and $B'$ are corresponding points in these ranges. Find also the values of $\lambda$ and $\mu$. Prove also conversely, that if $(ABCP) = (A'B'C'P')$ then the relation holds.*

Taking any origins we get

$$\lambda (x-a)(x'-b') + \mu (x'-c')(x-b) = (x-b)(x'-b'),$$

which is of the form $kxx' + lx + mx' + n = 0$.
Hence $P$ and $P'$ generate homographic ranges.
Take $P$ at $B$, then $x = b$, $\therefore \lambda (b-a)(x'-b') = 0$, $\therefore x' = b'$.
Hence $P'$ is at $B'$, i.e. $B$ and $B'$ correspond.
Again, let $P$ be at $C$ when $P'$ is at $C'$.
Put $x' = c'$. $\therefore \lambda (c-a) = (c-b)$, $\therefore \lambda = BC/AC$.
Let $P'$ be at $A'$ when $P$ is at $A$.
Put $x = a$. $\therefore \mu (a'-c') = a'-b'$, $\therefore \mu = B'A'/C'A'$.
Conversely, if $(ABCP) = (A'B'C'P')$, the relation

$$\frac{BC}{AC} \cdot \frac{AP}{BP} + \frac{B'A'}{C'A'} \cdot \frac{C'P'}{B'P'} = 1$$

is true; for it is of the form $xx' + lx + mx' + n = 0$, and it is satisfied by $(A, A')$, $(B, B')$ and $(C, C')$.

**Ex. 2.** *Treat the following relations in the same way—*
  (a)  $\lambda/BP + \mu \cdot C'P'/B'P' = 1;$
  (b)  $\lambda \cdot AP/IP + \mu \cdot C'P' = 1;$
  (c)  $\lambda/IP + \mu \cdot C'P' = 1;$
  (d)  $AP \cdot B'P' + \lambda \cdot CP + \mu \cdot C'P' = AC \cdot B'C';$
  (e)  $IP \cdot B'P' + \lambda \cdot CP + \mu = 0.$

Results—  (a)  $\lambda = BC$, $\mu = B'J'/C'J';$
  (b)  $\lambda = IC/AC$, $\mu = 1/C'A';$
  (c)  $\lambda = IC$, $\mu = 1/C'J';$
  (d)  $\lambda = -B'J'$, $\mu = -AI;$
  (e)  $\lambda = -B'J'$, $\mu = -IC \cdot B'C'$.

**Ex. 3.** Show that the following equations are satisfied by every two homographic ranges.

(a) $\dfrac{AP}{CP} \cdot \dfrac{B'P'}{D'P'} + \lambda \cdot \dfrac{AP}{CP} + \mu \cdot \dfrac{B'P'}{D'P'} + \nu = 0$;

(b) $\dfrac{AP}{AC} \cdot \dfrac{B'P'}{B'D'} + \lambda \cdot \dfrac{AP}{AC} + \mu \cdot \dfrac{B'P'}{B'D'} + \nu = 0$.

Each equation is of the required form, and $\lambda$, $\mu$, $\nu$ can be determined so that the equation shall be satisfied by any three pairs of points.

**Ex. 4.** Deduce in Ex. 3 definite formulae for $(ABCP) = (A'B'C'P')$, i.e. determine the values of $\lambda$, $\mu$, $\nu$.

**Ex. 5.** If $(ABCD) = (A'B'C'D')$, prove that

(a) $\dfrac{AB \cdot CD}{A'B'} + \dfrac{AC \cdot DB}{A'C'} + \dfrac{AD \cdot BC}{A'D'} = 0$;

(b) $\dfrac{AB \cdot CD}{A'B'} \cdot O'B' + \dfrac{AC \cdot DB}{A'C'} \cdot O'C' + \dfrac{AD \cdot BC}{A'D'} \cdot O'D' = 0$,

$O'$ being an arbitrary point on the line $A'B'$.

Take $D$ and $D'$ as variable points.

**Ex. 6.** If the pencil $V(ABCD)$ be homographic with the range $(A'B'C'D')$, show that

$$\dfrac{\sin AVB \cdot \sin CVD}{A'B'} + \dfrac{\sin AVC \cdot \sin DVB}{A'C'} + \dfrac{\sin AVD \cdot \sin BVC}{A'D'} = 0.$$

Use Ex. 5 (a).

**Ex. 7.** Show that $VP$, $V'P'$ generate homographic pencils if

(a) $\dfrac{\sin AVP}{\sin BVP} \cdot \dfrac{\sin BVC}{\sin AVC} + \dfrac{\sin C'V'P'}{\sin B'V'P'} \cdot \dfrac{\sin B'V'A'}{\sin C'V'A'} = 1$,

or (b) $\lambda \cot BVP + \mu \cot B'V'P' = 1$,

or (c) $\lambda \tan AVP + \mu \tan C'V'P' = 1$.

**Ex. 8.** If $VP$ and $V'P'$ generate two homographic pencils, and $AVP = \theta$ and $B'V'P' = \theta'$, $VA$ and $V'B'$ being any initial lines, show that

$$\tan \theta \cdot \tan \theta' + \lambda \tan \theta + \mu \tan \theta' + \nu = 0;$$

and conversely, if this relation be satisfied, then $VP$ and $V'P'$ generate homographic pencils.

Take transversals perpendicular to the initial lines, then

$$\tan \theta \propto x \quad \text{and} \quad \tan \theta' \propto x'.$$

## Common points of two homographic ranges on the same line.

**7.** Suppose corresponding points in two ranges on the same line to be connected by the relation

$$k \cdot UP \cdot V'P' + l \cdot UP + m \cdot V'P' + n = 0.$$

For the origins $U$ and $V'$ we can take the same point on the line, called $U$ or $V'$ according as it is considered to belong to one or the other range. The equation becomes

$$k \cdot UP \cdot UP' + l \cdot UP + m \cdot UP' + n = 0.$$

Now if $P$ correspond to itself, $P$ must coincide with $P'$. Hence the equation giving the self-corresponding or *common points* of the two ranges is
$$k \cdot UP^2 + (l+m) UP + n = 0.$$
Hence every two homographic ranges on the same line have two common points, real, coincident, or imaginary.

A graphic construction of the common points will be found in XVI. 6.

**Ex. 1.** *If $E$ and $F$ be the common points of the homographic ranges $(ABC...)$ and $(A'B'C'...)$, show that*
$$(EFAA') = (EFBB') = (EFCC') = \cdots.$$
For $(EF, AB) = (EF, A'B'), \therefore (EF, AA') = (EF, BB')$.

**Ex. 2.** *If $(EFABC...) = (EFA'B'C'...) = (EFA''B''C''...) = \cdots$, then $(EFAA'A''...) = (EFBB'B''...) = (EFCC'C''...) = \cdots$.*

**Ex. 3.** *If $(EF, PP')$ be constant, then $PP'$ generate homographic ranges of which $EF$ are the common points.*

**Ex. 4.** *If $ABC..., A'B'C'...$ be homographic ranges on the same line, and if $P'$, $Q$ be the points corresponding to the point $P (= Q')$ according as it is considered to belong to the first range or the second, show that $P'$, $Q$ generate homographic ranges whose common points are the same as those of the given ranges.*

The range generated by $P'$ is homographic with the range generated by $P$, i.e. by $Q'$, and this is homographic with the range generated by $Q$. Hence range $P'$ = range $Q$.

Again, suppose $P$ is a common point of the given ranges; then $P'$ coincides with $P$, i.e. $P'$ coincides with $Q'$; hence $P$ coincides with $Q$, i.e. $P'$ coincides with $Q$, i.e. $P$ is a common point of the derived ranges.

**Ex. 5.** *If $X$ be the fourth harmonic of $P$ for $P'$ and $Q$, then $PX$ is divided harmonically by the common points.*

Let the given homography be defined by
$$PA \cdot PA' + l \cdot PA + m \cdot PA' + n = 0.$$
Put $A = P$ and $A' = P', \therefore PP' = -n/m$. Put $A = Q, A' = Q' = P$,
$\therefore PQ = -n/l, \quad \therefore 2/PX = 1/PP' + 1/PQ = -(l+m)/n$.
Now $E$ and $F$ are given by $x^2 + (l+m) x + n = 0$,
$\therefore 1/PE + 1/PF = -(l+m)/n = 2/PX$,
$\therefore (PX, EF)$ is harmonic.

**Ex. 6.** *Construct the fourth harmonic of a given point for the (unknown) common points of two given homographic ranges.*

**Ex. 7.** *Show that $(EF, QP') = (EF, AA')^2$ in Ex. 4.*
For $(EF, AA')^2 = (EF, PP') \cdot (EF, QQ')$ where $P = Q'$.
This gives us another proof of Ex. 4, using Ex. 3.

**Ex. 8.** *If $ABA'B'$ be given collinear points, find a point $X$ in the same line, such that the compound ratio $AX \cdot A'X \div BX \cdot B'X$ may be a given quantity.*

$X$ is one of the common points of the homographic ranges determined by $AP/BP \div B'P'/A'P' =$ the given quantity.

**Ex. 9.** *Determine the point $X$, given the value of $AX \cdot A'X \div BX$.*

**8.** *If one of the common points of two homographic ranges $(ABC...)$ and $(A'B'C'...)$ on the same line be at infinity, then the points $ABC...$ divide the line in the same ratios as the points $A'B'C'$; and conversely.*

For if $\quad (AB, C\Omega) = (A'B', C'\Omega)$.

Then $\quad \dfrac{AC}{CB} \cdot \dfrac{\Omega B}{A\Omega} = \dfrac{A'C'}{C'B'} \cdot \dfrac{\Omega B'}{A'\Omega}$.

But $\quad \Omega B \div A\Omega = -1 = \Omega B' \div A'\Omega$;

$\quad \therefore AC : CB : : A'C' : C'B'$;

and similarly for any other pair of segments; i.e. the line is divided similarly by the two sets of points.

Conversely, if the line be divided similarly by the two sets of points.

Since $\quad AC : CB : : A'C' : C'B'$,

we have, retracing our steps, $\quad (AB, C\Omega) = (A'B', C'\Omega)$.

So $\quad (DB, C\Omega) = (D'B', C'\Omega)$, and so on.

Hence $\quad (\Omega ABC...) = (\Omega A'B'C'...)$,

i.e. $(ABC...)$ and $(A'B'C'...)$ are two homographic ranges with a common point at infinity.

*Or thus*—Let the homography be given by

$$kxx' + lx + mx' + n = 0.$$

The common points are given by $kx^2 + (l+m)x + n = 0$. If one of the common points be at infinity, then $k = 0$, i.e. the homography is given by $lx + mx' + n = 0$, i.e. the ranges are similar.

Conversely, if the ranges are similar, then

$$lx + mx' + n = 0,$$

i.e. $k = 0$, i.e. one of the common points is at infinity.

**Ex. 1.** *If in two homographic ranges on different lines the points at infinity correspond, the ranges are similar; and conversely.*

**Ex. 2.** *If one of the common points of two homographic ranges on the same line be at infinity, the other, $E$, is given by $EA : EA' : : BA : B'A'$.*

**Ex. 3.** *Show also that E is the meet with $AA'$ of the radical axis of any two circles through $AB'$ and $A'B$.*

**Ex. 4.** *If $AB/A'B' = BC/B'C' = \ldots = -1$, show that one common point is at infinity, and that the other bisects all the segments $AA'$, $BB'$, $CC'$, ....*

**Ex. 5.** *If each of the common points be at infinity, then all segments joining corresponding points are equal; and conversely.*

For if $F$ be at infinity, the ranges are divided proportionally, hence $AB/A'B' = EB/EB' = 1$, for $E$ is also at infinity. Conversely, if $AB = A'B'$, $BC = B'C'$, ..., the ranges are divided proportionally; hence $F$ is at infinity. And $E$ is given by $EB/EB' = 1$, hence $E$ is also at infinity.

Or thus. In this case the quadratic $kx^2 + (l+m)x + n = 0$ has both roots infinite. Hence $k = 0$ and $l + m = 0$. Hence the homography is given by $l(x - x') + n = 0$, i.e. $x - x' = $ constant, i.e. $AA'$ is constant. And conversely, if $AA'$ is constant, then $k = 0$ and $l + m = 0$. Hence both common points are at infinity.

### Common rays of two homographic pencils having the same vertex.

**9.** *In any two homographic pencils having the same vertex, two rays exist, each of which corresponds to itself.*

Let the pencils be $V(ABC\ldots) = V(A'B'C'\ldots)$. Suppose a line to cut the pencils in the ranges $(abc\ldots) = (a'b'c'\ldots)$, $a$ being on $VA$, and so on. Then if $VA$ and $VA'$ coincide, $a$ and $a'$ will coincide. Hence if $e$ and $f$ be the self-corresponding points of the ranges $(abc\ldots)$ and $(a'b'c'\ldots)$, $Ve$ and $Vf$ are the self-corresponding or *common rays* of the pencils $V(ABC\ldots)$ and $V(A'B'C'\ldots)$.

**Ex. 1.** *If $VP$ and $VP'$ be a pair of corresponding lines in two homographic pencils whose common lines are $VE$ and $VF$, show that*
$$\sin EVP/\sin FVP \div \sin EVP'/\sin FVP'$$
*is constant.*

**Ex. 2.** *Find a point on a given line through which shall pass a pair of corresponding lines of two given homographic pencils.*

Either of the common points of the homographic ranges determined on the line by the pencils.

**Ex. 3.** *If $VA$, $V'A'$ generate homographic pencils at $V$ and $V'$, show that in two positions $VA$ is parallel to $V'A'$; and that any transversal in either of these directions is cut by the two pencils proportionally.*

For without altering the directions of the rays, superpose $V'$ on $V$.

**Ex. 4.** *Two given homographic pencils $V(abc\ldots)$ and $V'(a'b'c'\ldots)$ meet a line in the points $ABC\ldots$ and $A'B'C'\ldots$; determine the position of the line so that $AB = A'B'$, $BC = B'C'$, $CD = C'D'$, &c.*

Suppose the line drawn. Since $(\Omega ABC\ldots) = (\Omega A'B'C'\ldots)$, the line must be parallel to one or other of the pairs of corresponding parallel

rays. Let it meet the other two corresponding parallel rays in $O$, $O'$. Draw $V'S$ parallel to the line to meet $VO$ in $S$. Then
$$SO = V'O', \quad OA = O'A', \text{ and } \angle SOA = \angle V'O'A'.$$
Hence $SA$ is parallel to $V'A'$.

Hence the construction—Take the corresponding rays $Vy$, $V'y'$ which are parallel, and also the corresponding rays $Vz$, $V'z'$ which are parallel. Let $Vz$ meet $V'y'$ in $S$, and through $S$ draw $SA$ parallel to $V'a'$ to meet $Va$ in $A$. Through $A$ draw $ABC... A'B'C'...$ parallel to $V'S$. This line satisfies the required condition.

For $Vy$, $V'y'$ meet the line in the same point $\Omega$ at infinity. Hence $(\Omega OAB) = (\Omega O'A'B')$. Hence $OA : OB :: O'A' : O'B'$. But $OA = O'A'$ by construction. Hence $OB = O'B'$. Hence $AB = A'B'$, and so on.

Hence there are two such lines, one parallel to each of the lines $Vy$, $Vz$.

**Ex. 5.** *Given any two homographic pencils, one can be moved parallel to itself so as to be in perspective with the other.*

**10.** *If $I$, $J'$ correspond to the points at infinity in two homographic ranges on the same line, and $O$ bisect $IJ'$, and $O'$ be the point corresponding to $O$, then the common points $E$, $F$ are given by*
$$OE^2 = OF^2 = OJ'. OO'.$$
For $\quad (O\Omega, IE) = (O'J', \Omega' E)$,
where $\Omega$ or $\Omega'$ is the point at infinity upon the line.
$$\therefore \quad \frac{OI}{I\Omega} \cdot \frac{E\Omega}{OE} = \frac{O'\Omega'}{\Omega'J'} \cdot \frac{EJ'}{O'E}.$$
But $\quad E\Omega \div I\Omega = 1$ and $O'\Omega' \div \Omega'J' = -1$,
$$\therefore \ OI \cdot O'E + OE \cdot EJ' = 0.$$
Take $O$ as origin, $\therefore \ OI(OE - OO') + OE(OJ' - OE) = 0$,
but $OI = -OJ'$, $\therefore \ -OJ'(OE - OO') + OE(OJ' - OE) = 0$,
$$\therefore \ OE^2 = OJ' \cdot OO'; \text{ so } OF^2 = OJ' \cdot OO'.$$

Hence *the two common points are equidistant from $O$*; therefore one is as far from $I$ as the other is from $J'$.

Notice that $(EF, O'J')$ is harmonic.

**Ex. 1.** *If $E$ and $F$ coincide, they both coincide with $O$.*
For $O$ bisects $EF$.

**Ex. 2.** *Show that the relation connecting two homographic ranges on the same line can be thrown into the form $EP \cdot FP + IP \cdot PP' = 0$.*

For this relation is of the required form, and it is satisfied by $(E, E)$ and $(F, F)$. Also putting the relation in the form
$$EP \cdot FP/PP' + IP = 0,$$
we see that it is satisfied by $(I, \Omega')$.

**Ex. 3.** *Prove the same for the relations*
(a) $EP \cdot FP' = EI \cdot PP'$;
(b) $OP \cdot OP' - OI \cdot PP' + OI \cdot OO' = 0$;
(c) $OP^2 + IP \cdot PP' + OI \cdot OO' = 0$.

**Ex. 4.** *If $E$ and $F$ coincide, $P$ and $P'$ are connected by the relation*
$$UP \cdot UP' - UJ' \cdot UP - UI \cdot UP' + UO^2 = 0.$$
For putting $P = P'$ in the general relation
$$UP \cdot UP' - UJ' \cdot UP - UI \cdot UP' + UU' \cdot UI = 0,$$
and noticing that $2 \cdot UO = UI + UJ'$, we get
$$UP^2 - 2 \cdot UO \cdot UP + UU' \cdot UI = 0.$$
And this is a perfect square, hence $UU' \cdot UI = UO^2$.

**Ex. 5.** *If $E$ and $F$ coincide, show that $P$ and $P'$ are connected by the relation $OP \cdot OP' = OI \cdot PP'$.*

It is of the required form, and is satisfied by $(I, \Omega')$, and by $(E, E)$ and $(F, F)$ since $E$ and $F$ coincide with $O$.

**Ex. 6.** *If $E$ and $F$ coincide, show also that*
(a) $(OP)^{-1} + (OP')^{-1} = (OA)^{-1} + (OA')^{-1}$;
(b) $OP \cdot OA/AP = OP' \cdot OA'/A'P'$;
(c) $OP^2 = PI \cdot PP'$.

**Ex. 7.** *Any two ranges whose common points coincide, can be placed in perspective with two ranges whose corresponding segments are equal.*

For join the two ranges to any point $V$ and consider the ranges on any line parallel to $VO$.

**11.** *If the common points be imaginary, then the ranges $(ABC...)$ and $(A'B'C'...)$ subtend at two points in the plane of the paper superposable pencils.*

For if $E$ and $F$ are imaginary, since $OE^2 = OJ' \cdot OO'$, we see that $OJ'$ and $OO'$ have different signs, i.e. $O$ lies between $O'$ and $J'$. On a perpendicular to the line $AA'$ through $O$ take $OU$, such that $OU^2 = OJ' \cdot O'O$. Two such points can be taken one on each side of the line $AA'$.

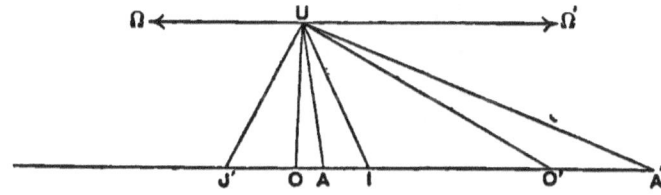

Then the pencils subtended at $U$ are superposable.

Since $I$ corresponds to the point $\Omega'$ at infinity, the ray

$U\Omega'$ is parallel to $AA'$; so the ray $U\Omega$ corresponding to $UJ'$ is parallel to $AA'$. Now since $UO^2 = J'O \cdot OO'$ it follows that $J'UO'$ is a right angle.

Hence  $\angle \Omega UJ' = \angle UJ'O = \angle OUO'$
$= \angle UIJ'$, since $J'O = OI$
$= \angle IU\Omega'$.

Hence the pencil $U(\Omega OI)$ can be superposed to the pencil $U(J'O'\Omega')$ by turning it through the angle $\Omega UJ'$. After the rotation three rays of the pencils $U(\Omega OIABC...)$ and $U(J'O'\Omega' A'B'C'...)$ coincide; hence every ray of one pencil coincides with the corresponding ray of the other pencil, i.e. the pencils are superposed.

Notice that the points $U$ give solutions of the problem— *Given, on one line, two homographic ranges $(ABC...)$ and $(A'B'C'...)$ of which the common points are imaginary, find a point at which the segments $AA'$, $BB'$, $CC'$, ... subtend equal angles.*

**Ex.** Determine a point at which three given collinear segments subtend equal angles.

**12.** *Two homographic pencils with the same vertex whose common rays are imaginary can be placed in perspective with two superposable pencils.*

For let any line cut the given pencils in $ABC...$ and $A'B'C'....$ In a plane not that of the pencils construct the point $U$ at which $AA$, $BB'$,... subtend equal angles. Take the vertex of projection on the line joining $U$ to the vertex $V$ of the given pencils; and take the plane of projection parallel to $UAA'$. Then the projection of $VA$ is parallel to $UA$, and of $VA'$ to $UA'$. Hence the projection of the angle $AVA'$ is equal to the angle $AUA'$; so for the other angles. Hence the angles $AVA'$, $BVB'$, $CVC'$,... project into equal angles.

# CHAPTER XI.

### ANHARMONIC PROPERTIES OF POINTS ON A CONIC.

1. WE have already shown in IX. 8 that the projection of a range of four points is homographic with the range, and in IX. 9 that the projection of a pencil of four lines is homographic with the pencil. We shall now proceed to investigate certain properties of a conic by proving the corresponding properties of the circle of which the conic is by definition the projection.

2. *Four fixed points on a conic subtend at a variable fifth point on the conic a constant cross ratio.*

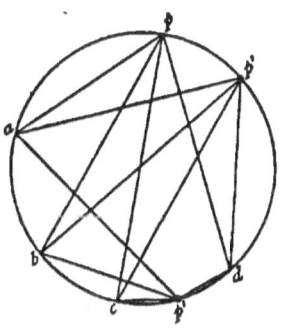

Let the four fixed points on the conic be $ABCD$ and the variable point $P$. Let $A, B, C, D, P$ be the projections of the points $a, b, c, d, p$ on the circle of which the conic is the projection. Now, in the circle, $abcd$ subtend the same cross ratio at every point on the circle. For take any two points $p$ and $p'$ on the circle. Then

$$p\,(ab,\,cd) = \frac{\sin apc}{\sin cpb} \div \frac{\sin apd}{\sin dpb} = \frac{\sin ap'c}{\sin cp'b} \div \frac{\sin ap'd}{\sin dp'b} = p'(ab,\,cd).$$

For in all cases the angle $apb$ is equal to the angle $ap'b$ or its supplement; and so for the other angles. Hence $P\,(ABCD) = p\,(abcd)$ by projection $= p'\,(abcd) = P'\,(ABCD)$

by projection. Hence $ABCD$ subtend the same cross ratio at every point $P$ on the conic.

The cross ratio subtended by the points $(AB, CD)$ on a conic at any point on the conic is called the *cross ratio of the points* $(AB, CD)$ *on the conic.*

Notice that, making $P$ coincide with $A$, the cross ratio of $(AB, CD)$ is equal to $A(AB, CD) = A(TB, CD)$, where $AT$ is the tangent to the conic at $A$.

**Ex. 1.** *Show that in a circle the pencils* $p(abcd)$ *and* $p'(abcd)$ *are superposable in all cases.*

This gives another proof of § 2.

**Ex. 2.** *A tangent to an ellipse meets the auxiliary circle in* $ZZ'$; *show that the cross ratio of the four points* $(AA', ZZ')$ *on the circle is* $(1-e) \div (1+e)$.

Consider the pencil at the point opposite to $Z'$.

**Ex. 3.** *Prove that the cross ratio* $(AB, CD)$ *of the four points* $A, B, C, D$ *on a circle is* $AC/CB \div AD/DB$, $AC$ *being the length of the line joining* $A$ *to* $C$.

For $\sin APC = AC \div 2R$.

**Ex. 4.** *Conjugate lines for a conic meet the conic in four points which subtend a harmonic pencil at every point on the conic.*

Consider the pencil at one of the four points.

Such points are called *harmonic points on the conic.*

**Ex. 5.** *If* $AA'$, $BB'$ *be pairs of harmonic points on a conic, show that* $AA'$ *and* $BB'$ *are conjugate lines for the conic.*

**Ex. 6.** *The chords* $AB$, $CD$ *of a conic are conjugate, and* $ACB$ *is a right angle; through* $D$ *is drawn the chord* $DP$ *meeting* $AB$ *in* $Q$; *show that* $CA$, $CB$ *are the bisectors of the angle* $PCQ$.

For $-1 = P(AB, CD) = P(AB, CQ) = C(AB, PQ)$.

**Ex. 7.** *Two conics* $\alpha$ *and* $\beta$ *touch at* $B$ *and* $C$. *Through* $A$, *the meet of the common tangents, is drawn a line meeting* $\alpha$ *in* $P, Q$. $BQ$, $BP$ *meet* $\beta$ *in* $V, U$. *Show that* $VU$ *passes through* $A$.

$(BC, UV) = B(BC, UV) = B(AC, PQ) = -1$.

**Ex. 8.** *If a variable circle cut a given arc of a given circle harmonically, it is orthogonal to the circle, which passes through the ends of the given arc and is orthogonal to the given circle.*

**Ex. 9.** *If* $AA'$, $BB'$ *be pairs of harmonic points on a circle, show that*
$$AA' . BB' = 2 . AB . A'B' = 2 . AB' . BA'.$$

By Ex. 3 we have $AB . A'B' = AB' . BA'$.

By Ptolemy's theorem we have
$$AA' . BB' = AB . A'B' + AB' . BA' = 2 . AB . A'B'.$$

**Ex. 10.** *Obtain the equation of a hyperbola referred to its asymptotes.*

Let $P, Q$ be any two points on the hyperbola, and $\Omega$, $\Omega'$ the points at infinity on the hyperbola. Then $\Omega(PQ\Omega\Omega') = \Omega'(PQ\Omega\Omega')$. Through $P$ and $Q$ draw $PL$, $QM$ parallel to one asymptote, and $PN$, $QR$ parallel to the other. Then $(LM\Omega\Omega') = (NR\Omega C)$ where $C$ is the centre of the conic.

Hence $\dfrac{CM}{CL} \div \dfrac{\Omega'M}{\Omega'L} = \dfrac{\Omega R}{\Omega N} \div \dfrac{CR}{CN}$, i.e. $CL \cdot CN = CM \cdot CR$.

**Ex. 11.** *Any diameter of a parabola meets the tangent at Q in T, the curve in P, and any chord $QQ'$ in R; show that $TP:PR::QR:RQ'$.*
For $Q(QP Q'\Omega) = \Omega(QPQ'\Omega)$.

**Ex. 12.** *A variable point P on a conic is joined to the fixed points L, M on the conic; show that the angle LPM is divided in a constant cross ratio by parallels through P to the asymptotes.*

**Ex. 13.** *Through four fixed points A, B, C, D is drawn a system of conics; show that the tangents at A, the tangents at B, the tangents at C, and the tangents at D form four homographic pencils.*
For $A(ABCD) = B(ABCD) = B(BADC)$.

**3. Pappus's theorem.** *If from any point P on a conic perpendiculars $a$, $\beta$, $\gamma$, $\delta$ be drawn on the lines AB, BC, CD, DA joining fixed points ABCD on the conic, then $a \cdot \gamma = k \cdot \beta \cdot \delta$, where k is independent of the position of P.*

For $P(AC, BD) = \sin APB \cdot \sin DPC \div \sin BPC \cdot \sin APD$.
But $PA \cdot PB \sin APB = a \cdot AB$, and so on.
Hence $a \cdot \gamma \cdot AB \cdot DC \div \beta \cdot \delta \cdot BC \cdot AD = P(AC, BD)$ is constant, i.e. $a\gamma = k \cdot \beta \cdot \delta$.

**Ex. 1.** *If the perpendiculars let fall from any point on a conic on the sides of an inscribed polygon of an even number of sides be called $1, 2, 3, \ldots, 2n$, show that $1 \cdot 3 \cdot 5 \ldots (2n-1) \div 2 \cdot 4 \cdot 6 \ldots 2n$ is constant.*

Suppose the theorem holds for $2n-2$ sides. Then
$$1 \cdot 3 \cdot 5 \ldots (2n-3) = k \cdot 2 \cdot 4 \cdot 6 \ldots (2n-4)x.$$
And by the above theorem $(2n-1)x = k'(2n-2)(2n)$. Multiplying,
$$1 \cdot 3 \cdot 5 \ldots (2n-1) = k'' \cdot 2 \cdot 4 \cdot 6 \ldots 2n.$$
Hence by Induction.

**Ex. 2.** *The product of the perpendiculars from any point on a conic on the sides of any inscribed polygon varies as the product of the perpendiculars on the tangents at the vertices.*

Make the alternate sides in Ex. 1 of zero length.

**Ex. 3.** *If the conic be a circle, the products are equal, in the theorem and in Ex. 1 and Ex. 2.* (See Ex. 3, § 2.)

**Ex. 4.** *The product of the perpendiculars from any point on a conic on two fixed tangents is proportional to the square of the perpendicular on the chord of contact.*

**Ex. 5.** *The product of the perpendiculars from any point on a hyperbola on two fixed lines parallel to the asymptotes is proportional to the perpendicular on the intercept on the curve.*
For $a \cdot \gamma \div \beta \cdot \delta = a' \cdot \gamma' \div \beta' \cdot \delta'$ and $a = a'$.

**Ex. 6.** *The product of the perpendiculars from any point on a parabola on two fixed diameters is proportional to the perpendicular on the intercept on the curve.*

128        *Anharmonic Properties of*        [CH.

**4.** *Any number of fixed points on a conic subtend homographic pencils at variable points on the conic.*

Let the fixed points be $A, B, C, D, \ldots$ and take two other points $P, Q$ on the conic; we have to prove that
$$P(ABCD\ldots) = Q(ABCD\ldots).$$
This follows at once from the fact that
$$P(ABCD) = Q(ABCD),$$
where $ABCD$ are any four of the fixed points.

**Ex. 1.** *$P, U, V$ are points on a hyperbola, $P$ being variable; show that the lines $PU$ and $PV$ intercept on either asymptote a constant length.*

Instead of the asymptote consider at first a chord $LM$ of the conic, and let $PU, PV$ cut $LM$ in $p$ and $p'$. Then $(p) = U(P) = V(P) = (p')$. And the common points of the homographic ranges $(p)$ and $(p')$ are seen, by taking $P$ at $L$ and $M$, to be $L$ and $M$. Hence in the given case the common points coincide at infinity; hence $pp'$ is constant.

**Ex. 2.** *Through a fixed point are drawn lines parallel to the rays of the pencils subtended at two points on a parabola by the other points on the parabola; show that corresponding lines cut off on a fixed diameter a constant length.*

Join the ranges determined on the line at infinity to the fixed point and proceed as above.

**Ex. 3.** *The fixed line $DA$ meets a fixed conic in $A$, and $EB$ touches at a fixed point $B$. A point $O$ is taken on the conic. Through $A$ is drawn a variable line meeting the conic again in $P$ and $EB$ in $Q$. $OP$ meets $DA$ in $U$ and $OQ$ meets $DA$ in $V$. Find the position of $O$ when $UV$ is of constant length.*

First take $EB$ to be a chord $BC$. Then
$$(U) = O(U) = O(P) = A(P) = (Q) = O(Q) = (V).$$
And the common points are where $OB$ and $OC$ meet $DA$. In the given case therefore these coincide. And they must be at infinity. Hence $OB$ is parallel to $DA$.

**5.** *The locus of the meets of corresponding rays of two homographic pencils, at different vertices and not in perspective, is a conic which passes through the vertices.*

Let the pencils be $O(PQR\ldots)$ and $V(PQR\ldots)$. Then we have to prove that the locus of the points $PQR\ldots$ is a conic through $O$ and $V$. Since the pencils are not in perspective, corresponding to the ray $VO$ in the $V$ pencil, we shall have some ray $OU$, say, in the $O$ pencil which does not coincide with $VO$. Draw any circle touching $OU$ at $O$. Let this circle cut $OV$ in $V'$, $OP$ in $P'$, and so on.

Now $V(OPQ\ldots) =$ by hypothesis $O(UPQ\ldots)$
$$= O(UP'Q'\ldots) = O(OP'Q'\ldots) = V'(OP'Q'\ldots).$$

from the circle. Hence the two pencils $V(OPQ...)$ and $V'(OP'Q'...)$ are homographic. And they have a common ray, viz. $VV'O$. Hence they are in perspective. Hence

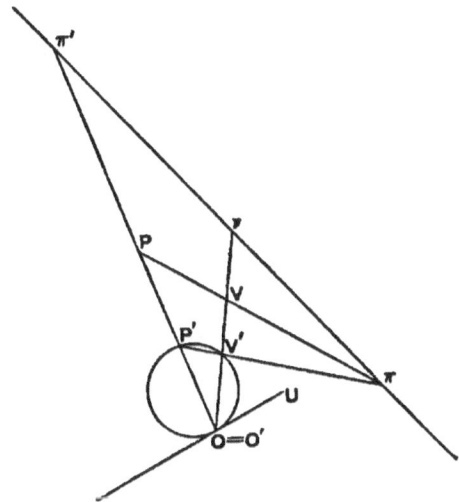

all the points $(VP\ ;\ V'P')$, $(VQ\ ;\ V'Q')$,... lie on a line, viz. the axis of perspective. Let $(VP\ ;\ V'P')$ be called $\pi$; and let the axis meet $OV$ in $v$ and $OP$ in $\pi'$; so for $Q$, $R$,....

Now rotate the figure of the circle out of the original plane about the axis $\pi\pi'$...; and let $O'$ be the new position of $O$. Then the triangles $OPV$ and $O'P'V'$ are coaxal; for $OP$ and $O'P'$ meet in $\pi'$, and $OV$ and $O'V'$ meet in $v$, and $PV$ and $P'V'$ meet in $\pi$. Hence these triangles are copolar, i. e. $OO'$, $PP'$, $VV'$ meet in a point. Hence $PP'$ passes through a fixed point, viz. the meet of $OO'$, $VV'$. Hence the figure $OVPQR...$ is the projection of the figure $O'V'P'Q'R'...$. But the latter figure is a circle; hence the locus of $P$ is the projection of a circle, i. e. is a conic. Also, since the circle passes through $V'$ and $O'$, the conic passes through $V$ and $O$.

Notice that if the pencils are in perspective, the locus degenerates into a conic consisting of the axis of perspective and the join of the vertices.

**6.** *One, and only one, conic can be drawn through five given points.*

Let the five points be $A, B, C, D, E$. Take $A$ and $B$ as vertices. Through $A$ draw any ray $AP$, and let $BQ$ be such that $A(CDEP) = B(CDEQ)$. Then the rays $AP$ and $BQ$ generate homographic ranges of which $AC$ and $BC$, $AD$ and $BD$, $AE$ and $BE$ are corresponding rays. Hence the locus of the meet $R$ of the rays $AP$ and $BQ$ is a conic through $ABCDE$. Hence a conic can be drawn through $ABCDE$.

Also only one conic can be drawn through $ABCDE$. For the other point $R$, in which any ray $AP$ cuts a conic through $ABCDE$, is given by the relation $A(CDER) = B(CDER)$. Hence every ray through $A$ cuts all conics through $ABCDE$ in the same point, i.e. all the conics coincide.

*The locus of points at which four given points subtend a constant cross ratio is a conic through the given points.*

Let the points $ABCD$ subtend the same cross ratio at $E, P, Q, R\ldots$. Then, taking $E$ and $P$ as vertices, since

$$E(ABCD) = P(ABCD),$$

we know that $ABCDEP$ lie on a conic. Hence the locus of $P$ is the conic drawn through the five fixed points $A, B, C, D, E$.

**7.** *Every two conics cut in four points.*

Two conics cannot cut in more than four points; for if they have five points in common, they must coincide. Also we see that two equal ellipses laid across one another cut in four points. Hence we conclude that if two conics do not apparently cut in four points, some of the meets are imaginary or coincident. (See also XXVII. 4.)

*Through four given points can be drawn an infinite number of conics.*

For we can draw a conic through the four given points and any fifth point.

*All conics through four given points have a common self-conjugate triangle;* viz. the harmonic triangle of the quadrangle formed by the points.

**Ex. 1.** *Any four points A, B, C, D are taken, and M is the middle point of AC; BQ a parallel to AC cuts DM in Q, and DP a parallel to AC cuts BM in P; show that ABCDPQ lie on a conic.*
   For $P(AC, BD) = Q(AC, BD) = -1$ on $AC$.

**Ex. 2.** *Through four given points can be drawn one and only one rectangular hyperbola.*
   For a fifth point is the orthocentre of any of the other three. An exception is when this orthocentre coincides with the fourth point, when an infinite number of rectangular hyperbolas can be drawn through the four points.

**Ex. 3.** *Given in position two pairs of conjugate diameters of a conic and a point P on the conic, to construct it.*
   Through $P$ draw parallels to a pair of conjugate diameters; this gives two more points on the conic. Proceeding similarly with the other pair, we have five points on the conic.

**Ex. 4.** *If P, Q, A, B, C, D be six points on a conic; show that the meets of PA and QB, of PB and QA, of PC and QD, and of PD and QC lie on a conic through PQ.*   For $P(BCAD) = Q(BCAD) = Q(ADBC)$.

**Ex. 5.** *The sides PQ, QR, RP of a triangle inscribed in a conic meet a diameter in Z, X, Y, and W, U, V are the reflexions of these points in the centre; show that PU, QV, RW meet on the conic.*
   Let the diameter be $LM$. Let $PU$, $QV$ meet in $N$. Then
$$P(LMRN) = Q(LMRN),$$
for $(LMYU) = (LMXV)$, since $(LMYU)$ and $(MLVX)$ are superposable.

**Ex. 6.** *If a conic coincide with its reciprocal, it must coincide also with the base conic, or have double contact with it.*
   For let the conic $a$ and the base conic $\Gamma$ meet in the point $P$. Then the reciprocal of $P$ touches $\Gamma$, and therefore $a$ at $P$. Hence $a$ and $\Gamma$ touch at $P$; so they touch at every common point.

**Ex. 7.** *In the case of Ex. 6 when a and $\Gamma$ have double contact, if R be the point where the reciprocal of any point Q on a touches a, then QR passes through the pole of the chord of contact of a and $\Gamma$.*
   Let the tangent at $R$ meet the chord of contact $BC$ in $L$. Let $A$ be the pole of $BC$. Let $AR$ cut $a$ in $Q'$. Then $Q'$ is the reciprocal of $RL$. Let $AR$ cut $BC$ in $M$. Then since $AR$ is the polar of $L$ for $a$, hence $(LM, CB) = -1$. Hence $AR$ is the polar of $L$ for $\Gamma$. Hence $AR, RL$ are conjugate for $\Gamma$. Hence the reciprocal of $RL$ lies on $RA$; and also by hypothesis on $a$. Hence $Q'$ is the reciprocal of $RL$. Hence $QR$ passes through $A$.

**8.** *A, B, C, D are fixed points. CD meets AP in M and BP in N; find the locus of P, given that the ratio CM : DN is constant. Discuss the locus when AB and CD are parallel.*

Since $CM = k \cdot DN$, $M$ and $N$ generate homographic ranges on $CD$ (see X. 8). Hence
$$A(P_1 P_2 \ldots) = A(M_1 M_2 \ldots) = B(N_1 N_2 \ldots) = B(P_1 P_2 \ldots).$$
Hence the locus of $P$ is a conic through $A$ and $B$.

K 2

If $AB$ and $CD$ be parallel, it follows from elementary geometry that the locus is the line dividing $CD$ and $AB$ is the given ratio.

**Ex. 1.** *The locus of the vertex of a triangle, whose base is fixed, and whose sides cut off a constant length from a given line, is a conic, which is a rectangular hyperbola, when the constant length is equal to the length of the base.*

For in this case $AM$ and $BN$ are parallel when $AM$ is parallel to either of the bisectors of the angles between the given lines.

**Ex. 2.** *A triangle $ABC$ is such that $B$ and $C$ move on fixed lines $OL$ and $OM$, whilst its sides $BC, CA, AB$ pass through fixed points $P$, $Q$, $R$; show that the locus of $A$ is a conic passing through $R$, $Q$, $O$ and through the meet of $PQ$ and $OL$ and through the meet of $PR$ and $OM$.*

**Ex. 3.** *All but one of the vertices of a polygon move on fixed lines and each side passes through a fixed point; find the locus of the remaining vertex.*

**Ex. 4.** *The locus of $Q$ is a line. The angles $QOP$ and $QO'P$ are given, $O$ and $O'$ being fixed points. Show that the locus of $P$ is a conic.*

**Ex. 5.** *$A$, $A'$ are fixed points on a circle and the arc $PP'$ moves round the circle; show that the locus of the intersection of $AP$, $A'P'$ is a conic.*

For $A(P...) = A(P'...)$ since $\angle PAP'$ is given
$= A'(P'...)$.

**Ex. 6.** *$A$ and $M$ are fixed points, $P$ is a variable point moving on a fixed line $l$, $QM$ at right angles to $PM$ meets $PA$ in $Q$; show that the locus of $Q$ is a conic. If $l$ meet the circle on $AM$ as diameter in $B$ and $C$, show that the asymptotes of the conic are parallel to $AB$, $AC$.*

**Ex. 7.** *$A$ and $B$ are fixed points, and $P$ and $Q$ are points such that the angles $PAQ$ and $PBQ$ are constant; if $P$ describe a conic through $A$ and $B$, so will $Q$.*

**Ex. 8.** *$(PQR...)$ and $(P'Q'R'...)$ are two homographic ranges on the lines $OA$, $OB$; if the parallelogram $POP'V$ be constructed, show that the locus of $V$ is a conic.*

Viz. a conic through the points at infinity on $OA$ and $OA'$.

**Ex. 9.** *All but one of the vertices of a polygon move on fixed lines, and each side subtends a fixed angle at a fixed point; find the locus of the remaining vertex.*

**Ex. 10.** *$PCP'$ and $DCD'$ are fixed conjugate diameters of an ellipse. On $CP$ and $CD$ are taken $X$ and $Y$ such that $PX.DY = 2CP.CD$. Show that $DX$ and $PY$ meet on the given ellipse.*

For $X$ and $Y$ generate homographic ranges of which $P$ and $D$ are the vanishing points. To get the constant, take $X$ at $P'$; then $Y$ is at $C$.

**Ex. 11.** *$EF$, $FD$, $DE$ pass through the fixed points $A$, $B$, $C$. The centroid of $DEF$ is fixed at $G$. $AG$ is produced to $H$, so that $GH = 2.AG$. Show that the locus of $D$ is a conic through $BCGH$.*

For $D(GH, BC) = -1$ on $EF$.

**Ex. 12.** *$Q$ moves on a fixed line, $PQ$ passes through a fixed point, the angle $QAP$ is constant, and $A$ is a fixed point. Find the locus of $P$.*

**Ex. 13.** *A variable line $PQ$ passes through a fixed point $D$ and meets the fixed* lines $AB$ and $AC$ in $P$ and $Q$. Through $P$ and $Q$ are drawn $PR$ and $QR$ in given directions. Show that the locus of $R$ is a hyperbola with asymptotes in the given directions; and find where the locus meets $AB$ and $AC$.

**9.** *The locus of the meets of corresponding rays of two pencils whose corresponding angles are equal but measured in opposite directions is a rectangular hyperbola with the vertices of the pencils at the ends of a diameter.*

The locus is clearly the locus of the meets of corresponding rays of two homographic pencils, i.e. is a conic through the vertices of the pencils.

Let $OP$ be one of the rays of the pencil at $O$ and $O'P'$ the corresponding ray of the pencil at $O'$. Through $O$ draw $Op'$ parallel to $O'P'$. Then clearly all the angles $POp'$ have the same bisector. Now draw this bisector $OL$ and its perpendicular $OM$, and the parallels $O'L'$ and $O'M'$. Then $OL$ and $O'L'$ correspond and are parallel, hence their meet is at infinity; hence $OL$ is parallel to an asymptote of the conic. Similarly $OM$ is parallel to an asymptote of the conic. Hence the conic is a rectangular hyperbola.

Again, the ray corresponding to $OO'$, viz. the tangent at $O'$, is parallel to the reflexion in $OL$ of $OO'$; and the ray corresponding to $O'O$, viz. the tangent at $O$, is the reflexion in $OL$ of $O'O$. Hence the tangents at $O$ and $O'$ are parallel, i. e. $OO'$ is a diameter.

**Ex. 1.** *The point of trisection of a given arc of a circle may be constructed as one of the meets of the arc with a rectangular hyperbola.*

Let $AB$ be the arc and $BT$ the tangent at $B$. Let $C$ be the centre of the circle. Make the angle $ACP$ equal to the angle $TBP$. Then if $P$ is on the arc we have $\angle BCP = 2 \angle ACP$. If $P$ is not on the arc, the locus of $P$ is a rectangular hyperbola; and if $Q$ be that meet of the circle and the rectangular hyperbola which lies between $A$ and $B$, $Q$ trisects the arc $AB$.

The other meets trisect the other arc $AB$ and the arc supplementary to $AB$.

**Ex. 2.** *The locus of the points of contact of parallel tangents to a system of confocal conics is a rectangular hyperbola through the foci. Prove this, and obtain the reciprocal property of coaxal circles.*

**10.** Converse of Pappus's theorem. *If a point move so that its perpendicular distances $a$, $\beta$, $\gamma$, $\delta$ from four fixed lines $AB$, $BC$, $CD$, $DA$ are connected by the relation $a \cdot \gamma = k \cdot \beta \cdot \delta$, then the locus of $P$ is a conic through $ABCD$.*

# 134 *Anharmonic Properties of Points on a Conic.*

For $\dfrac{a \cdot \gamma}{\beta \cdot \delta} \cdot \dfrac{AB}{BC} \dfrac{DC}{AD}$ is constant. Hence, reasoning as in § 3, we see that $P(AC, BD)$ is constant.

**Ex. 1.** *Given two pairs of lines which are conjugate for a circle, the locus of the centre of the circle is a rectangular hyperbola.*

Let $AB$, $CD$ be conjugate, and also $BC$, $AD$. Assume $O$ to be a position of the centre. From $O$ drop $OP$ perpendicular to $DC$ to meet $AB$ in $P'$. Then $P'$ is the pole of $CD$, hence $OP \cdot OP' = $ (radius)$^2$. So if $OQ$, perpendicular to $AD$, meet $BC$ in $Q'$, we have $OP \cdot OP' = OQ \cdot OQ'$. Also $OP = \gamma$, $OP' \propto a$, $OQ = \delta$, $OQ' \propto \beta$. Hence $a \cdot \gamma \propto \beta \cdot \delta$. Hence the locus of $O$ is a conic through $ABCD$. Also the orthocentre of $ADC$ gives $OP \cdot OP' = OQ \cdot OQ'$. Hence the conic is a r. h.

**Ex. 2.** *The locus of the foci of conics inscribed in a parallelogram is a r. h. circumscribing the parallelogram.*

Here $a \cdot \gamma = \beta \cdot \delta$.

**11.** *The projection of a conic is a conic.*

We have to prove that any projection of a conic can be placed in perspective with a circle. Now every projection of a conic is such that all the points on it subtend homographic pencils at two points on it; for this is true in the conic which was projected and is a projective property. Hence the projection is the locus of the meets of two homographic pencils and is therefore a conic.

# CHAPTER XII.

ANHARMONIC PROPERTIES OF TANGENTS OF A CONIC.

**1.** *Four fixed tangents of a conic cut any variable fifth tangent of the conic in a constant cross ratio.*

Consider first the circle of which the conic is the projection. Let the fixed tangents of the conic be the projections of the tangents at $ABCD$ of the circle, and let the variable tangent of the conic be the projection of the variable tangent at $P$ of the circle. Let the tangent at $A$ cut the tangent at $P$ in $a$, and so on.

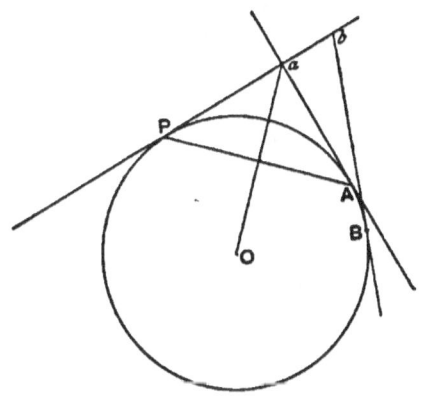

Then if $O$ be the centre of the circle, $Oa$ is perpendicular to $PA$. Hence the pencils $O(abcd)$ and $P(ABCD)$ are superposable and therefore homographic. But $P(ABCD)$ is independent of the position of $P$ on the circle. Hence $O(abcd)$, i.e. $(abcd)$, is independent of the position of the variable tangent of the circle. Hence the proposition is true for a circle; and being a projective theorem, it follows at once for the conic by projection.

The constant cross ratio $(ab, cd)$ determined on a variable tangent by four fixed tangents is called a *cross ratio of the four tangents*.

Notice that the point where a tangent cuts itself is its point of contact; for as two tangents approach, their meet approaches the point of contact of each.

Similarly *any number of tangents of a conic determine on two other tangents of the conic two ranges which are homographic.*

Notice that we have in the above proof incidentally shown that *the range determined on any tangent of a conic by several other tangents of the conic is homographic with the pencil subtended at any point on the conic by the points of contact of the other tangents.*

**Ex. 1.** *Show that the angle aOb is the same for every position of the variable tangent.*

This gives us another proof of the proposition of § 1.

**Ex. 2.** *A variable tangent of a conic meets at $Q$ and $Q'$ the tangents at the ends $P$, $P'$ of a fixed diameter of the conic; show that $PQ \cdot P'Q' = CD^2$, $CD$ being the semi-diameter conjugate to $CP$.*

For $P$ and $P'$ are the vanishing points of the ranges determined by $Q$ and $Q'$ on the tangents at $P$ and $P'$. Hence $PQ \cdot P'Q'$ is constant. To get the constant in the ellipse, take $QQ'$ parallel to $PP'$. To get the constant in the hyperbola, take an asymptote as $QQ'$. Then
$$PQ = P'Q' = CD.$$

**Ex. 3.** *If the joins of the ends $PP'$ of a diameter to a point on the conic cut the tangents at $P$ and $P'$ in $Q$ and $Q'$, show that $PQ \cdot P'Q' = 4 \cdot CD^2$.*

**Ex. 4.** *If $R$ and $R'$ be the meets of these joins and $DD'$, then $CR \cdot CR' = CD^2$, and $R$ and $R'$ are conjugate points.*

**Ex. 5.** *A variable tangent to a conic meets the adjacent sides $AB$, $BC$ of the parallelogram $ABCD$ circumscribed to the conic in $P$ and $Q$; show that $AP \cdot CQ$ is constant.*

**Ex. 6.** *A variable tangent cuts the asymptotes of a hyperbola in $T$ and $T'$; show that $CT \cdot CT'$ is constant, $C$ being the centre.*

**Ex. 7.** *Deduce the equation of a hyperbola referred to its asymptotes, viz. $xy = $ constant.*

**Ex. 8.** *$B$ and $C$ are the points of contact of tangents from $A$ to a conic. A variable tangent meets $AB$ in $P$ and $AC$ in $Q$. Show that the locus of $(BQ; CP)$ is a conic touching the given conic at $B$ and $C$.*

For $B(R) = (Q) = (P) = C(R)$. Also when $P$ approaches $B$, $R$ approaches $B$.

**Ex. 9.** *The two pairs of tangents from a pair of conjugate points meet any tangent in two pairs of harmonic points.*

Such pairs of tangents are called *harmonic pairs of tangents*.

**Ex. 10.** *If $AA'$, $BB'$ be pairs of harmonic points on a conic, show that the four tangents at $ABA'B'$ cut any fifth tangent in a harmonic range.*

**Ex. 11.** *On a fixed tangent of a conic are taken two fixed points $AB$ and also two variable points $QR$, such that $(AB, QR) = -1$; show that the locus of the meet of the other tangents from $Q$ and $R$ is the join of the points of contact of the other tangents from $A$ and $B$.*

**2.** *If $AB$, $BC$, $CD$, $DA$ touch a conic, and $p$, $q$, $r$, $s$ be the perpendiculars from $A$, $B$, $C$, $D$ on a variable tangent of the conic, then $p \cdot r = k \cdot q \cdot s$.*

Let two variable tangents cut $BC$ in $P$, $P'$ and $AD$ in $Q$, $Q'$.

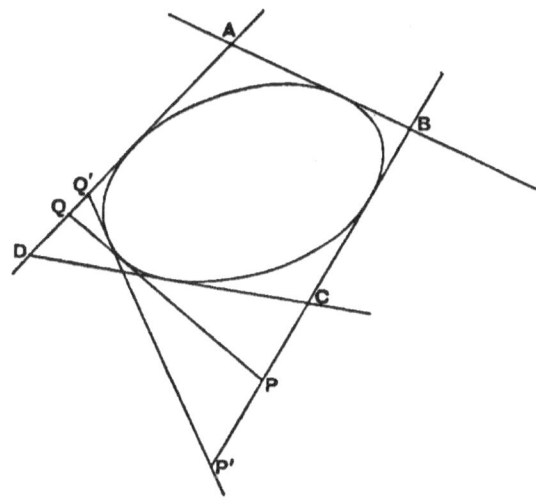

Then $(BC, PP') = (AD, QQ')$.

$$\therefore \frac{BP}{PC} \cdot \frac{P'C}{BP'} = \frac{AQ}{QD} \cdot \frac{Q'D}{AQ'};$$

$\therefore BP \cdot QD \div PC \cdot AQ$ is constant.

But $\dfrac{BP}{PC} = \dfrac{q}{r}$ and $\dfrac{AQ}{QD} = \dfrac{p}{s}$,

$\therefore p \cdot r \div q \cdot s$ is constant.

**Ex. 1.** *Extend the theorem to a $2n$-sided circumscribed polygon.*

**Ex. 2.** *Deduce a theorem concerning a n-sided circumscribed polygon.*

**Ex. 3.** *If the conic be a circle, show that $p \cdot r + q \cdot s$ is equal to*
$$OA \cdot OC + OB \cdot OD,$$
*$O$ being the centre.*
For $\sin AOQ = \sin BOP$.

**Ex. 4.** *If the conic be a parabola, then $p \cdot r = q \cdot s$.*
For taking the line at infinity as tangent, $k = p' \cdot r' \div q' \cdot s' = 1$.

**Ex. 5.** *Show that for any conic the $k$ of $p \cdot r = k \cdot q \cdot s$ is the cross ratio of the four tangents divided by the cross ratio of the pencil formed by four lines drawn parallel to them through any vertex.*

Let $PQ$ meet $AB$ in $M$ and $CD$ in $N$. Then
$$MQ \div \sin MAQ = AQ \div \sin AMQ;$$ and so on.
Hence the ratio of cross ratios corresponding to $(MN, QP)$ is
$$AQ \cdot PC \div QD \cdot BP = p \cdot r \div q \cdot s.$$

**Ex. 6.** *The lines $AB$ $BC$, $CD$, $DA$ touch a conic; one tangent meets $AB$, $CD$ in $M$, $N$ and another tangent meets $AD$, $BC$ in $P$, $Q$; show that*
$$AM \cdot BQ \cdot CN \cdot DP = AP \cdot BM \cdot CQ \cdot DN.$$

**Ex. 7.** *The sides $BC$, $CA$, $AB$ of a triangle touch a conic at $P$, $Q$, $R$; show that if $t$ be any tangent*
(i) $(P, t) \cdot (A, t) \propto (B, t) \cdot (C, t)$ ;
(ii) $(R, t) \cdot (Q, t) \propto (A, t)^2$.

**3.** *Deduce, from the theorem* $a \cdot \gamma = k \cdot \beta \cdot \delta$ *of XI.* 3, *the theorem* $p \cdot r = k \cdot q \cdot s$ *by Reciprocation.*

Call the sides of the inscribed figure in XI. 3 $a, b, c, d$; and let the reciprocals of $a, b, c, d$ be the points $A, B, C, D$ of a four-sided figure circumscribing a conic; then $p$, the reciprocal of $P$, touches this conic.

The given theorem $a \cdot \gamma = k \cdot \beta \cdot \delta$ asserts that
$$(P, a) \cdot (P, c) \div (P, b) \cdot (P, d)$$
is constant.

But by Salmon's theorem $OP/(P, a) = OA/(A, p)$, and so on.

Hence, dividing by $OP^2$, we see that
$$\frac{(A, p)}{OA} \cdot \frac{(C, p)}{OC} \div \frac{(B, p)}{OB} \cdot \frac{(D, p)}{OD}$$
is constant.

Now $O$ is a fixed point, hence
$$(A, p) \cdot (C, p) \div (B, p) \cdot (D, p)$$
is constant, i.e. $p \cdot r \div q \cdot s$ is constant.

**Ex. 1.** *Given any fixed point $O$ and any conic, two lines $s$ and $h$ can be found such that $OP^2 \div (P, s) \cdot (P, h)$ is constant, $P$ being a variable point on the conic.*

Viz. the lines corresponding to the foci of a reciprocal of the conic for $O$.

**Ex. 2.** *$AA'$, $BB'$ $CC'$ are the three pairs of opposite vertices of a quadrilateral circumscribed to a parabola whose focus is $S$; show that*
$$SA \cdot SA' = SB \cdot SB' = SC \cdot SC'.$$

Take the four-sided figure whose vertices are $AB'A'B$. Then $p \cdot r = q \cdot s$. Hence in the reciprocal circle we have
$$SA \cdot SA' \cdot a \cdot \gamma = SB \cdot SB' \cdot \beta \cdot \delta.$$
But $k = 1$ in the circle. Hence $SA \cdot SA' = SB \cdot SB'$, $= SC \cdot SC'$ similarly.

XII.]  *Tangents of a Conic.*  139

**Ex. 3.** *If the tangents at ABCD ... to a circle meet in LMN ..., then, t being any tangent and O the centre of the circle,*
$$\text{II}(A, t) : \text{II}(L, t) :: \text{II } OA : \text{II } OL,$$
II *denoting a product.*
  For II $(T, a)$ = II $(T, l)$ in a circle.

**4.** *The lines joining corresponding points of two homographic ranges which are on different axes and not in perspective touch a conic which touches the axes.*

Let the ranges be $(PQR...)$ and $(P'Q'R'...)$ on the axes $OP$ and $OP'$. Since they are not in perspective, the point which corresponds in the range $(P'Q'R'...)$ to $O$ will be some point $O'$ not coinciding with $O$.
Draw any circle touching $OP'$ at $O'$, and from $O$ and $P'$ draw the second tangents to this circle, meeting in $p$.
Then the range $(P)$ = range $(P')$ by hypothesis = range $(p)$ from the circle. Hence the ranges $(P)$ and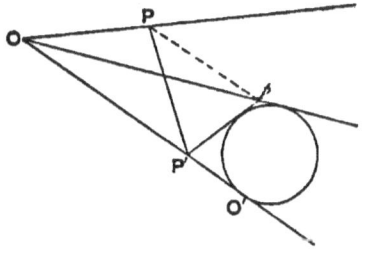
$(p)$ are homographic. Also when $P'$ coincides with $O'$, both $P$ and $p$ coincide with $O$. Hence the ranges are in perspective.

Now rotate the figure of the circle out of the original plane about the axis $OO'$. Then the ranges $(P)$ and $(p)$ are still in perspective. Hence all the lines $Pp, Qq, Rr, ...$ meet in a point, say $V$. Hence, taking $V$ as vertex of projection, $p$ projects into $P$, and therefore the line $P'p$ into the line $P'P$. Hence, since $P'p$ in all positions touches a circle, $P'P$ in all positions touches the projection of a circle, i.e. a conic. Also, since the circle touches $Op$ and $OP'$, the conic touches the projections of these lines, viz. $OP$ and $OP'$.

Notice that if the ranges be in perspective the envelope of $PP'$ degenerates into the centre of perspective and the meet of the axes.

**5.** *One, and only one, conic can be drawn touching five given lines.*

*The envelope of a line which cuts four given lines in a given cross ratio is a conic touching the given lines.*

These propositions can be proved like the reciprocal propositions in XI. 6 or they may be deduced from these by Reciprocation.

**6.** *Every two conics have four common tangents.*

Two conics cannot have more than four common tangents; for if they had five, they would coincide. Also we see that two equal ellipses laid across one another have four common tangents. Hence we conclude that if two conics have not apparently four common tangents, some of the tangents are imaginary, or coincident. (See also XXVII. 4.)

*Touching four given lines can be drawn an infinite number of conics.*

For we can draw a conic touching the four given lines and any fifth line.

*All the conics which touch four given lines have a common self-conjugate triangle,* viz. the harmonic triangle of the quadrilateral formed by the common tangents.

**Ex. 1.** *Given two homographic ranges $ABC...$ and $A'B'C'...$ on different lines; show that two points can be found at each of which the segments $AA'$, $BB'$, $CC'$, ... subtend the same angle.*

Viz. the foci of the touching conic.

**Ex. 2.** *There are also two points at which $AA'$, $BB'$, $CC'$.... subtend angles having the same bisectors.*

Let the enveloped conic touch the lines in $P$ and $Q$. The required points are the meets of $PQ$ with the director; as may be shown by reciprocating for one of these meets.

**Ex. 3.** *The vertices $A$, $B$, $C$ of a triangle lie on the fixed lines $MN$, $NL$, $LM$. and the sides $BA$, $AC$ pass through the fixed points $W$ and $V$; show that the envelope of $BC$ is a conic touching the five lines $LM$, $LN$, $VW$, $NV$, $MW$.*

**Ex. 4.** *All but one of the sides of a polygon pass through fixed points and each vertex moves on a fixed line; find the envelope of the remaining side.*

**Ex. 5.** *From the variable point $O$ situated on a fixed line are drawn the lines $OA$, $OB$, $OC$ to the fixed points $A$, $B$, $C$, meeting $BC$, $CA$, $AB$ in $X$, $Y$, $Z$; $BC$, $YZ$ meet in $X'$, $CA$, $ZX$ meet in $Y'$, and $AB$, $XY$ meet in $Z'$. Show that the line $X'Y'Z'$ envelopes a conic which touches each side of the triangle at the fourth harmonic of the fixed line for the side.*

By a previous example $X'Y'Z'$ are collinear. Also

$(O) = A(O) = (X) = (X')$ since $(BC, XX')$ is harmonic

$= (Y')$ similarly.

Hence $X'Y'$ envelopes a conic. Let the locus of $O$ meet $BC$ in $P$. Then when $O$ coincides with $P$, $X$ coincides with $P$, $X'$ coincides with $P'$ where $(PP', BC) = -1$, $Y$ and $Y'$ coincide with $C$, and $Z$ and $Z'$ coincide with $B$. Hence $BC$ touches at $P'$.

**Ex. 6.** *Reciprocate the previous example.*

**Ex. 7.** *The vertices BC of a triangle lie on given lines and the vertex A lies on a conic on which also lie fixed points VW through which the sides CA, AB pass. Show that the envelope of BC is a conic touching the given lines.*

**Ex. 8.** *The side BC of a triangle touches a conic, and the vertices B and C move on fixed tangents of this conic, whilst the sides AB, AC pass through fixed points; show that the locus of A is a conic through the fixed points.*

**Ex. 9.** *If $e\,(ab, cd)$ mean the cross ratio determined on the line $e$ by the lines $a, b, c, d$; show that*
$$e\,(ab, cd) \cdot c\,(ab, de) \cdot d\,(ab, ec) = 1,$$
*where $a, b, c, d, e$ are any five lines.*
Estimate the cross ratios on any tangent to the conic touching $abcde$.

**Ex. 10.** *Show that the problem—'To find a line on which five given lines, no three of which are concurrent, shall determine a range homographic with a given range'—has four solutions.*

**Ex. 11.** *Given in position two pairs of conjugate diameters of a conic and a tangent, construct the conic.*
Construct the parallel tangent (which is equidistant from the centre). Let these tangents cut a pair of conjugate diameters in $LL'$ and $MM'$. Then $LM$ and $L'M'$ also touch the conic. Proceeding similarly with the other pair, we have seven tangents.

**Ex. 12.** *Given in position a pair of conjugate diameters and two tangents, construct the conic.*

**Ex. 13.** *Prove the converse of § 2.*

**7.** *If two quadrangles have the same harmonic points, then their eight vertices lie on a conic; as a particular case, if any three of the vertices are collinear, the eight vertices lie on two lines.*

Let $ABCD$, $A'B'C'D'$ be the two given quadrangles, and $UVW$ the common harmonic triangle.

If no three of the eight vertices lie on a line, we can draw a conic through any five, say $A'$, $B'$, $C'$, $D'$ and $A$. Then from the inscribed quadrangle $A'B'C'D'$ we see that $UVW$ is a self-conjugate triangle for this conic. Also by hypothesis $UVW$ is the harmonic triangle of the quadrangle $ABCD$. Hence (see figure of V. 9) $B$ is such that $(WANB)$ is harmonic; hence $B$ is on the conic, for $A$ is on the conic, and $W$ is the pole of $UV$; similarly $C$ and $D$ are on the conic.

Hence $ABCDA'B'C'D'$ lie on a conic.

If three of the vertices lie on a line, say $ACD'$, then since $B'D'$ passes through $V$ we see that $B'$ also lies on $AC$. Again, $BD$ and also $A'C'$ form with $AC$ or $B'D'$ a pair harmonic with $VU$ and $VW$. Hence $BD$ and $A'C'$ coincide. Hence the eight vertices lie on two lines, i.e. on a conic.

**Ex. 1.** *Prove that two quadrilaterals which have the same harmonic triangle are such that the eight sides touch a conic (which may be two points).*

**Ex. 2.** *A conic can be drawn through the eight points of contact of two conics inscribed in the same quadrilateral.*

**Ex. 3.** *The eight tangents at the four meets of any two conics touch the same conic.*

**8.** *Any number of tangents of a parabola determine on two other tangents of the parabola two ranges which are similar.*

Let the two ranges be $(PQR...)$ and $(P'Q'R'...)$. Let $\Omega$ and $\Omega'$ be the two points at infinity upon the lines $PQ$ and $P'Q'$. Then since the line at infinity touches the parabola, the line $\Omega\Omega'$ is a tangent. Hence the two ranges $(\Omega PQR...)$ and $(\Omega'P'Q'R'...)$ are homographic; also the points at infinity $\Omega\Omega'$ correspond. Hence the ranges are similar.

Conversely, *the lines joining corresponding points of two similar ranges which are on different axes and not in perspective touch a parabola which touches the axes.*

For if the ranges $(PQR...)$ and $(P'Q'R'...)$ are similar, the ranges $(\Omega PQR...)$ and $(\Omega'P'Q'R'...)$ are homographic. Hence the lines $\Omega\Omega'$, $PP'$, $QQ'$, ... all touch a conic which touches $PQ$ and $P'Q'$. And this conic is a parabola since $\Omega\Omega'$ touches it.

**Ex. 1.** *One and only one parabola can be drawn touching four given lines.*

**Ex. 2.** *The envelope of a line which cuts three given lines in a constant ratio is a parabola.*

**Ex. 3.** *Every two parabolas have three finite common tangents.*

**Ex. 4.** *Touching three given lines can be drawn an infinite number of parabolas.*

**Ex. 5.** *$TP$, $TP'$ touch a parabola at $P$ and $P'$, and cut a third tangent in $Q$, $Q'$; show that $QP:TP::TQ':TP'$.*

For $(QT, P\Omega) = (Q'P', T\Omega')$, considering the ranges determined on the two tangents $TP$, $TP'$ by the four tangents $QQ'$, $TP'$, $PT$, $\Omega\Omega'$.

**Ex. 6.** *If $QQ'$ touch at $R$, then $PQ/QT = QR/RQ'$.*

**Ex. 7.** *Through the fixed points A, B is drawn a variable circle meeting fixed lines through A in P, Q ; show that PQ envelopes a parabola.*

**Ex. 8.** *The envelope of the axes of conics which touch two given lines at given points is a parabola.*

Let $TP$, $TP'$ be the fixed tangents. Then

$$PG : P'G' = Pg : P'g' = CD : CD' = TP : TP', \text{ which is constant.}$$

**Ex. 9.** *The normals at the points P and P' on a conic, the chord PP' and the axes of the conic touch a parabola.*

**Ex. 10.** *Determine a line which shall meet given lines $AA'$, $BB'$, $CC'$ in points P, Q, R such that $AP = BQ = CR$.*

On $AA'$, $BB'$ take $X$, $Y$ such that $AX = BY$, and construct a parabola $a$ touching $AA'$, $BB'$, $AB$, $XY$. On $CC'$ take $Z$ such that $BY = CZ$, and construct a parabola $\beta$ touching $BB'$, $CC'$, $BC$, $YZ$. Let $PQR$ be either of the remaining two common tangents of the parabolas. Then $PQR$ is one position of the required line. For $(A\Omega, XP) = (B\Omega, YQ) = (C\Omega, ZR)$ (the $\Omega$s being different). Hence

$$AX + AP = BY + BQ = CZ + CR, \text{ i.e. } AP = BQ = CR.$$

**Ex. 11.** *The ends PQ of a segment move on fixed lines, and the orthogonal projection of PQ on a fixed line is of constant length; show that the envelope of PQ is a parabola whose axis is in the direction of the projecting lines.*

Let $pq$ be the projection of $PQ$. Then range $(P)$ is similar to range $(p)$, which is equal to range $(q)$, which is similar to range $(Q)$. Also when $pq$ approaches infinity, $PQ$ approaches being perpendicular to $pq$.

**Ex. 12.** *From points P on one line are drawn perpendiculars PQ, PR on two other lines, show that QR touches a parabola.*

**Ex. 13.** *If through any point parallels be drawn to the tangents of a parabola, a pencil is constructed homographic with the range determined by the tangents on any tangent.*

**Ex. 14.** *If through points of a range on a given line there be drawn lines parallel to the corresponding rays of a pencil, which is homographic with the given range, these lines will touch a parabola.*

**Ex. 15.** *If all the tangents of a parabola be turned through the same angle and in the same direction about the points where they meet a tangent, they will still touch a parabola.*

**Ex. 16.** *If the angle OPQ be constant, O being a fixed point and P moving on a fixed line, show that PQ envelopes a parabola.*

# CHAPTER XIII.

### POLES AND POLARS.  RECIPROCATION.

**1.** *A RANGE formed by any number of points on a given line is homographic with the pencil formed by the polars of these points for a conic.*

Consider the circle of which the conic is the projection. Let $A, B, \ldots$ on the line $p$ be the points in the figure of the circle which project into the points on the given line in the figure of the conic.

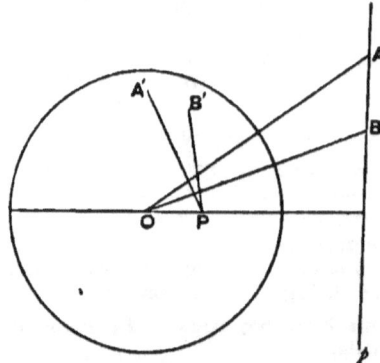

Now since $A, B, \ldots$ lie on $p$, the polars $PA'$, $PB'$, ... all pass through $P$, the pole of $p$. Also $PA'$ is perpendicular to $OA$, $O$ being the centre of the circle. Hence the pencil $P(A'B'\ldots)$ is superposable to and therefore homographic with the pencil $O(AB\ldots)$, and is therefore homographic with the range $(AB\ldots)$. Hence the proposition is true for a circle; and being a projective theorem, it is true for the conic by projection.

Taking the base conic as the given conic, the theorem becomes—

*The reciprocal of a range of points is a pencil of lines which is homographic with the given range.*

# Poles and Polars. Reciprocation.

**Ex. 1.** *Through a fixed point O is drawn a variable line cutting a fixed line in Q' and a fixed conic in PP'. If (PP', QQ') be harmonic, show that the locus of Q is a conic passing through O, through the pole of the fixed line, through the meets of this line with the conic, and through the feet of the tangents from O.*

For $O(Q) = (Q') = V(Q)$, $V$ being the pole of the locus of $Q'$.

**Ex. 2.** *Obtain the reciprocal theorem to that of example 1.*

**Ex. 3.** *If on fixed lines OL and OL' points PP' be taken which are conjugate for a fixed conic, show that PP' envelopes a conic which touches OL, OL' and also the four tangents to the fixed conic at its meets with OL and OL'.*

The join of $P'$ to the pole of $OL$ is the polar of $P$.

**Ex. 4.** *If OL, OL' be conjugate lines, then the envelope degenerates into two points; also if O be on the conic.*

**Ex. 5.** *Two vertices of a triangle self-conjugate for a given conic move on fixed lines; show that the locus of the third vertex is a conic passing through the intersections of the given lines with the given conic and through the poles of the given lines for the given conic.*

**Ex. 6.** *AA' are a pair of opposite vertices of a quadrilateral whose sides touch a conic at L, M. N, R. Through A and A' are drawn conjugate lines meeting in P. Show that the locus of P is the conic AA'LMNR.*

**Ex. 7.** *AP, AQ, harmonic with two fixed lines through A, meet a conic in P. Q; show that the envelope of PQ is a conic touching the fixed lines at points on the polar of A, and touching the tangents to the conic at the points where the fixed lines meet it.*

For $PQ$ meets the fixed lines in conjugate points.

**Ex. 8.** *Through a fixed point O is drawn a variable line, and PY is the perpendicular on this line from its pole P for a fixed conic; show that PY envelopes a parabola, which touches the polar of O, and also touches the tangents at the feet of the normals from O.*

Let $PY$ cut the line at infinity in $Q$. Through any point $V$ draw $Vq$ parallel to $PY$; then $Vq$ passes through $Q$. Hence

$$(Q_1 Q_2 \ldots) = V(q_1 q_2 \ldots) = O(Y_1 Y_2 \ldots)$$

[corresponding rays being perpendicular] $= (P_1 P_2 \ldots)$. Hence $PQ$, i.e. $PY$, envelopes a conic touching $P_1 P_2$ and $Q_1 Q_2$, i.e. the polar of $O$ and the line at infinity. This parabola touches the tangent at $R$, a foot of a normal from $O$; for if $OY$ be $OR$, then $PY$ is the tangent at $R$.

**Ex. 9.** *If instead of being perpendicular to the variable line, PY make a given angle with it; show that PY envelopes a parabola, which touches the polar of O, and also touches the tangents at the points where the tangents make the above angle with the radii from O.*

**Ex. 10.** *If the given angle be the angle between the polar of O and the conjugate diameter, the envelope reduces to a point; and the locus of Y is a circle.*

For when $P$ is at infinity, $Q$ coincides with it.

**Ex. 11.** *If through every point on a line, there be drawn the chord of a conic which is bisected at this point, the envelope of these chords is a parabola which touches the line.*

Consider the pencil of diameters.

**Ex. 12.** *Through points PQ... on the line l are drawn the lines PP', QQ',... parallel to the polars of P, Q,... for a conic; show that PP', QQ',... touch a parabola which touches l.*

L

**Ex. 13.** *The reciprocals of the four points $A, B, P, Q$ are the four lines $a, b, p, q$; show that*

$$\frac{(P, a)}{(P, b)} + \frac{(A, p)}{(B, p)} = \frac{(Q, a)}{(Q, b)} + \frac{(A, q)}{(B, q)}.$$

Let $PQ$ cut $a$ in $L$ and $b$ in $M$; also let $AB$ cut $p$ in $N$ and $q$ in $U$. Then we have to prove that $(PQLM) = (ABNU)$; and this is true, for the polars of $P, Q, L, M$ are $ON, OU, OA, OB$, if $O$ be the meet of $p$ and $q$.

**Ex. 14.** *Show that*

$$\frac{(P, a)}{(P, b)} + \frac{(A, p)}{(B, p)} = \frac{(B, a)}{(A, b)} = \frac{(C, a)}{(C, b)}.$$

Take $Q$ successively at $B$ and at the centre $C$.

**2.** *The reciprocal of a conic for a conic is a conic.*

We may define the original conic as the locus of a point $P$ such that $P(ABCD) = E(ABCD)$, where $A, B, C, D, E$ are fixed points on the conic. Let the reciprocals of the points $A, B, C, D, E, P$ be the lines $a, b, c, d, e, p$. Now the reciprocal of the pencil $P(ABCD)$ is the range of points determined on the line $p$ by the lines $a, b, c, d$. Hence this range is homographic with $P(ABCD)$. So the range of points determined on $e$ by $a, b, c, d$ is homographic with $E(ABCD)$, i.e. with $P(ABCD)$, i.e. with the range of points determined on $p$ by $a, b, c, d$. Hence the reciprocal of the given conic, viz. the envelope of $p$, the reciprocal of $P$, is the envelope of a line which cuts four given lines $a, b, c, d$ in a constant cross ratio. Hence the reciprocal is a conic touching $a, b, c, d, e$.

**3.** *The reciprocal of a pole and polar for a conic is a polar and pole for the reciprocal conic.*

Let $P$ be the pole and $e$ its polar. Through $P$ draw any line $r$ cutting $e$ in $P'$ and the conic in $Q, Q'$. Then $(PP', QQ')$ is harmonic. Let the reciprocals of $P, e, r, P', Q, Q'$ be $p, E, R, p', q, q'$. Then on a fixed line $p$ is taken a variable point $R$, and from $R$ are drawn the tangents $q, q'$ to the reciprocal conic, and the line $p'$ is taken such that $(pp', qq')$ is harmonic. We are given that $p'$ always passes through $E$, and we have to prove that $E$ is the pole of $p$. But this is obvious, for $p$ and $p'$ are conjugate in all positions of $p'$, since $(pp', qq') = -1$. Hence $p'$ always passes through the pole of $p$, i.e. $E$ is the pole of $p$.

XIII.] *Poles and Polars. Reciprocation.* 147

**Ex. 1.** *The reciprocal of a triangle self-conjugate for a conic is a triangle self-conjugate for the reciprocal conic.*

**Ex. 2.** *A triangle self-conjugate for the base conic reciprocates into itself.*

**Ex. 3.** *A conic, its reciprocal, and the base conic have a common self-conjugate triangle.*

Viz. the common self-conjugate triangle of the given conic and the base conic.

**4. Given any two conics, a base conic can be found for which they are reciprocal.**

Of the two given conics $a$ and $\beta$, let $P$ be a common point, $q$ a common tangent, and $UVW$ the common self-conjugate triangle. Describe by XXV. 12 the conic $\Gamma$ for which $UVW$ is a self-conjugate triangle and $P$ is the pole of $q$. Then $\Gamma$ is the required base conic.

For let $a'$ be the reciprocal of $a$ for $\Gamma$. Then since $P$ is on $a$, its reciprocal $q$ touches $a'$. Again, since $q$ touches $a$, its reciprocal $P$ is on $a'$. Also since $UVW$ is self-conjugate for $a$ and $\Gamma$, it is self-conjugate for $a'$. Hence $a$, $a'$ and $\beta$ pass through $P$, touch $q$, and have $UVW$ as a self-conjugate triangle.

Now by V. 9 to be given a point and a self-conjugate triangle is equivalent to being given four points. Hence $a$, $a'$ and $\beta$ pass through the same four points and touch the same line. But by XXI. 3, Ex. 4, two, and only two, conics can satisfy these conditions. Hence $a'$ coincides with $a$ or $\beta$.

Now if the meets of the conics are distinct, $a'$ cannot coincide with $a$. For let $q$ touch $a$ at $R$. Then, by XI. 7, Ex. 6 and 7, $a$ and $\Gamma$ have double contact, and $PR$ passes through the common pole $A$ of the chord of contact $BC$. Now $A$ is the pole of $BC$ for $a$ and $\Gamma$. Hence $A$ must be $U$ or $V$ or $W$. Hence $PR$ passes through $U$ or $V$ or $W$. Hence $PR$ is a common chord of $a$ and $\beta$, i.e. $R$ is a common point; which is impossible unless $a$ and $\beta$ touch.

Hence $a'$ does not coincide with $a$. Hence $a'$ coincides with $\beta$. Hence $a$ and $\beta$ are reciprocal for $\Gamma$.

If two or more of the common points of $a$ and $\beta$ coincide, this may be taken as the limit of a case when no two coincide; and the proposition still holds.

Note that there are four base conics. For we may take any one of the four common tangents as the reciprocal of $P$. Then as the conics are reciprocal, each of the common points will have, as polar, one of the common tangents.

The above construction is imaginary unless the conics have a real common point and also a real common tangent.

**Ex.** *The cross ratio of the four common points of two conics for one of the conics is equal to the cross ratio of the four common tangents for the other conic.*

**5.** *Reciprocate—a segment divided in a given ratio.*

Let $AC$ be divided in $B$. Let $l$ be the line $AB$ and $i$ the line at infinity, and let $\Omega$ be the meet of $l$ and $i$. The reciprocals of the points $ABC\Omega$ on the line $l$ are the lines $abc\omega$ through the point $L$. Also the reciprocal of $i$ is the centre $O$ of the base conic. Hence $AB \div BC = -(AC, B\Omega)$ of the given range of points $= -(ac, b\omega)$ of the reciprocal pencil, where $\omega$ is the join of $L$ to $O$.

As a particular case *the middle point of a segment $AC$ reciprocates into the fourth harmonic for $a$ and $c$ of the join of $ac$ to the centre of the base conic.*

**Ex.** *Reciprocate the theorem—*

'*The locus of the centres of conics inscribed in a given quadrilateral is a line which bisects each of the three diagonals.*'

# CHAPTER XIV.

### PROPERTIES OF TWO TRIANGLES.

**1.** *If the vertices of two triangles lie on a conic, the sides touch a conic; and conversely.*

Let the vertices $ABC$, $A'B'C'$ of the two triangles lie on a conic. Let $AB$, $AC$ meet $B'C'$ in $L$, $M$; let $A'B'$, $A'C'$ meet $BC$ in $L'$, $M'$. Then
$$(C'LMB') = A(C'BCB')$$
$$= A'(C'BCB') = (M'BCL').$$
Hence the six lines $C'M'$, $LB$, $MC$, $B'L'$, $B'C'$, $BC$ touch a conic; i.e. $C'A'$, $AB$, $AC$, $B'A'$, $B'C'$, $BC$ touch a conic; i.e. the sides of the triangles touch a conic.

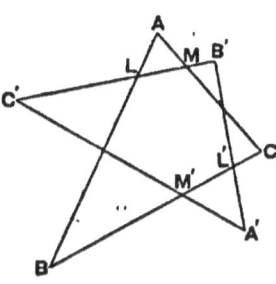

Let the sides touch a conic. Then
$$A(C'BCB') = (C'LMB') = (M'BCL') = A'(C'BCB').$$
Hence the six points $C'$, $B$, $C$, $B'$, $A$, $A'$ lie on a conic; i.e. the vertices lie on a conic.

**Ex. 1.** *If two triangles circumscribe the same conic, then a conic drawn through five of the vertices will pass through the sixth also.*

**Ex. 2.** *If two triangles be inscribed in the same conic, then a conic drawn to touch five of the sides will touch the sixth also.*

**Ex. 3.** *If two conics be such that one triangle can be drawn which is circumscribed to one conic and inscribed in the other, then an infinite number of such triangles can be drawn.*

For suppose $ABC$ to be circumscribed to $\beta$ and inscribed in $\gamma$. Draw any tangent to $\beta$ cutting $\gamma$ in $B'$ and $C'$. From $B'$ and $C'$ draw the other tangents to $\beta$ meeting in $A'$. Then, since $ABC$, $A'B'C'$ are circumscribed to $\beta$, the vertices $ABCA'B'C'$ lie on one conic; hence $A'$ lies on $\gamma$. Hence $A'B'C'$ satisfies the required conditions.

**Ex. 4.** *If BC be the points of contact of tangents from A, and B'C' be the points of contact of tangents from A' to a conic; show that the triangles ABC, A'B'C' are inscriptible in a conic, and circumscriptible to a conic.*

Let $AB$, $AC$ cut $B'C'$ in $L$, $M$; let $A'B'$, $A'C'$ cut $BC$ in $L'$, $M'$. Then $(LB'C'M)$ of poles $= A'(BL'M'C)$ of polars. Hence $(LB'C'M) = (BL'M'C)$. Hence the triangles are circumscriptible, and therefore inscriptible.

**Ex. 5.** *If O be the centre of the conic circumscribing ABC, A'B'C' (of Ex. 4), and if BC and B'C' meet in D, show that DO bisects AA'.*

For $D$ is the pole of $AA'$ for the new conic as well as for the given conic.

**Ex. 6.** *A conic is drawn through a fixed point A and through the points of contact B, C of tangents from A to a circle, so as to touch the circle at a variable point P. Show that the curvatures of all the conics at the points P are equal.*

In Ex. 4 let $A'B'C'$ coincide in $P$. Then the circle of curvature of the conic at $P$ is the circum-circle of $A'B'C'$, whose radius is one-half of that of the given circle.

**Ex. 7.** *Through a point O on a conic is drawn a line cutting the conic in p and the sides of an inscribed triangle in a, b, c; show that (abcp) is constant.*

Draw another line $O\ a'b'c'p'$ and consider the triangles $ABC$, $Opp'$.

**2.** *If two triangles be self-conjugate for a conic, the six vertices lie on a conic, and the six sides touch a conic; conversely, if the six vertices lie on a conic, or if the six sides touch a conic, the triangles are self-conjugate for a conic.*

In the figure of § 1, let $ABC$, $A'B'C'$ be self-conjugate for a conic. Then the polar of $C'$ is $A'B'$, the polar of $L$ where $B'C'$ and $AB$ meet is $A'C$, the polar of $M$ where $B'C'$ and $AC$ meet is $A'B$, and the polar of $B'$ is $A'C'$. Hence

$$(C'LMB') = A'(B'CBC') = (L'CBM') = (M'BCL').$$

Hence the six sides $C'M'$, $LB$, $MC$, $B'L'$, $B'C'$, $BC$ touch a conic; and hence the six vertices lie on a conic.

If the two triangles are inscriptible in a conic $\gamma$, describe by XXV. 12 a conic $a$ such that $ABC$ is self-conjugate for $a$, and that $A'$ is the pole of $B'C'$ for $a$. Let the polar of $B'$ for $a$ cut $B'C'$ in $C''$. Then $ABC$ and $A'B'C''$ are self-conjugate for $a$; hence $ABCA'B'C''$ lie on a conic. But this conic is $\gamma$, for the points $ABCA'B'$ lie on both conics. Hence $B'C'$ cuts $\gamma$ in three points unless $C'$ and $C''$ coincide. Hence $C'$ and $C''$ coincide. Hence $ABC$, $A'B'C'$ are self-conjugate for a conic, viz. for the conic $a$.

If the two triangles are circumscribed to a conic, they are also inscribed in a conic, and the above proof applies.

## Properties of two Triangles.

**Ex. 1.** *If two triangles be self-conjugate for a conic α, then a conic β drawn to touch five of the sides will touch the sixth also, and a conic γ drawn to pass through five of the vertices will pass through the sixth also; and γ and β are reciprocal for α.*

**Ex. 2.** *Through the centre of a conic and the vertices of a triangle self-conjugate for the conic can be drawn a hyperbola with its asymptotes parallel to any pair of conjugate diameters of the conic.*

For, adding the line at infinity, we have two self-conjugate triangles.

**Ex. 3.** *If two conics be such that one triangle can be circumscribed to one conic which is self-conjugate for the other conic, then an infinite number of such triangles can be drawn.*

Let $ABC$ be the given triangle touching conic $\beta$ and self-conjugate for conic $a$. Take any tangent $B'C'$ of $\beta$, and take its pole $A'$ for $a$; draw from $A'$ one tangent $A'B'$ to $\beta$, and take $C'$, the pole of $A'B'$ for $a$. Then, since $ABC$, $A'B'C'$ are self-conjugate for $a$, the sides touch a conic. But five sides touch $\beta$; hence the sixth side $C'A'$ touches $\beta$. Hence $A'B'C'$ satisfies the required conditions.

**Ex. 4.** *If two conics be such that one triangle can be inscribed in one conic which is self-conjugate for the other conic, then an infinite number of such triangles can be drawn.*

**Ex. 5.** *An infinite number of triangles can be described having the same circumscribing, nine-point, and polar circles as a given triangle.*

For the nine-point circle is given when the circum-circle and the polar circle are given, being half the circum-circle, taking the orthocentre as centre of similitude.

**Ex. 6.** Gaskin's theorem. *The circum-circle of any triangle self-conjugate for a conic is orthogonal to the director circle of the conic.* (See also XXIII. 5, Ex. 9.)

Let the two circles meet in $T$. Let the polar $PP'$ of $T$ for the conic meet the circum-circle in $QQ'$. Then, as in Ex. 4, since $T$ is the pole of $QQ'$, it follows that $TQQ'$ is a self-conjugate triangle for the conic. Hence $QQ'$ are conjugate points for the conic; hence if $CT$ meet $PP'$ in $V$, we have $VQ \cdot VQ' = VP^2$, for $V$ bisects $PP'$. Also $PTP'$ is a right angle. Hence $VQ \cdot VQ' = VT^2$; i.e. $CT$ touches the circum-circle. Hence the circles are orthogonal.

**Ex. 7.** *Two conics $\beta$ and $a$ are such that triangles can be circumscribed to $\beta$ which are self-conjugate for $a$; find the locus of the point from which the pairs of tangents to $a$ and $\beta$ are harmonic.*

From $P$, any point on the locus, draw tangents $PT$ and $PT'$ to $\beta$. These tangents are conjugate for $a$, for they are harmonic for the tangents to $a$. Hence the pole of $PT$, viz. $Q$, lies on $PT'$, and the pole of $PT'$, viz. $R$, lies on $PT$. Hence the triangle $PQR$ is self-conjugate for $a$. Let $ABC$ be a triangle self-conjugate for $a$ and circumscribed to $\beta$. Then since the two triangles $ABC$, $PQR$ are self-conjugate for the same conic, their sides touch a conic, i.e. $QR$ always touches $\beta$. Hence $P$, the pole of $QR$ for $a$, always lies on the reciprocal of $\beta$ for $a$.

**Ex. 8.** *If two conics $\gamma$ and $a$ are such that triangles can be inscribed in $\gamma$ which are self-conjugate for $a$, find the envelope of a line which cuts $a$ and $\gamma$ in pairs of harmonic points.*

**Ex. 9.** *If Q and R be the points of contact of the tangents from P to any conic a, and any conic γ be drawn to pass through P and to touch QR at Q, then triangles can be inscribed in γ which are self-conjugate for a.*

For $PQQ$ is such a triangle, $QQ$ being $QR$.

**Ex. 10.** *If Q and R be the points of contact of the tangents from P to any conic a, and any conic β be drawn to touch PQ at P and to touch QR, then triangles can be circumscribed to β which are self-conjugate for a.*

For $PQQ$ is such a triangle, $QQ$ being $QR$.

**Ex. 11.** *If triangles can be circumscribed to β which are self-conjugate for a, then triangles can be inscribed in a which are self-conjugate for β; and conversely.*

For we can reciprocate $a$ into $β$.

**Ex. 12.** *The triangle ABC is inscribed in the conic a, and the triangle DEF is self-conjugate for a. Show that a conic β can be found such that DEF is circumscribed to β and ABC is self-conjugate for β.*

Viz. that conic inscribed in $DEF$ for which $A$ is the pole of $BC$.

**Ex. 13.** *The centre of the circle circumscribing a triangle which is self-conjugate for a parabola is on the directrix.*

Consider the triangle $O\Omega\Omega'$ where $O\Omega$, $O\Omega'$ are the tangents to the parabola from the centre of the circle.

**Ex. 14.** *The conic a is drawn touching the lines PQ, PR at Q, R; the conic β is drawn touching the lines QP, QR at P, R; show that (i) triangles can be inscribed in a which are self-conjugate for β, (ii) triangles can be inscribed in β which are self-conjugate for a, (iii) triangles can be circumscribed to a which are self-conjugate for β, (iv) triangles can be circumscribed to β which are self-conjugate for a, (v) triangles can be inscribed in a and circumscribed to β, (vi) triangles can be inscribed in β and circumscribed to a.*

On $RP$ and $RQ$ take $L$, $L'$ consecutive to $R$; on $PR$, $QR$ take $M$, $M'$ consecutive to $P$, $Q$; on $QP$, $PQ$ take $N$, $N'$ consecutive to $Q$, $P$. Then consider the triangles (i) $QRL$, (ii) $PRL'$, (iii) $QPM$, (iv) $PQM'$, (v) $RQN$, (vi) $RPN'$.

**Ex. 15.** *If a triangle can be drawn inscribed in a and circumscribed to β and also a triangle self-conjugate for a and circumscribed to β, then the conics a and β are related as in Ex. 14.*

At $R$, one of the meets of $a$ and $β$, draw $RQ$ touching $β$ and meeting $a$ again in $Q$; draw the tangent at $Q$, and on it take $N$ consecutive to $Q$. Then by the first datum $QN$ touches $β$, at $P$ say. Then by the second datum $QR$ is the polar of $P$ for $a$, i.e. $PR$ touches $a$ at $R$.

Similarly many other converses of Ex. 14 can be proved.

**Ex. 16.** *The centre of a circle touching the sides of a triangle self-conjugate for a rectangular hyperbola is on the r. h.*

For triangles can be inscribed in the r. h. which are self-conjugate for the circle. Now one triangle self-conjugate for the circle is $O\Omega\Omega'$, and two of its vertices $\Omega\Omega'$ lie (at infinity) on the r. h.; hence $O$, the centre of the circle, lies on the r. h.

**Ex. 17.** *Given a triangle self-conjugate for a r. h., we know four points on the r. h.*

Viz. the centres of the touching circles.

**Ex. 18.** *Given a self-conjugate triangle of a conic and a point on the director, show that four tangents are known, viz. the directrices of the four conics which can be drawn to circumscribe the triangle and to have the point as corresponding focus.*
Reciprocate for the point.

**Ex. 19.** *The necessary and sufficient condition that triangles can be circumscribed to a circle which are self-conjugate for a r. h. is that the centre of the circle shall be on the r. h.*

**Ex. 20.** *An instance of Ex. 14 is a rectangular hyperbola which passes through the vertices of a triangle and also through the centre of a circle touching the sides.*
This follows from Ex. 15 and Ex. 19.

**Ex. 21.** *If two conics $\beta$ and $\gamma$ be so situated that one triangle can be circumscribed to $\beta$ so as to be inscribed in $\gamma$, then an infinite number of such triangles can be drawn, and all of these will be self-conjugate for a third conic $\alpha$; also the two conics $\beta$ and $\gamma$ are reciprocal for $\alpha$.*

The first part has been proved. To prove the third part, notice that $ABC$, $A'B'C'$ are self-conjugate for a conic $\alpha$. Define $\gamma$ by $ABCA'B'$; then since the polars of these points for $\alpha$, viz. $BC$, $CA$, $AB$, $B'C'$, $C'A'$ touch $\beta$, it follows that $\beta$ is the reciprocal of $\gamma$ for $\alpha$.
Again, take any point $A''$ on $\gamma$, and let $B''$ be one of the points in which the polar of $A''$ for $\alpha$ (which touches $\beta$) cuts $\gamma$. Let the polar of $B''$ for $\alpha$ (which touches $\beta$ and passes through $A''$) cut the polar of $A''$ in $C''$. Then the triangle $A''B''C''$ is self-conjugate for $\alpha$. Hence, since two sides touch $\beta$ and two vertices are on $\gamma$, it is circumscribed to $\beta$ and inscribed in $\gamma$.

**Ex. 22.** *Prove by this article that 'The orthocentre of a triangle inscribed in a rectangular hyperbola lies on the r. h.'*
The given triangle and the triangle formed by the orthocentre and the points at infinity on the r. h. are self-conjugate for the polar circle.

**3.** The *two triangles $ABC$, $A'B'C'$ are said to be reciprocal for a conic if $A$ be the pole of $B'C'$, $B$ of $C'A'$, $C$ of $A'B'$, $A'$ of $BC$, $B'$ of $CA$ and $C'$ of $AB$ for the conic.*

*Two triangles which are reciprocal for a conic are homologous; and conversely, if two triangles be homologous they are reciprocal for a conic.*

Let the triangles $ABC$, $A'B'C'$ be reciprocal for a conic; then they are homologous. For let $BC$ and $B'C'$ meet in $U$, and let $AA'$ meet $BC$ in $L$ and $B'C'$ in $L'$. Then the polar of $B$ is $A'C'$, the polar of $C$ is $A'B'$, the polar of $U$ where $BC$ and $B'C'$ meet is $A'A$, the polar of $L$ where $BC$ and $A'A$ meet is $A'U$. Hence $(LBCU)$ of poles $= A'(UC'B'L')$. Hence $(LBCU) = (L'B'C'U)$; hence the ranges $(LBCU)$ and $(L'B'C'U)$ are in perspective. Hence $LL'$, $BB'$, $CC'$

meet in a point, i.e. the triangles $ABC$, $A'B'C'$ are homologous.

Let the triangles $ABC$, $A'B'C'$ be homologous, then they are reciprocal for a conic. For let $BC$ and $A'C'$ meet in $M$.

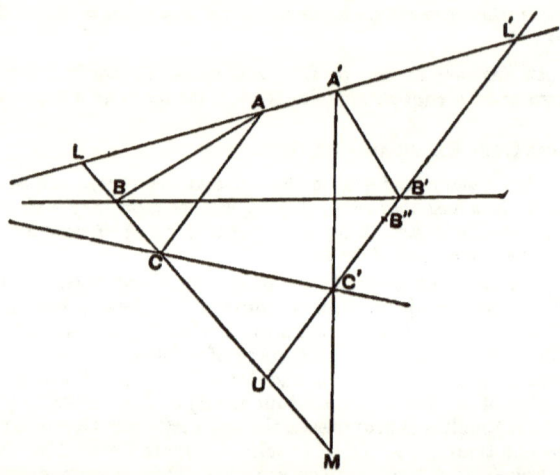

By XXV. 12 describe a conic such that the triangle $A'BM$ is self-conjugate for it, and that $A$ is the pole of $B'C'$.

Then $A'$ is the pole of $BC$, $B$ is the pole of $A'C'$, and $A$ is the pole of $B'C'$. Hence $C'$ is the pole of $AB$. Now let the polar of $C$ cut $C'B'$ in $B''$. Then the triangles $ABC$ and $A'B''C'$ are reciprocal and therefore homologous. Hence $AA'$, $BB''$, $CC'$ meet in a point. But $AA'$, $BB'$, $CC'$ meet in a point. Hence $B'$ and $B''$ coincide, i.e. the triangles $ABC$, $A'B'C'$ are reciprocal for the above conic.

Given a triangle $ABC$ and a conic $a$, we can describe the reciprocal triangle $A'B'C'$, and then determine the centre $O$ and axis $s$ of perspective of the triangles $ABC$, $A'B'C'$. It is convenient to call $O$ *the pole* and $s$ *the polar of the triangle $ABC$ for the conic $a$*.

**Ex. 1.** *If two triangles be reciprocal for a conic, show that the centre of homology of the triangles is the pole of the axis of homology for this conic.*

**Ex. 2.** *$BC$, $CA$, $AB$ meet any conic in $XX'$, $YY'$, $ZZ'$, and the conic meets $AX$ again in $L$, $AX'$ in $L'$, $BY$ in $M$, $BY'$ in $M'$, $CZ$ in $N$, $CZ'$ in $N'$. Show that $LL'$, $MM'$, $NN'$, meet $BC$, $CA$, $AB$ on a line.*

Viz. on the axis of homology of $ABC$ and its reciprocal for the conic.

## Properties of two Triangles.

**Ex. 3.** *Any triangle inscribed in a conic and the triangle formed by the tangents at the vertices are homologous.*

**Ex. 4.** Hesse's theorem. *If the opposite vertices $AA'$ and the opposite vertices $BB'$ of a complete quadrilateral be conjugate for the same conic, then the opposite vertices $CC'$ are also conjugate for this conic.* (See also XX. 1, Ex. 11.)

Let the triangle reciprocal to the triangle $ABC$ for the conic be $PQR$. Then $QR$ passes through $A'$, since $A$ and $A'$ are conjugate. So $RP$ passes through $B'$. Hence $PQ$ passes through $C'$; for the triangles $ABC$ and $PQR$ are homologous. Hence $C$ and $C'$ are conjugate.

**Ex. 5.** *If two pairs of opposite sides of a complete quadrangle be conjugate for the same conic, then the third pair is also conjugate for this conic.*

**Ex. 6.** *The points $PP'$, $QQ'$, $RR'$ divide harmonically the diagonals $AA'$, $BB'$, $CC'$ of a quadrilateral; show that the six points $P, P', Q, Q', R, R'$ lie on a conic.*

# CHAPTER XV.

## PASCAL'S THEOREM AND BRIANCHON'S THEOREM.

### Pascal's Theorem.

**1.** *The meets of opposite sides of a hexagon (six-point) inscribed in a conic are collinear.*

Let the six points be $A, B, C, D, E, F$. Let the opposite sides $AB, DE$ meet in $M$, and the opposite sides $BC, EF$

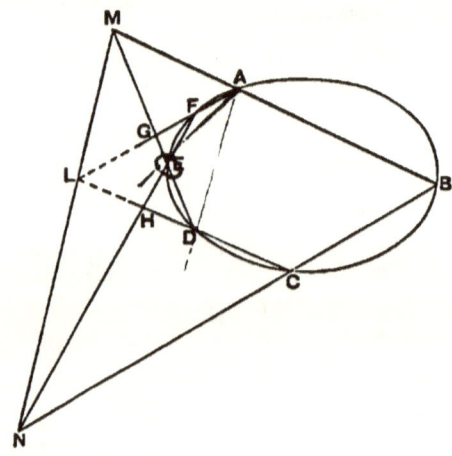

meet in $N$. Let $AF$ meet $MD$ in $G$, and let $CD$ meet $NF$ in $H$. Then we have to show that $MN, FG, HD$ are concurrent. This is true if $(EMGD) = (ENFH)$, for the ranges, having a common point, will be in perspective; i.e. if
$$A(EBFD) = C(EBFD),$$
which is true. Hence the meet $M$ of $AB, DE$, the meet $N$ of $BC, EF$, and the meet $L$ of $CD, FA$ are collinear.

## Pascal's Theorem and Brianchon's Theorem. 157

Conversely, *if the meets of opposite sides of a hexagon (six-point) are collinear, the six vertices lie on a conic.*
For if $LMN$ are collinear, we have $(EMGD) = (ENFH)$. Hence $A(EBFD) = C(EBFD)$. Hence $A, B, C, D, E, F$ lie on the same conic.

The line $LMN$ is called the *Pascal line* of the six-point $ABCDEF$. Observe that for every different order of the points $A, B, C, D, E, F$ we get a different Pascal line.

Notice that if two consecutive points, e.g. $B$ and $C$, coincide, the side $BC$ becomes the tangent at $B$ or $C$.

**Ex. 1.** *If $AD$, $BE$, $CF$ meet in a point, the Pascal line is the polar of this point.*

**Ex. 2.** *The triangles $ABC$, $A'B'C'$ are homologous. $BC$ meets $A'B'$ in $Y$ and $A'C'$ in $Z'$, $CA$ meets $B'C'$ in $Z$ and $B'A'$ in $X'$, and $AB$ meets $C'A'$ in $X$ and $C'B'$ in $Y'$. Show that*

$$BY \cdot BZ' \cdot CZ \cdot CX' \cdot AX \cdot AY' = CY \cdot CZ' \cdot AZ \cdot AX' \cdot BX \cdot BY'.$$

For $XY'ZX'YZ'$ lie on a conic.

**Ex. 3.** *In every hexagon inscribed in a conic, the two triangles formed by taking alternate sides are homologous.*

**Ex. 4.** *Six points on a conic determine 60 hexagons inscribed in the conic.*

**Ex. 5.** *The 60 Pascal lines belonging to six given points on a conic intersect three by three.*

Let the homologous triangles of any one hexagon be $XYZ$, $X'Y'Z'$. Then $XX'$, $YY'$, $ZZ'$ meet in a point. Also $XX'$ is the Pascal line of $CDEBAF$, $YY'$ of $ABCFED$, $ZZ'$ of $BCDAFE$.

**Ex. 6.** *Two triangles are inscribed in a conic. The sides of the one meet the sides of the other in nine points. Show that any join of two of these nine points is a Pascal line of the six vertices of the triangles, unless it is one of the sides of the triangles.*

**Ex. 7.** *$ABC$ is a triangle inscribed in a circle. $P$ is any point on this circle. A perpendicular at $P$ to $PA$ meets $BC$ in $D$, to $PB$ meets $CA$ in $E$, and to $PC$ meets $AB$ in $F$. Show that $DEF$ is a line passing through the centre of the circle.*

Call the centre of the circle $O$. Let $PD$, $PE$ meet the circle in $A'$, $B'$. Then $AA'PB'BC$ proves that $ODE$ are concurrent.

**Ex. 8.** *Reciprocate Ex. 7, (i) for the circle itself, (ii) for any circle.*

**Ex. 9.** *If $AOA'$, $BOB'$, $COC'$, $POP'$ be chords of a conic, show that the meets of $PA$, $B'C'$, of $PB$, $C'A'$, of $PC$, $A'B'$, of $P'A'$, $BC$, of $P'B'$, $CA$ and of $P'C'$, $AB$ all lie on the same line through $O$.*

Use $(BCC'P'A'A)$, $(B'C'CPAA')$, $(BAPP'C'B')$.

**Ex. 10.** *Taking the conic as a circle and $O$ as its centre, deduce by reciprocating for $P$ the theorem—The orthocentre of a triangle about a parabola is on the directrix.*

**Ex. 11.** $A, B, C, D, E$ are any five points. $EA, BC$ meet in $A'$; $AB, CD$ meet in $B'$; $BC, DE$ meet in $C'$; $CD, EA$ meet in $D'$; $DE, AB$ meet in $E'$; and $AD, BC$ meet in $F$. Show that $FB'$ touches the conic through $A'B'C'D'E'$.

**Ex. 12.** $AA', BB', CC'$ are the diagonals of a complete quadrilateral, $A'B'C'$ being collinear points. $AO$ meets $BC$ in $M$, $CO$ meets $AB$ in $L$, $LM$ meets $B'C'$ in $N$ and $AC$ in $P$. If $PB$ and $ON$ meet in $R$, show that $R$ is the remaining intersection of the conics $OBB'AA'$ and $OBB'CC'$, and that $OR$ is the tangent at $O$ to the conic $OCC'AA'$.

Consider the hexagons $ORBA'B'A$, $ORBC'B'C$, and $OOCA'C'A$.

**Ex. 13.** $ABC, A'B'C'$ are coaxal triangles; $AC$ and $A'B'$ meet in $P$, $AB$ and $A'C'$ meet in $Q$; show that $BCB'C'PQ$ are on a conic.

**Ex. 14.** The chord $QQ'$ of a conic is parallel to the tangent at $P$, and the chord $PP'$ is parallel to the tangent at $Q$; show that $PQ$ and $P'Q'$ are parallel.

Consider $PPP'Q'QQ$.

**Ex. 15.** The tangents at the vertices of a triangle inscribed in a conic meet the opposite sides in three collinear points.

**Ex. 16.** $PQ, PR$ are chords of a parabola. $PR$ meets the diameter through $Q$ in $V$, and $PQ$ meets the diameter through $R$ in $U$; show that $UV$ is parallel to the tangent at $P$.

Consider $PPR\Omega\Omega Q$, where $\Omega$ is the point at infinity on the parabola.

**Ex. 17.** Deduce by Reciprocation a property of a circle.

**2.** Since Pascal's theorem is true for a hyperbola however near the hyperbola approaches two lines, it is true for two lines, the six points being situated in any manner on the two lines.

But each case may be proved as in § 1.

**Ex. 1.** If any four-sided figure be divided into two others by a line, the three meets of the internal diagonals are collinear.

Let the four-sided figure $ABCD$ be divided into two others $ABFE$, $EFCD$. Now apply Pascal's theorem to $ACEBDF$.

**Ex. 2.** $P, Q, R$ are fixed points on the sides $MN, NL, LM$ of a triangle. $A$ is taken on $MN$, $AQ$ meets $LM$ in $B$, $BP$ meets $NL$ in $C$, $CR$ meets $MN$ in $A'$, $A'Q$ meets $LM$ in $B'$, $B'P$ meets $NL$ in $C'$; show that $C'A$ passes through $R$.

Consider the hexagon $BQCA'PB'$.

**Ex. 3.** On the fixed lines $LM, MN, NL$ are taken the fixed points $C, A, B$. On $BC$ is taken the variable point $P$; $NP$ meets $CA$ in $Q$, and $MP$ meets $BA$ in $R$. Show that $RLQ$ are collinear.

Consider $ACMPNB$.

**3.** *If $OQ$ and $OR$ be the tangents of a conic at $Q$ and $R$, and if $P$ be any point on the conic, then $PQ$ and $PR$ cut any line through $O$ in points which are conjugate for the conic.*

Let $PQ$ and $PR$ cut any line through $O$ in $F$ and $G$. Let $FR$ and $GQ$ meet in $U$. Consider the six-point $PQQURR$.

Then since the meets of opposite sides are collinear, the six points lie on a conic. But five points lie on the given conic; hence the sixth point $U$ also lies on the given conic. Hence $F$ and $G$ are two harmonic points of the inscribed quadrangle $PQUR$. Hence $F$ and $G$ are conjugate points.

Conversely, *if any two conjugate points lying on a line through O be joined to the points of contact of the tangents from O, then the joining lines meet on the conic.*

Let $F$ and $G$ be conjugate points on a line through $O$. Join $FQ$ cutting the conic again in $P$, and join $PR$ cutting $FG$ in $G'$. Then $F$ and $G'$ are conjugate, and also $F$ and $G$. Hence $G'$ coincides with $G$; i.e. $FQ$ and $GR$ meet on the conic. So $FR$ and $GQ$ meet on the conic.

**Ex. 1.** *If $PP'$ be conjugate points for a central conic, and $QQ'$ be the ends of the diameter which bisects chords parallel to $PP'$; show that $PQ$, $P'Q'$ cut on the conic, and so do $PQ'$, $P'Q$.*

**Ex. 2.** *If $R$ and $R'$ be conjugate points lying on a diameter of a hyperbola, show that parallels to the asymptotes through $R$ and $R'$ cut again on the curve.*

**Ex. 3.** *The diameter bisecting the chord $QQ'$ of a parabola cuts the curve in $P$, and $RR'$ are points on this diameter equidistant from $P$; show that the other lines joining $QQ'RR'$ meet on the curve.*

**Ex. 4.** *If $F$ and $G$ be conjugate points on $PQ$ and $PR$, then $FG$ and $QR$ are conjugate lines.*

**Ex. 5.** *The lines $BC$, $CA$, $AB$ touch a conic at $A'$, $B'$, $C'$. Show that an infinite number of triangles can be drawn which are inscribed in $A'B'C'$ and circumscribed to $ABC$. Show also that each of these triangles is self-conjugate for the conic.*

Through $B$ draw any line meeting $A'B'$ in $\gamma$, and $B'C'$ in $a$. Let $A\gamma$ meet $C'A'$ in $\beta$. Then $\gamma$ and $a$ are conjugate, and $a$ lies on $B'C'$; hence $a$ is the pole of $\beta\gamma$. So $\beta$ is the pole of $\gamma a$. Hence $a\beta\gamma$ is self-conjugate. Hence $a$, $\beta$ are conjugate. Hence $a\beta$ passes through $C$.

## Brianchon's Theorem.

**4.** *The joins of opposite vertices of a hexagon (six-side) circumscribing a conic are concurrent.*

Let the six sides be $AB$, $BC$, $CD$, $DE$, $EF$, $FA$. Let the four tangents $AB$, $CD$, $DE$, $EF$ meet the tangents $FA$, $BC$ in $ASPF$ and $BCQT$. Then $(ASPF) = (BCQT)$. Hence $D(ASPF) = E(BCQT)$. But the rays $DP$, $EQ$ coincide. Hence $(DA \,;\, EB)$ and $(DS \,;\, EC)$ and $(DF \,;\, ET)$ are collinear; i.e. $(DA \,;\, EB)$, and $C$ and $F$ are collinear; i.e. $DA$, $EB$, $CF$ are concurrent.

Conversely, *if the joins of opposite vertices of a hexagon (six-side) are concurrent, the six sides touch a conic.*

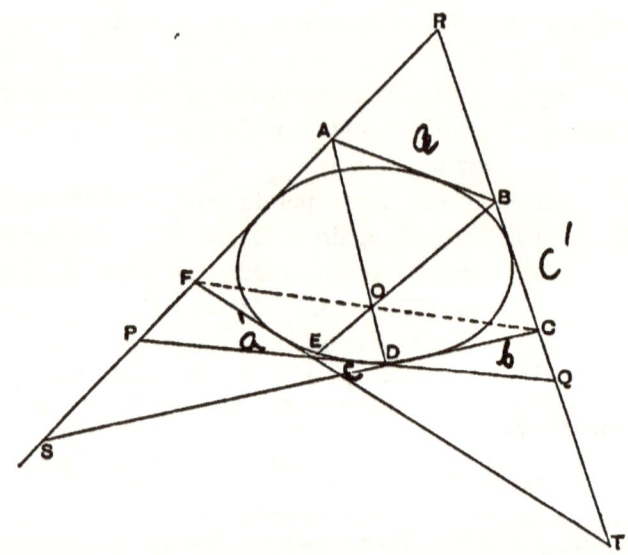

For if $DA$, $EB$, $CF$ are concurrent, we have
$$D(ASPF) = E(BCQT),$$
hence $(ASPF) = (BCQT)$; hence the six lines $AB$, $BC$, $CD$, $DE$, $EF$, $FA$ touch the same conic.

The point $O$ is called the *Brianchon point* of the hexagon $ABCDEFA$.

Notice that when two of the sides, e.g. $CD$ and $DE$, coincide, the point $D$ becomes the point of contact of either $CD$ or $DE$.

**Ex. 1.** *In every hexagon circumscribed to a conic, the two triangles formed by taking alternate vertices are homologous.*

**Ex. 2.** *Six tangents to a conic determine 60 hexagons circumscribed to the conic.*

**Ex. 3.** *The 60 Brianchon points belonging to six given tangents to a conic are collinear three by three.*

Reciprocate.

**Ex. 4.** *The hexagon formed by the six lines in order obtained by joining alternate pairs of vertices of a Brianchon hexagon is a Pascal hexagon.*

For the triangles are coaxal.

XV.] *Brianchon's Theorem.* 161

**Ex. 5.** Reciprocate Ex. 4.

**Ex. 6.** *Three angles have collinear vertices. Show that their six legs intersect in twelve other points which can be divided in four ways into a Pascal hexagon and a Brianchon hexagon.*

**Ex. 7.** *If two triangles be the reciprocals of one another for a conic $a$, the meets of non-corresponding sides lie on a conic $\beta$, and the joins of non-corresponding vertices touch a conic $\gamma$; and $\beta$ and $\gamma$ are reciprocals for $a$. If one triangle be inscribed in the other, the three conics coincide.*

**Ex. 8.** Steiner's theorem. *The orthocentre of a triangle circumscribing a parabola is on the directrix.*

Let $ABC$ be the triangle. Through $Z$, the meet of $BC$ and the directrix, draw the other tangent $Z\Omega$ where $\Omega$ is at infinity. Through $Z'$, the meet of $CA$ and the directrix, draw the other tangent $Z'\Omega'$ where $\Omega'$ is at infinity. From the circumscribing six-side $ABZ\Omega\Omega'Z'A$ we conclude that $ZZ'$, $B\Omega'$ and $A\Omega$ meet in a point. Now $ZZ'$ is the directrix; $B\Omega'$ is a parallel through $B$ to $Z'\Omega'$, i.e. $B\Omega'$ is the perpendicular from $B$ on $CA$; so $A\Omega$ is the perpendicular from $A$ on $BC$. Hence these two perpendiculars meet on the directrix; i.e. the orthocentre is on the directrix.

**Ex. 9.** *The orthocentres of the four triangles formed by taking three out of four given lines are collinear.*

**Ex. 10.** *$ABCDE$ is a pentagon circumscribing a parabola; show that the parallel through $A$ to $CD$, and the parallel through $B$ to $DE$ meet on $CE$.*

**Ex. 11.** *$ABCDA$ is a quadrilateral circumscribing a parabola; show that the parallel through $A$ to $CD$ and the parallel through $C$ to $DA$ meet on the diameter through $B$.*

**Ex. 12.** *The lines $AB$, $BC$, $CD$, $DA$ touch a conic in $L$, $M$, $N$, $R$; show that $AC$, $BD$, $LN$, $MR$ are concurrent.*

Consider $ALBCNDA$ and $ABMCDRA$.

**Ex. 13.** *The lines $BC$, $CA$, $AB$ touch a conic at $L$, $M$, $N$; show that $AL$, $BM$, $CN$ are concurrent*

**Ex. 14.** *The line $C'B'A$ touches a conic in $P$, $ACB$ touches in $P'$, $B'CA'$ touches in $Q$ and $C'BA'$ in $Q'$. Show that $A'P'$, $AQ$ meet on $CC'$, and so do $A'P$, $AQ'$.*

**Ex. 15.** *If two triangles be inscribed in a conic, their sides touch a conic.*

Consider the Pascal hexagon $ABC'A'B'C$, and the Brianchon hexagon $BC$, $CA$, $A'C'$, $C'B'$, $B'A'$, $AB$.

**Ex. 16.** *If two triangles be circumscribed to a conic, their vertices lie on a conic.*

**Ex. 17.** *If $AB$ and $AC$ touch a conic at $B$ and $C$, and $A'B'$ and $A'C'$ touch the same conic at $B'$ and $C'$, then $ABCA'B'C'$ lie on a conic and the six sides touch a conic.*

The proof is like that of Ex. 15.

5. *If $OQ$ and $OR$ be the tangents of a conic at $Q$ and $R$, and if any tangent meet $OQ$, $OR$ in $K$, $L$; then the joins of $K$ and $L$ to any point $E$ on $QR$ are conjugate lines.*

Let $LE$ cut $OQ$ in $M$, and let $KE$ cut $OR$ in $N$. Consider the six-side $KL$, $LR$, $RN$, $NM$, $MQ$, $QK$. Since $ML$, $QR$,

M

## 162 *Pascal's Theorem and Brianchon's Theorem.*

$KN$ meet in a point, the six sides touch a conic. But five sides touch the given conic; hence the sixth side $MN$ also touches the given conic. Hence $ML$, $KN$, being two harmonic lines of the circumscribed quadrilateral $KLNM$, are conjugate lines.

Conversely, *if through any point $E$ on $QR$ any two conjugate lines be drawn cutting $OQ$ in $M, K$ and $OR$ in $L, N$, then $MN$ and $KL$ touch the conic.*

For if $KL$ does not touch, let $KL'$ touch. Then $EL$ and $EL'$ are both conjugate to $EK$. Hence $L$ and $L'$ coincide. Hence $KL$ touches; so $MN$ touches.

**Ex. 1.** *Parallel to a diameter of a conic are drawn a pair of conjugate lines; show that the diagonals of the parallelogram formed by these lines and the tangents at the ends of the diameter touch the conic.*

**Ex. 2.** *Two parallel lines which are conjugate for a hyperbola meet the asymptotes in points such that the other lines joining them touch the curve.*

**Ex. 3.** *If the tangents of a parabola at $P$ and $Q$ cut in $T$, and on the diameter through $P$ there be taken any point $R$; show that $RT$ is conjugate to the parallel through $R$ to the tangent at $Q$.*

**Ex. 4.** *Through a point on the chord of contact $PQ$ of the tangents from $T$ to a parabola are drawn parallels to $TP$ and $TQ$ meeting $TQ$ and $TP$ in $R$ and $U$; show that $RU$ touches the parabola.*

# CHAPTER XVI.

## HOMOGRAPHIC RANGES ON A CONIC.

**1.** Two systems of points $ABC...$ and $A'B'C'...$ on a conic are said to be *homographic ranges on the conic* when the pencils $P(ABC...)$ and $Q(A'B'C'...)$ are homographic, $P$ and $Q$ being points on the conic. Hence two ranges on a conic which are homographic subtend, at any points on the conic, pencils which are homographic.

To construct homographic ranges on a conic, take two homographic pencils at points $P$ and $Q$ on the conic; the rays of these pencils will determine on the conic two homographic ranges. Given one of these pencils, three rays of the other pencil may be taken arbitrarily. Hence given a range of points on a conic, in constructing a homographic range on the conic, three points may be taken arbitrarily.

**2.** *If $(ABC...)$ and $(A'B'C'...)$ be two homographic ranges on a conic, then the meet of $AB'$ and $A'B$, of $BC'$ and $B'C$, and generally of $PQ'$ and $P'Q$, where $PP'$, $QQ'$ are any two pairs of corresponding points, all lie on a line* (called the *homographic axis*).

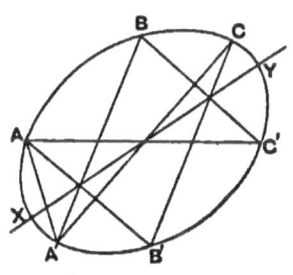

First consider all the meets which belong to $A$ and $A'$. These all lie on a line. For $A(A'B'C'...) = A'(ABC...)$. Hence all the meets $(AB'; A'B)$, $(AC'; A'C)$, $(AD'; A'D)$, ... lie on an axis. So all the meets which belong to $B$ and $B'$ lie on an axis. So for $CC'$, $DD'$, ....

We have now to prove that all these axes are the same. The inscribed six-point $AB'CA'BC'$ shows that the meets $(AB'; A'B)$, $(B'C; BC')$, $(CA'; C'A)$ are collinear. Now $(AB'; A'B)$ and $(CA'; C'A)$ determine the axis of $AA'$; so $(AB'; A'B)$ and $(B'C; BC')$ determine the axis of $BB'$. Hence the axes of $AA'$ and $BB'$ coincide; i.e. every two axes, and therefore all the axes, coincide. Hence all the cross meets $(PQ'; P'Q)$ lie on the same line.

**3.** *Given three pairs of corresponding points $ABC$, $A'B'C'$ of two homographic ranges on a conic, to construct the point $D'$ corresponding to $D$.*

The meets $(AB'; A'B)$ and $(AC'; A'C)$ give the homographic axis; and we know that $(AD'; A'D)$ is on the homographic axis. Hence the construction—Let $A'D$ cut the homographic axis in $\delta$, join $A\delta$, cutting the conic again in the required point $D'$.

**4.** *Two homographic ranges on a conic have two common points, viz. the points where the homographic axis cuts the conic.*

Let the homographic axis cut the conic in $X$ and $Y$. To get the point $X'$ corresponding to $X$, we join $A'$ to $X$ cutting $XY$ in $X$ and then join $AX$ cutting the conic again in $X'$. Hence $X'$ is $X$. So $Y'$ is $Y$.

And there can be no common point other than $X$ and $Y$. For if $D$ and $D'$ coincide, then each coincides with $\delta$. Hence $D$, $D'$ and $\delta$ must be at $X$ or $Y$.

**5.** *Reciprocally, two homographic sets of tangents to a conic can be formed by dividing two tangents homographically in $ABC...$ and $A'B'C'...$; then the second tangents from $ABC...$ will form a set of tangents homographic with the second tangents from $A'B'C'....$*

For any tangent will cut the two sets in homographic ranges.

Again, *all the cross joins will pass through a point called the homographic pole; and the tangents from the homographic pole* will be the self-corresponding lines in the two sets of homographic tangents.

This follows by Reciprocation from the previous articles.

**Ex. 1.** *The points of contact of two homographic sets of tangents are homographic ranges; and conversely, the tangents at points of two homographic ranges on a conic form homographic sets of tangents.*

**Ex. 2.** *If $OO'$ be fixed points on a conic and $AA'$ variable points on the conic, such that $(OO', AA')$ is constant; show that $A$ and $A'$ generate homographic ranges on the conic of which $O$ and $O'$ are the common points.*

**Ex. 3.** *If the lines joining a fixed point $P$ on a conic to the corresponding points $AA'$ of two homographic ranges on the conic cut the homographic axis in $aa'$, show that $aa'$ generate homographic ranges, and that the ranges obtained by varying $P$ are identical.*

For $(XY, aa')$ is constant and independent of the position of $P$ on the conic.

**Ex. 4.** *A conic is drawn through the common points of two homographic ranges $AB..., A'B'...$ on the same line. $P$ is any point on the conic, and $PA, PA'$ cut the conic again in $a, a'$. Show that $aa'$ generate homographic ranges on the conic, and that the ranges obtained by varying $P$ are identical.*

**Ex. 5.** *Reciprocate Examples 3 and 4.*

**Ex. 6.** *The pencils $A(PQR...)$ and $A'(PQR...)$ are homographic. A line meets $AP$ in $p$, $A'P$ in $p'$, and so on. Show that there are two positions of the line such that $pp' = qq' = rr' = ....$*

Viz. the asymptotes of the conic through $AA'PQR...$.

**Ex. 7.** *The joins of corresponding points of two homographic ranges on a conic touch a conic having double contact with the given conic at the common points of the given ranges.*

Let $AA'$ cut $XY$ in $L$, the tangent at $X$ in $a$, and the tangent at $Y$ in $a'$; let $BB'$ cut $XY$ in $M$, the tangent at $X$ in $b$, and the tangent at $Y$ in $b'$. Let $AB', A'B$ cut $XY$ in $K$. Then

$$(ALA'a) = X(ALA'a) = (AYA'X) = (BYB'X)$$

[since $X, Y$ are the common points]

$$= Y(BYB'X) = (Bb'B'M) = (B'MBb').$$

Now $AB', LM$ and $A'B$ meet in $K$. Hence $ab'$ passes through $K$. So $a'b$ passes through $K$. Hence $XY, ab', a'b$ are concurrent. Hence, by Brianchon, a conic touching the conic at $X$ and at $Y$ and touching $AA'$ will also touch $BB'$, and similarly $CC'$, etc. (See also XXIX. 10.)

**6.** *Given a conic and a ruler, construct the common points of two homographic ranges on the same line.*

Let the ranges be $ABC...$ and $A'B'C'....$ Take any point $p$ on the conic, and let $pA, pA', pB, pB', ...$ cut the conic again in $a, a', b, b', ....$ The ranges $abc...$ and $a'b'c'...$ on the conic are homographic; for

$$(abc...) = p(abc...) = (ABC...) = (A'B'C'...)$$
$$= p(A'B'C'...) = (a'b'c'...).$$

## Homographic Ranges on a Conic.

Now determine the homographic axis of the ranges $(abc...)$ and $(a'b'c'...)$ by connecting the cross meets $(ab'; a'b)$, etc.; and let this axis cut the conic in $x$ and $y$. Then if $px$ and $py$ cut $AB$ in $X$ and $Y$, $X$ and $Y$ are the common points of the ranges $ABC...$ and $A'B'C'...$.

For
$$(XYABC...) = p(XYABC...) = (xyabc...) = (xya'b'c'...)$$
$$= p(xya'b'c'...) = (XYA'B'C'...);$$

i.e. $XY$ correspond to themselves in the ranges $ABC...$ and $A'B'C'...$.

*Given a conic and a ruler, construct the common rays of two homographic pencils having the same vertex.*

Join the vertex to the common points of the ranges determined by the pencils on any line.

# CHAPTER XVII.

RANGES IN INVOLUTION.

1. If we take pairs of corresponding points, viz. $AA'$, $BB'$, $CC'$, $DD'$, $EE'$, ... on a line, such that a cross ratio of any four of these points (say $AD'$, $C'E$) is equal to the corresponding cross ratio of the corresponding points (viz. $A'D$, $CE'$), then the pairs of points $AA'$, $BB'$, $CC'$, ... are said to be in involution or to form an involution range.

Or more briefly — If the ranges $(AA'BB'CC'...)$ and $(A'AB'BC'C...)$ are homographic, then the pairs of points $AA'$, $BB'$, $CC'$, ... are in involution.

To avoid the use of the vague word 'conjugate' let us call each of a pair of corresponding points, $AA'$ say, the *mate* of the other, so that $A$ is the mate of $A'$ and $A'$ is the mate of $A$. Let us call $AA'$ together a pair of the involution.

There is no good notation for involution. The notation we have used above implies that $A$ and $B$ are related to one another in a way in which $A$ and $B'$ are not related; and this is not true. If we use the notation $AB$, $CD$, $EF$, ... for pairs of points in involution, this objection disappears; but there is now nothing to tell us that $A$ and $B$ are corresponding points.

2. The following is the fundamental proposition in the subject and enables us to recognise a range in involution.

*If two homographic ranges, viz.*

$$(AA'BCD...) \text{ and } (A'AB'C'D'...),$$

*be such that to one point $A$ corresponds the same point, viz. $A'$,* whichever range $A$ is supposed to belong to, the same is true of every other point, and the pairs of corresponding points $AA'$, $BB'$, $CC'$, $DD'$, ... are in involution.

We have to prove that
$$(AA'BB'CC'DD'...) = (A'AB'BC'CD'D...),$$
given that $(AA'BCD...) = (A'AB'C'D'...).$

Now if $P$ be considered to belong to the first range, its mate $P'$ in the second range is determined by the equation
$$(AA'BP) = (A'AB'P').$$

Let $P$ be $B'$, then the mate $P'$ of $B'$ is given by the equation $(AA'BB') = (A'AB'P')$. Now we have identically $(AA'BB') = (A'AB'B)$. Hence $P'$ is $B$. Hence $B$ has the same mate, viz. $B'$, whichever range it is considered to belong to. Again, we may consider the homography to be determined by the equation $(AA'CP) = (A'AC'P')$; hence, as before, $C$ has the same mate in both ranges. Similarly every point has the same mate in both ranges, i.e.
$$(AA'BB'CC'...) = (A'AB'BC'C...).$$

The commonest case of this proposition is—

*If* $(AA'BC) = (A'AB'C')$;
*then* $AA'$, $BB'$, $CC'$ *are in involution.*

*Two pairs of points determine an involution.*

For the pairs of points $PP'$ which satisfy the relation $(AA'BP) = (A'AB'P')$ are in involution.

**Ex. 1.** *If* $(CB, AA')$ *and* $(C'B', AA')$ *be harmonic, then* $(AA', BB', CC')$ *are in involution.*

**Ex. 2.** *If* $(CA, A'B') = (AB, A'C') = -1$, *then* $(AA', BB', CC')$ *is an involution.*

**Ex. 3.** *If* $(AA', BC) = (BB', CA) = (CC', AB) = -1$, *show that* $(A'A, B'C') = (B'B, C'A') = (C'C, A'B') = -1$, *and that* $(AA', BC, B'C')$, $(BB', AC, A'C)$ *and* $(CC', AB', A'B)$ *are involutions.*

Project the range so that $A$ goes to infinity.

**Ex. 4.** *If* $(AB, XX') = (CD, XX')$, *where $A$, $B$, $C$, $D$ are fixed points on the same line, then $X$ and $X'$ generate homographic ranges.*

For $(AB, XX') = (DC, X'X)$, hence $(AD, BC, XX')$ is an involution. Hence $(ADBX) = (DACX')$.

**Ex. 5.** *ABC and $A'B'C'$ are homologous triangles. $BC$ and $B'C'$ meet in $X$,*

*CA* and *C'A'* meet in *Y*, and *AB* and *A'B'* meet in *Z*. *OAA'*, *OBB'*, *OCC'* meet the line *XYZ* in *X'*, *Y'*, *Z'*. Show that $(XX', YY', ZZ')$ is an involution. For $(XX'Y'Z') = O(XABC) = A(XOBC) = (XX'ZY) = (X'XYZ)$.

**3.** *To construct with the ruler only the mate of a given point in a given involution.*

Let the involution be determined by the two pairs $AA'$, $BB'$. Take any vertex $V$, and let $VA$, $VA'$, $VB$, $VB'$, &c. cut any line in $a$, $a'$, $b$, $b'$, &c. Then the ranges $AA'BB'...$ and $a'ab'b...$ are homographic; for $(a'ab'b...) = (A'AB'B...)$ by projection through $V = (AA'BB'...)$ by invo‐

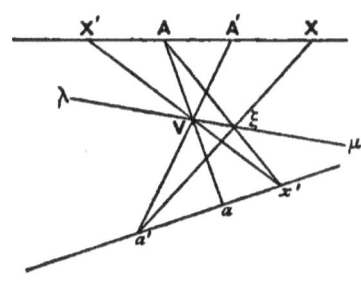

lution. Construct the homographic axis $\lambda\mu$ of these ranges. We observe that $V$ is on $\lambda\mu$, being the cross meet $(Aa\,; A'a')$. Take any point $X$ on $AA'$. Let $Xa'$ cut $\lambda\mu$ in $\xi$. Let $A\xi$ cut $aa'$ in $x'$. Let $Vx'$ cut $AA'$ in $X'$. Then $X'$ is the mate of $X$ in the given involution. For
$(XAA'BB'...) = (x'a'ab'b...)$ by the homographic axis
$= (X'A'AB'B...)$ by projection through $V$.

Hence $(XAA'BB'...) = (X'A'AB'B...)$. Hence $X'$ is the mate of $X$ in the involution.

**4.** *If $AA'$, $BB'$, $CC'$ be three pairs of points in involution, the following relations are true, viz.*

$$AB'.BC'.CA' = -A'B.B'C.C'A,$$
$$AB'.BC.C'A' = -A'B.B'C'.CA,$$
$$AB.B'C'.CA' = -A'B'.BC.C'A,$$
$$AB.B'C.C'A' = -A'B'.BC'.CA.$$

Take any one of the relations, viz.
$$AB.B'C'.CA' = -A'B'.BC.C'A.$$
This is true if $AB/BC \div AC'/\mathbf{1} = -A'B'/B'C' \div A'C/\mathbf{1}$,
i.e. if $AB/BC \div AC'/C'C = A'B'/B'C' \div A'C/CC'$,
i.e. if $(AC, BC') = (A'C', B'C)$.

And this is true; hence the relation in question is true.

Similarly the other relations can be proved.

Conversely, *if any one of these relations be true, then $AA'$, $BB'$, $CC'$ are in involution.*

For suppose $AB \cdot B'C' \cdot CA' = -A'B' \cdot BC \cdot C'A$; then as above $(AC, BC') = (A'C', B'C)$; hence $AA'$, $BB'$, $CC'$ are in involution.

Remark that, given one of these relations, the others follow at once. For in the definition of involution, there is no distinction made between two corresponding points. Hence in any relation connecting the points, we may interchange $A$ and $A'$, or $B$ and $B'$, or $C$ and $C'$, or we may make any of these interchanges simultaneously.

To obtain the second relation from the first, we interchange $C$ and $C'$, to obtain the third we interchange $B$ and $B'$, to obtain the fourth we interchange $B$ and $B'$ and $C$ and $C'$ simultaneously.

**Ex. 1.** If $(AA', BB', CC')$ be in involution, then
$$(A'A, BC) \cdot (B'B, CA) \cdot (C'C, AB) = -1.$$

**Ex. 2.** Circles of a coaxal system whose centres are $A, B, C$ touch the sides of a triangle $LMN$ in $P, Q, R$, and circles of the same system whose centres are $A', B', C'$ pass through the vertices of the triangle; if $PQR$ be a line, then $(AA', BB', CC')$ is an involution.

For $LR^2 : LQ^2 :: A'C : A'B$.

**5.** *If $AA'$, $BB'$, $CC'$, ... be in involution, and if any fixed pair of corresponding points $UU'$ be taken as origins, and if $PP'$ be any variable pair of corresponding points, then*
$$UP \cdot UP' \div U'P \cdot U'P'$$
*is constant.*

It will be sufficient to prove that
$$UP \cdot UP' \div U'P \cdot U'P' = UA \cdot UA' \div U'A \cdot U'A',$$
where $AA'$ is a fixed pair of corresponding points. This is true if $PU/UA \div PU'/U'A = P'U'/U'A' \div P'U/UA'$, i.e. if $(PA, UU') = (P'A', U'U)$. And this is true; hence the relation in question is true.

Particular cases of this theorem are—
$$AB \cdot AB' \div A'B \cdot A'B' = AC \cdot AC' \div A'C \cdot A'C',$$
$$CA \cdot CA' \div C'A \cdot C'A' = CD \cdot CD' \div C'D \cdot C'D'.$$

Conversely, *if $UU'$ be fixed points, and if $PP'$ be variable points such that $UP \cdot UP' \div U'P \cdot U'P'$ is constant; then $PP'$ generate an involution in which $UU'$ are corresponding points.*

For take any point $A$ and let $A'$ be the position of $P'$ when $P$ is at $A$. Then
$$UP \cdot UP' \div U'P \cdot U'P' = UA \cdot UA' \div U'A \cdot U'A';$$
hence $(PA, UU') = (P'A', U'U)$, i.e. $P$ and $P'$ are corresponding points in the involution determined by the two pairs $AA', UU'$.

**6.** *In an involution range, if any two of the segments $AA'$, $BB'$, ... bounded by corresponding points overlap, then every two overlap; and if any two do not overlap, then no two overlap.*

For suppose $AA'$ and $BB'$ overlap, then any two others $CC'$ and $DD'$ overlap.

<u>　　A　　　　B　　A'　　B'</u>

For $\dfrac{AB \cdot AB'}{A'B \cdot A'B'} = \dfrac{AC \cdot AC'}{A'C \cdot A'C'}.$

But since $AA'$ and $BB'$ overlap, the sign of
$$AB \cdot AB' \div A'B \cdot A'B'$$
is $-$. Hence the sign of $AC \cdot AC' \div A'C \cdot A'C'$ is $-$. Hence $AA'$ and $CC'$ overlap; for if $AA'$ and $CC'$ do not overlap, the sign of this expression is $+$, as we see from the figures—

<u>　A　　A'　　C　　C'.　　A　　C　　C'　　A'</u>

We have shown that if $AA'$ and $BB'$ overlap, then $AA'$ and $CC'$ overlap. Hence, since $CC'$ and $AA'$ overlap, it follows that $CC'$ and $DD'$ overlap, i.e. that every two such segments overlap.

Conversely, suppose $AA'$ and $BB'$ do not overlap, then $CC'$ and $DD'$ do not overlap; for if they do overlap, by the first part of the proof it follows that $AA'$ and $BB'$ overlap.

**7.** The *centre of an involution range* is the point corresponding to the point at infinity.

*If $O$ be the centre of the involution of which $P$ and $P'$ are a* pair of corresponding points, then $OP \cdot OP'$ is constant; and, conversely, if a pair of points $P$ and $P'$ be taken on a line, such that $OP \cdot OP'$ is constant, then $P$ and $P'$ generate an involution range of which $O$ is the centre.

Let $O$ be the centre of the involution range $(AA', BB', PP', ...)$. Then $\Omega'$ being the point at infinity upon the line, we have by definition

$$(O\Omega'AA'BB'PP') = (\Omega'OA'AB'BP'P);$$
$$\therefore (O\Omega', AP) = (\Omega'O, A'P');$$
$$\therefore OA/A\Omega' \div OP/P\Omega' = \Omega'A'/A'O \div \Omega'P'/P'O,$$
and $A\Omega' = P\Omega'$ and $\Omega'A' = \Omega'P'$;
$$\therefore OP \cdot OP' = OA \cdot OA', \text{ which is constant.}$$

Conversely, if $OP \cdot OP'$ be constant, let $A'$ be the position of $P'$ when $P$ is at $A$. Then we have $OP \cdot OP' = OA \cdot OA'$. Hence by writing the above steps backward we get

$$(O\Omega'AP) = (\Omega'OA'P'),$$

where $\Omega'$ is the point at infinity on the line. Hence $P$ and $P'$ are a pair of corresponding points in the involution determined by $(O\Omega', AA')$, i.e. $P$ and $P'$ generate an involution of which $O$ is the centre.

**Ex. 1.** *If $O$ be the centre of the involution $(AA', BB', CC', ...)$, show that*
$$AB \cdot AB' \div A'B \cdot A'B' = AO \div A'O.$$

To prove this, make the relation projective by introducing infinite segments in such a manner that the same letters occur on each side of the relation. We get
$$AB \cdot AB' \div A'B \cdot A'B' = AO \cdot A\Omega' \div A'O \cdot A'\Omega',$$
and this is a particular case of the theorem
$$AB \cdot AB' \div A'B \cdot A'B' = AC \cdot AC' \div A'C \cdot A'C'.$$

**Ex. 2.** *Show that $OA : OB :: AB' : BA'$; and deduce three other relations by interchanging corresponding points.*

**Ex. 3.** *If $a$ bisect $AA'$ and $\beta$ bisect $BB'$, show that*
  (a) $2 \cdot AO \cdot a\beta = AB \cdot AB'$;
  (b) $4 \cdot aO \cdot a\beta = AB \cdot AB' + A'B \cdot A'B'$;
  (c) $2 \cdot AA' \cdot a\beta = AB \cdot AB' - A'B \cdot A'B'$.
For if $O$ be origin, then $aa' = bb'$.

**Ex. 4.** *If $R$ bisect $CC'$ and $R'$ be the mate of $R$, then $RC^2 = RR' \cdot RO$.*

**Ex. 5.** *Any two homographic ranges, whether on the same line or not, can be placed in two ways so as to be in involution.*

Viz. by placing $I$ on $J'$ and placing $A$ and $A'$ on the same or opposite sides of $I$.

**Ex. 6.** *Of the two involutions one is overlapping and the other not.*

**Ex. 7.** *Any line through the radical centre of three circles cuts them in a range in involution.*

**8.** A point on the line of an involution range which coincides with its mate is called a *double point* (or focus) of the involution.

*An involution range has two, and only two, double points; and the segment joining the double points is bisected by the centre and divides the segment joining any pair of corresponding points harmonically.*

If $AA'$, $PP'$ be two pairs of corresponding points of an involution whose centre is $O$, we have seen that
$$OP \cdot OP' = OA \cdot OA'.$$
Suppose $P$ and $P'$ coincide in $E$. Then $OE^2 = OA \cdot OA'$, hence $OE = \pm \sqrt{OA \cdot OA'}$. Hence there are two double points, $E$ and $F$ say, which are equidistant from $O$. Also, since $OE^2 = OF^2 = OA \cdot OA'$ and $O$ bisects $EF$, it follows that $(AA', EF)$ is harmonic, i.e. $EF$ divides the segment joining any two corresponding points harmonically.

Notice that *the centre is always real*, being the mate of the point at infinity. But the double points will be imaginary when $OA \cdot OA'$ is negative, i.e. when $O$ lies between $A$ and $A'$. *The double points cannot coincide*, for then each coincides with $O$, in which case $OA \cdot OA' = OE^2 = 0$; i.e. $A$ or $A'$ coincides with $O$, and $A'$ or $A$ is anywhere, i.e. half the points are at $O$ and half are indeterminate, i.e. the involution is nugatory.

**9.** *The double points of an overlapping involution are imaginary and those of a non-overlapping involution are real.*

Take $O$ the centre of the involution. Then
$$OA \cdot OA' = OB \cdot OB' = \cdots = OE^2 = OF^2.$$
Now in an overlapping involution $O\Omega'$ and $AA'$ overlap, i.e. $O$ lies between $A$ and $A'$. Hence $OA \cdot OA'$ is negative, i.e. $OE^2$ and $OF^2$ are negative, i.e. $E$ and $F$ are imaginary.

174    *Ranges in Involution.*    [CH.

Similarly in a non-overlapping involution, $OE^2$ and $OF^2$ are positive. i.e. $E$ and $F$ are real.

An overlapping involution is sometimes called a *negative involution* and a non-overlapping involution is called a *positive involution*.

**Ex. 1.** *If $E$ and $F$ divide harmonically the segments $AA'$, $BB'$, $CC'$, ..., show that $(AA', BB', CC', ...)$ is an involution.*

Bisect the segment $EF$ at $O$. Then
$$OA \cdot OA' = OB \cdot OB' = \cdots = OE^2.$$

**Ex. 2.** *If $E$ and $F$ be the double points of $(AA', BB', CC', ...)$, show that*
$$AB \cdot AB' \div A'B \cdot A'B' = AE^2 \div A'E^2.$$

**Ex. 3.** *Also* $\dfrac{AB \cdot AB'}{A'B \cdot A'B'} = -\dfrac{AE \cdot AF}{A'E \cdot A'F}.$

**Ex. 4.** *Also $AB' \cdot BE \cdot EA' = -A'B \cdot B'E \cdot EA$.*

**Ex. 5.** *Also $EF^2 \cdot \alpha\beta^2 = AB \cdot AB' \cdot A'B \cdot A'B'$.*

For $EF = 2 \cdot OE = 2e$ if $aa' = e^2$.

**Ex. 6.** *Also $4 \cdot a\beta \cdot aE = [\sqrt{AB \cdot AB'} + \sqrt{A'B \cdot A'B'}]^2.$*

✗ **Ex. 7.** *If the segments $AA'$, $BB'$, ... joining corresponding points have the same middle point, show that $AA'$, $BB'$, ... form an involution; and find the centre and double points.*

$\Omega'$ the point at infinity and $E$ the middle point are harmonic with every segment $AA'$. Hence $\Omega'$, $E$ are the double points and $\Omega'$ is the centre.

**Ex. 8.** *If $AA'$, $BB'$ be pairs of points in an involution, one of whose double points is at infinity, then $AB = -A'B'$.*

For $E$ the other double point must bisect $AA'$ and $BB'$.

✗ **Ex. 9.** *If any two segments $AA'$, $BB'$ joining corresponding points in an involution have the same middle point, then all such segments have the same middle point.*

For the other double point must be $\Omega'$.

**Ex. 10.** *If $AP \cdot AP' = A'P \cdot A'P'$, show that the points $P$ and $P'$ form an involution in which $A$ and $A'$ are corresponding points; and find the centre and double points.*

**Ex. 11.** *If any transversal through $V$ (the internal vertex of the harmonic triangle of a quadrilateral circumscribing a conic) cut the sides in $AA'$, $BB'$ and the conic in $PP'$; show that $(AA', BB', PP')$ is an involution, the double points being $V$ and the meet of $UW$ with the transversal.*

**Ex. 12.** *Through a given point $O$ draw a line meeting two conics (or two pairs of lines) in points $AA'$, $BB'$ such that $(OAA'BB') = (OA'AB'B)$.*

Join $O$ to the meet of the polars of $O$.

**Ex. 13.** *If $ABC...$, $A'B'C'...$ be two homographic ranges on the same line, and if the mates of $P$ ($= Q'$) be $P'$ and $Q$, we know that the ranges $A$ and $A'$ and the ranges $P'$ and $Q$ have the same common points ($E$, $F$ say); show that $P$ has the same fourth harmonic for $P'Q$ and for $EF$. (See X. 7. Ex. 4.)*

We have only to prove that $P(=Q')$ is one of the double points of the involution determined by $P'Q$, $EF$.
Now $(PQEF) = (P'Q'EF)$ from the first homography
$= (P'PEF) = (PP'FE)$.
Hence $P'Q$, $EF$, $PP$ are in involution, i.e. $P$ is a double point of the involution.

**Ex. 14.** *With the same data, $P$ and the fourth harmonic of $P$ for $P'Q$ generate an involution.*

**10.** *A system of coaxal circles is cut by any transversal in pairs of points in involution.*

For if the transversal cut the circles in $AA'$, $BB'$, $CC'$, ... and the radical axis in $O$, then

$$OA \cdot OA' = OB \cdot OB' = OC \cdot OC' = \ldots .$$

Hence $AA'$, $BB'$, $CC'$, ... form an involution of which $O$ is the centre.

**Ex. 1.** *Give a geometrical construction for the double points of the involution determined on a line by a system of coaxal circles.*

**Ex. 2.** *A line touches two circles in $A$ and $A'$ and cuts a coaxal circle in $B$ and $B'$; show that $(AA', BB')$ is harmonic.*

**Ex. 3.** *Of the involution determined by a system of coaxal circles on the line of centres, the limiting points are the double points.*

**Ex. 4.** *If a line meet three circles in three pairs of points in involution, then either the circles are coaxal or the line passes through their radical centre.*

**Ex. 5.** *If each of the sides of a triangle meet three circles in pairs of points in involution, the circles are coaxal.*

**Ex. 6.** *The three circles drawn through a given point $V$, one coaxal with the circles $\alpha$ and $\beta$, one coaxal with the circles $\beta$ and $\gamma$, and one coaxal with the circles $\gamma$ and $\alpha$, are coaxal.*

Let two of the circles cut again in $V'$, and consider the involution on $VV'$.

**Ex. 7.** *Two circles $\alpha$ and $\beta$ are drawn having the radical axis $p$ with the circle $\gamma$, and $\delta$ and $\epsilon$ are drawn having the radical axis $q$ with $\gamma$; show that the meets of $\alpha\delta$ and of $\beta\epsilon$ are concyclic.*

Consider the involution on the radical axis of $\alpha$ and $\delta$.

**11.** *If $EF$ be the double points of an involution of which $AA'$ and $BB'$ are any two pairs of corresponding points, then $(AB', A'B, EF)$ are in involution, and so are $(AB, A'B', EF)$.*

For $(AB', A'B, EF)$ are in involution if
$$(ABEF) = (B'A'FE), \text{ i.e. } = (A'B'EF);$$
and this is true, for $E$ corresponds to itself and so does $F$.

Similarly $(AB, A'B', EF)$ are in involution.

**Ex. 1.** *Prove the following construction for the double points of the involution $AA'$, $BB'$, $CC'$,... viz.—Take any point $P$ and let the circles through $PAB'$ and $PA'B$ cut in $Q$; so let the circles through $PAB$ and $PA'B'$ cut in $R$; then the circle through $PQR$ cuts $AA'$ in the required double points.*

For if the circle through $PQR$ cut $AA'$ in $EF$, then from the radical axis $PQ$ we see that $(AB', A'B, EF)$ are in involution; hence
$$(ABEF) = (B'A'FE) = (A'B'EF).$$
So from the radical axis $PR$, we get $(AB'EF) = (A'BEF)$.
Hence $$(ABEFB') = (A'B'EFB).$$
Hence $EF$ are the double points of the involution determined by $AA'$ and $BB'$.

**Ex. 2.** *If $E$ and $F$ be the limiting points of the circles on the collinear segments $AA'$, $BB'$ as diameters, show that the circles on $AB$, $A'B'$, and $EF$ as diameters are coaxal.*

**Ex. 3.** *If $E$, $F$ be the common points of the two homographic ranges $(ABC...)$ and $(A'B'C'...)$, then $AB'$, $A'B$, $EF$ are in involution.*

**Ex. 4.** *Prove the following construction for the common points of the two homographic ranges $(ABC...)$ and $(A'B'C'...)$.—Take any point $P$ and let the circles $PAB'$ and $PA'B$ cut in $Q$, and let the circles $PAC'$ and $PA'C$ cut in $R$; then the circle $PQR$ will cut $AA'$ in the required points.*

**Ex. 5.** *Given two pairs of points $AA'$, $BB'$ of two homographic ranges and one common point, construct the other.*

**12.** *If $AA'$, $BB'$, $CC'$ be pairs of points in involution, and if $P$, $Q$, $R$ be the middle points of $AA'$, $BB'$, $CC'$, show that*
$$A'A^2 . QR + B'B^2 . RP + C'C^2 . PQ + 4PQ . QR . RP = 0;$$
*and if $U$ be any point on the same line, then*
$$(AU^2 + A'U^2) QR + (BU^2 + B'U^2) RP + (CU^2 + C'U^2) PQ$$
$$= -4PQ . QR . RP.$$

Take the centre of the involution as origin and use abridged notation; then if $OA' = a_1$, and so on,
$$A'A^2 = (a - a_1)^2 = a^2 + a_1^2 - 2aa_1 = (a + a_1)^2 - 4aa_1.$$
But $\qquad a + a_1 = 2p$ and $QR = r - q$,
and $\qquad aa_1 = bb_1 = cc_1 = \lambda$, say;
$$\therefore A'A^2 . QR = (4p^2 - 4\lambda)(r - q);$$
$$\therefore \Sigma(A'A^2 . QR) = 4 \Sigma p^2 (r-q) - 4\lambda \Sigma (r-q)$$
$$= -4(q-p)(r-q)(p-r)$$
$$= -4PQ . QR . RP.$$
Again, if $x$ be the distance of $U$ from the origin
$$AU^2 = (x - a)^2:$$

Hence $\Sigma\{(AU^2 + A'U^2)QR\}$
$= \Sigma\{[2x^2 - 2x(a+a_1) + a^2 + a_1^2](r-q)\}$
$= 2x^2 \Sigma(r-q) - x \Sigma p(r-q)$
$\quad + \Sigma\{a^2 + a_1^2 - 2aa_1 + 2\lambda\}(r-q)$
$= \Sigma A'A^2 \cdot QR$
$= -4PQ \cdot QR \cdot RP$ by the former part.

**Ex. 1.** *With the same notation, show that*
$AB \cdot AB'/AC \cdot AC' = PQ/PR$.

**Ex. 2.** *Also if $E$ be a double point, then*
$A'A^2/PE - B'B^2/QE = 4PQ$.

**Ex. 3.** *Also, $X$ being any point on the same line,*
$XA \cdot XA' \cdot QR + XB \cdot XB' \cdot RP + XC \cdot XC' \cdot PQ = 0$.

**Ex. 4.** *Also* $XA \cdot XA' \cdot EF + XE^2 \cdot FP + XF^2 \cdot PE = 0$.

**Ex. 5.** *Also* $XA \cdot XA' - XB \cdot XB' + 2PQ \cdot XO = 0$.

**Ex. 6.** *Also* $RC^2 \cdot PQ = RA \cdot RA' \cdot QR + RB \cdot RB' \cdot RP$.

**Ex. 7.** *Given two collinear segments $AA'$, $BB'$, determine a point $C$ such that* $CA \cdot CA' : CB \cdot CB' :: \lambda : 1$.

Determine the point $R$ from the relation $RP : RQ :: \lambda : 1$. Through any point $V$ draw the two circles $VAA'$, $VBB'$ cutting again in $V'$. Draw the circle through $VV'$, having its centre on the perpendicular to $AA'$ through $R$. This circle will cut $AA'$ in the required points (see Ex. 1).

**13.** Take any point $V$ on the line of the involution. Then $OA = VA - VO = x - r$, say; so $OA' = x' - r$.

$\therefore OA \cdot OA' =$ constant gives $(x-r)(x'-r) =$ constant.

Hence *the distances of pairs of points in an involution from any point on the line satisfy the relation $kxx' + l(x+x') + n = 0$, where $k$, $l$ and $n$ are constants.*

Conversely, *if this relation be satisfied, the pairs of points form an involution.*

For $kxx' + l(x+x') + n = 0$ can be thrown into the form $(x-r)(x'-r) =$ constant; which is the same as
$OA \cdot OA' =$ constant.

*Or thus.* If $(AA', BB', CC', \ldots)$ be in involution, then $(AA'BB'CC', \ldots)$ is homographic with $(A'AB'BC'C, \ldots)$. Hence corresponding points in the two ranges are connected by a relation of the form $xx' + lx + mx' + n = 0$. Also, as there is no distinction in an involution between $A$ and $A'$, we must have $l = m$. Conversely, if $xx' + l(x+x') + n = 0$, $A$ and $A'$ generate homographic ranges in which $A$ and $A'$ are interchangeable. Hence $A$ and $A'$ generate an involution.

**Ex. 1.** *Show that $P$, $P'$ determine an involution if*
$$AP \cdot B'P' + \lambda \cdot AP + \mu \cdot B'P' + \nu = 0,$$
*provided* $\lambda - \mu = AB'$.

**Ex. 2.** *Show that $P$, $P'$ determine an involution if*
$$2 \cdot AP \cdot BP' = AB \cdot PP';$$
*and that $A$ and $B$ are the double points.*

**Ex. 3.** *Show that $P$, $P'$ determine an involution if $AP + B'P' = \nu$; and that one double point is at infinity. Find also the second double point.*

**14.** *If $(AA', BB', CC')$ be pairs of points of an involution,*

then $\qquad \dfrac{CA}{C'A'} \cdot BB' + \dfrac{CB}{C'B'} \cdot B'A + \dfrac{CB'}{C'B} \cdot AB = 0.$

We have to prove that
$$CA \cdot BB' \cdot C'B' \cdot C'B + CB \cdot B'A \cdot C'A' \cdot C'B$$
$$+ CB' \cdot AB \cdot C'A' \cdot C'B' = 0.$$

This relation is of the first order in $A$ and in $A'$. Consider the points $X$, $X'$ connected by the relation
$$CX \cdot BB' \cdot C'B' \cdot C'B + CB \cdot B'X \cdot C'X' \cdot C'B$$
$$+ CB' \cdot XB \cdot C'X' \cdot C'B' = 0.$$

Reducing to any origin, this relation assumes the form
$$xx' + lx + mx' + n = 0.$$

Hence $X$ and $X'$ generate homographic ranges.

Now the relation is satisfied by $X = C$ and $X' = C'$, and by $X = B$ and $X' = B'$, and by $X = B'$ and $X' = B$. Hence the homography is that determined by $(CBB') = (C'B'B)$, i.e. is the given involution. Hence the above relation between $X$ and $X'$ is satisfied by any corresponding pair of points of the involution. Hence the relation is satisfied if we replace $X$, $X'$ by $A$, $A'$.

**Ex. 1.** *Show that* $AA' = \dfrac{AB \cdot A'C}{BC} + \dfrac{AB' \cdot A'C'}{B'C'}.$

**Ex. 2.** *Also* $\dfrac{AB \cdot A'C}{AC \cdot A'B} + \dfrac{AA' \cdot B'C'}{AB' \cdot A'C'} = 1.$

**Ex. 3.** *Also* $\dfrac{CA}{C'A'} \cdot BC' + \dfrac{CB}{C'B'} \cdot C'A = AB.$

XVII.]     *Ranges in Involution.*     179

**Ex. 4.** *Also* $\dfrac{CA}{C'A'} \cdot BB' + \dfrac{CB}{C'B'} \cdot B'A + \dfrac{CB'}{C'B} \cdot AB = 0.$

**Ex. 5.** *Also* $AB = \dfrac{AC \cdot AC'}{AB'} - \dfrac{BC \cdot BC'}{BA'}.$

**Ex. 6.** *Also, P being any point on the same line,*
$\dfrac{CA}{C'A'} \cdot BB' \cdot PA' + \dfrac{CB}{C'B'} \cdot B'A \cdot PB' + \dfrac{CB'}{C'B} \cdot AB \cdot PB = 0.$

**Ex. 7.** *Also* $\dfrac{CA}{C'A'} \cdot BC' \cdot PA' + \dfrac{CB}{C'B'} \cdot C'A \cdot PB' = AB \cdot PC.$

# CHAPTER XVIII.

### PENCILS IN INVOLUTION.

**1.** The pencil of lines $VA$, $VA'$, $VB$, $VB'$, $VC$, $VC'$, ... is said to form a pencil in involution if
$$V(AA'BB'CC'...) = V(A'AB'BC'C...).$$
*Any transversal cuts an involution pencil in an involution range; and, conversely, the pencil joining any involution range to any point is in involution.*

Let a transversal cut an involution pencil in the pairs of points $AA'$, $BB'$, $CC'$, .... Then, since
$$V(AA'BB'CC'...) = V(A'AB'BC'C...),$$
we have $V(AA'BC) = V(A'AB'C')$; hence
$$(AA'BC) = (A'AB'C').$$
Hence $C$, $C'$ are a pair in the involution determined by the pairs $AA'$, $BB'$. Similarly for any other pair of points in which the transversal is cut by a pair of lines of the involution pencil.

Conversely, if $(AA'BB'CC'...) = (A'AB'BC'C...)$, we have $(AA'BC) = (A'AB'C')$; hence $V(AA'BC) = V(A'AB'C')$. Hence $VC$, $VC'$ are a pair of rays in the involution pencil determined by $V(AA', BB')$. So for any other pair of corresponding rays.

**Ex. 1.** *If $V$ be any point on the homographic axis of the two homographic ranges $(ABC...) = (A'B'C'...)$ on different lines; show that*
$$V(AA', BB', CC',...)$$
*is an involution pencil.*

Let $X'Y$ be the mates of the point $X(=Y')$ where $AB$ and $A'B'$ meet. Then $V$ is on $X'Y$. Hence
$$V(XX'ABC...) = V(XYABC...)$$

## Pencils in Involution. 181

$= V(X'Y'A'B'C'...)$ by homography $= V(X'XA'B'C'...)$.
Hence $V(XX', AA', BB', ...)$ is an involution.

**Ex. 2.** *Reciprocate Ex.* 1.

**Ex. 3.** *Two homographic pencils have their vertices at infinity. Show that any line through their homographic pole determines an involution of which the pole is the centre.*

**Ex. 4.** *Any two homographic pencils can be placed in two ways so as to be in involution.*

Let the pencils be $V(ABC...) = V'(A'B'C'...)$. First, superpose the pencils so that $V$ is on $V'$ and $VA$ on $V'A'$. This can be done in two ways. Let $VX (= V'X')$ be the other common line of the two pencils $V(ABC...) = V(AB'C'...)$. Then in the original figure $AVX = A'V'X'$. Second, place $V$ on $V'$ and $VA$ on $V'X'$ and $VX$ on $V'A'$. The two pencils are now in involution; for $VA (= V'X')$ has the same mate, viz. $V'A' (= VX)$ whichever pencil it is supposed to belong to.

If the vertices are at infinity, place the pencils so that all the rays are parallel. Let any line now cut them in the homographic ranges $(abc...) = (a'b'c'...)$. Now slide $(a'b'c'...)$ along $(abc...)$ until the two ranges are in involution (either by Ex. 5. of XVII. 7, or by a construction similar to the above).

**2.** *A pencil of rays in involution has two double rays (i.e. rays each of which coincides with its corresponding ray), and the double rays divide harmonically the angle between every pair of rays.*

Let any transversal cut the pencil in the involution $(AA', BB', CC', ...)$, and let $E, F$ be the double points of this involution. Then the ray corresponding to $VE$ in the pencil is clearly $VE$ itself. Hence $VE$ is a double ray. So $VF$ is a double ray. Also $(AEA'F)$ is a harmonic range; hence $V(AEA'F)$ is a harmonic pencil. Similarly $VE, VF$ divide each of the angles $BVB', CVC', ...$ harmonically.

There is nothing in an involution pencil which is analogous to the centre of an involution range. In fact the point at infinity in the range $AA', BB', CC', ...$ will project into a finite point on another transversal, and $O$ will project into the mate of this finite point.

If, however, $V$ is at infinity, i.e. if the rays of the pencil are parallel, then all sections of the pencil are similar, and there is a central ray which is the locus of the centres of all the involution ranges determined on transversals.

**Ex. 1.** *If the angles $AVA', BVB', CVC', ...$ be divided harmonically by the same pair of lines, the pencil $V(AA', BB', CC', ...)$ is in involution.*

**Ex. 2.** *If the angles be bisected by the same line, then the pencil is in involution.*

**Ex. 3.** *If the double rays of a pencil in involution be perpendicular, they bisect all the angles bounded by corresponding rays.*

✶ **Ex. 4.** *If two angles $AVA'$, $BVB'$ bounded by corresponding rays of a pencil in involution have the same bisectors, then all such angles have the same bisectors.*

**Ex. 5.** *Find the locus of a point at which every segment $(AB)$ of an involution subtends the same angle as the corresponding segment $(A'B')$.*

The circle on $EF$ as diameter.

**Ex. 6.** *Through any point $O$ are drawn chords $AA'$, $BB'$, $CC'$, ... of a conic; show that $AA'$, $BB'$, $CC'$ subtend an involution pencil at any point of the polar of $O$.*

**Ex. 7.** Reciprocate Ex. 6.

**Ex. 8.** *If $ABA'B'$ be four points on a conic, and if through any point $O$ on the external side $UW$ of the harmonic triangle of $ABA'B'$ there be drawn two tangents $OT$ and $OT'$ to the conic; show that $O(AA', BB', TT')$ is a pencil in involution, the double lines being $OU$ and $OV$.*

**Ex. 9.** *Through a fixed point $O$ is drawn a variable line to cut the sides of a given triangle in $A'B'C'$; find the locus of the point $P$ such that*
$$(PB', A'C') = -1.$$
Now $B(AC, B'P) = -1$, $\therefore$ $BB'$ and $BP$ generate an involution
$$\therefore B(P) = B(B') = O(B') = O(P),$$
$\therefore$ the locus is a conic through $B$ and $O$.

**3.** *If $AVA', BVB', CVC', ...$ be all right angles, then the pencil $V(AA', BB', CC', ...)$ is in involution.*

We have to show that
$$V(AA'BB'CC'...) = V(A'AB'BC'C...).$$

Produce $AV$ to $a$, $BV$ to $b$, and so on.

Then if we place $VA$ on $VA'$, $VA'$ will fall on $Va$, and so on. Hence the two pencils
$$V(AA'BB'...) \text{ and } V(A'aB'b...)$$
are superposable and therefore homographic. But
$$V(A'aB'b...)$$
is homographic with $V(A'AB'B...)$; hence $V(AA'BB'...)$ and $V(A'AB'B...)$ are homographic.

*Otherwise:*—From the vertex $V$ drop the perpendicular $VO$ on any transversal $AA'BB'...$. Then, since $AVA'$ is a right angle, we have $VO^2 = AO.OA'$.

Hence $OA.OA' = OB.OB' = OC.OC' = \cdots$.

Hence $(AA', BB', CC', ...)$ is an involution range.
Hence $V(AA', BB', CC', ...)$ is an involution pencil.

**Ex.** *If through the centre of an overlapping involution $(AA', BB', ...)$, there be drawn $VO$ perpendicular to $AA'$ and such that $VO^2 = AO \cdot OA'$, then any four points of the involution subtend at $V$ a pencil superposable to that subtended by their mates.*

4. *In every involution pencil, there is one pair of corresponding rays which is orthogonal; and if more than one pair be orthogonal, then every pair is orthogonal.* (See also XX. 6.)

Take any transversal cutting the pencil in the involution $(AA', BB', CC', ...)$. Through the vertex $V$ draw the circles $VAA', VBB'$ cutting again in $V'$. Let $VV'$ cut $AA'$ in $O$. Then $OA \cdot OA' = OV \cdot OV' = OB \cdot OB'$. Hence $O$ is the centre of the involution. Hence
$$OC \cdot OC' = OA \cdot OA' = OV \cdot OV'.$$
Hence the four points $V, V', C, C'$ are concyclic.

In this way, we prove that all the circles $VAA', VBB', VCC', ...$ pass through $V'$. Also every circle through $VV'$ cuts $AA'$ in a pair of points $PP'$ of the involution; for
$$OP \cdot OP' = OV \cdot OV' = OA \cdot OA'.$$
Let the line bisecting $VV'$ at right angles cut $AA'$ in $Q$. Describe a circle with $Q$ as centre and with $QV$ as radius, cutting $AA'$ in $PP'$. Then $P, P'$ are a pair in the involution, and $PVP'$ is a right angle.

This construction fails in only one case, viz. when $VV'$ is perpendicular to $AA'$. In this case, the orthogonal pair are $VV'$ and the perpendicular to $VV'$ through $V$.

Also if two pairs are orthogonal, every pair is orthogonal. For suppose $AVA', BVB'$ are right angles. Then the centres of the circles $AVA'$ and $BVB'$ are on $AA'$. Hence $AA'$ bisects $VV'$ orthogonally. Hence the centres of all the circles $AVA', BVB', CVC', ...$ are on $AA'$. Hence all the angles $AVA', BVB', CVC', ...$ are right angles.

**Ex. 1.** *Show that a given line $VX$ through the vertex always bisects one of the angles $AVA', BVB', ...$ of an involution; and if it bisect two of the angles, it bisects all. Discuss the case when $VX$ is perpendicular to one of the double rays.*

Take $AA'$ perpendicular to $VX$, and take the centre of the circle on $VX$.

**Ex. 2.** *Show that the pencils*
$$V(AA', BB', CC', \ldots) \text{ and } V'(A'A, B'B, C'C, \ldots) \text{ of § 4}$$
*are superposable.*

For $\angle AVB'$ is equal to $\angle A'V'B$ or its supplement.

**Ex. 3.** *Given two homographic pencils, we can always find in the first pencil rays $VA$, $VB$, and in the second pencil corresponding rays $V'A'$, $V'B'$, such that both $AVB$ and $A'V'B'$ are right angles. Can more than one such pair exist?*

**Ex. 4.** $(AA', BB', CC', \ldots)$ *is an involution. Show that the circles*
$$PAA', PBB', PCC', \ldots,$$
*where $P$ is any point, are coaxal.*

**Ex. 5.** *Deduce a construction for the mate of a given point in the involution.*

**Ex. 6.** *Also given $AA'$ and $BB'$, and the middle point of $CC'$, construct $C$ and $C'$.*

**Ex. 7.** *Given two segments $AA'$, $BB'$ of an involution, construct geometrically the centre $O$.*

**Ex. 8.** *Given a segment $AA'$ of an involution and the centre $O$, construct the mate of $C$.*

**Ex. 9.** *Given two involutions $(AA', BB', \ldots)$ and $(aa', bb', \ldots)$ on the same line : find two points which correspond to one another in both involutions.*

**Ex. 10.** *If any two circles be drawn through $AA'$ and $BB'$, their radical axis passes through $O$.*

**Ex. 11.** *If $A$, $A'$ generate an involution range, and $QA$ be perpendicular to $PA$ and $QA'$ to $PA'$, show that if $P$ be a fixed point, then $Q$ generates a line.*

For the locus of the centres of the coaxal circles $PAA'$ is a line.

**5.** *Every overlapping pencil in involution can be projected into an orthogonal involution.*

Let any transversal cut the pencil in the overlapping involution of points $(AA', BB', CC', \ldots)$. On $AA'$, $BB'$ as diameters describe circles. Since $AA'$, $BB'$ overlap, the circles will cut in two real points $U$ and $U'$. Now, since in the pencil in involution $U(AA', BB', CC', \ldots)$ two pairs of rays, viz. $UA$, $UA'$ and $UB$, $UB'$, are orthogonal, hence every pair is orthogonal.

Rotate $U$ about $AA'$ out of the plane of the paper. With any vertex $W$ on the line joining $U$ to the vertex $V$ of the given pencil, project the given pencil on to any plane parallel to the plane $UAA'$. Then $VA$ projects into a line parallel to $UA$, and $VA'$ projects into a line parallel to $UA'$; hence $AVA'$ projects into a right angle; similarly $BVB'$, $CVC'$, ... project into right angles.

**Ex. 1.** $(AA', BB', CC', \ldots)$ is an involution. Show that the circles on $AA', BB', CC', \ldots$ as diameters are coaxal.

**Ex. 2.** Show also that $AA', BB', CC', \ldots$ subtend right angles at two points in the plane. When are these points real?

**6.** If $P, Q, R$ be the fourth harmonics of the point $X$ for the segments $AA', BB', CC'$ of an involution range, then

$$\frac{PA^2}{XA^2} \cdot \frac{QR}{XP} + \frac{QB^2}{XB^2} \cdot \frac{RP}{XQ} + \frac{RC^2}{XC^2} \cdot \frac{PQ}{XR} + \frac{PQ \cdot QR \cdot RP}{XP \cdot XQ \cdot XR} = 0.$$

Join the points to any vertex $V$; and cut the pencil so formed by a transversal $aa', bb', cc', p, q, r, x$. Then since the given relation is homogeneous in each point, as in proving Carnot's Theorem, we see that the relation is also true of the projections $aa'$, &c. of the given points. Now take $aa'$ parallel to $VX$. Then $x$ is at infinity. Hence

$$\frac{xa^2 \cdot xp}{xb^2 \cdot xq} = 1, \quad \frac{xa^2 \cdot xp}{xc^2 \cdot xr} = 1, \quad \frac{xa^2 \cdot xp}{xp \cdot xq \cdot xr} = 1.$$

Hence the given relation is true if

$$pa^2 \cdot qr + qb^2 \cdot rp + rc^2 \cdot pq + pq \cdot qr \cdot rp = 0.$$

But now $p, q, r$ bisect $aa', bb', cc'$; hence this relation is true by XVII. 12.

If in addition to the above notation $P_1, Q_1, R_1$ bisect $AA', BB', CC'$, show that in Examples 1-6

**Ex. 1.** $\dfrac{AB \cdot AB'}{XB \cdot XB'} + \dfrac{AC \cdot AC'}{XC \cdot XC'} = \dfrac{PQ}{PR} \div \dfrac{XQ}{XR}$.

**Ex. 2.** $AB \cdot AB' \div AC \cdot AC' = PQ \cdot XQ_1 \div PR \cdot XR_1$.
For $XQ \cdot XQ_1 = XB \cdot XB'$, &c.

**Ex. 3.** $\dfrac{YA \cdot YA'}{XA \cdot XA'} \cdot QR \cdot XP + \dfrac{YB \cdot YB'}{XB \cdot XB'} \cdot RP \cdot XQ$
$\qquad + \dfrac{YC \cdot YC'}{XC \cdot XC'} \cdot PQ \cdot XR = 0,$

where $Y$ is any point on the same line.

**Ex. 4.** $YA \cdot YA' \cdot \dfrac{QR}{XP_1} + YB \cdot YB' \cdot \dfrac{RP}{XQ_1} + YC \cdot YC' \cdot \dfrac{PQ}{XR_1} = 0$.

**Ex. 5.** $\dfrac{QR \cdot XP}{XA \cdot XA'} + \dfrac{RP \cdot XQ}{XB \cdot XB'} + \dfrac{PQ \cdot XR}{XC \cdot XC'} = 0$.

Take $Y$ at infinity.

**Ex. 6.** $\dfrac{QR}{XP_1} + \dfrac{RP}{XQ_1} + \dfrac{PQ}{XR_1} = 0$.

## Pencils in Involution.

**Ex. 7.** If $(CP, AA') = (CQ, BB') = (PP', BB') = (QQ', AA')$
$$= (CC', P'Q') = -1,$$
then $(AA', BB', CC')$ are in involution.

Project $C$ to infinity, and use the relation
$$2(pp' + bb') = (p+p')(b+b'), \text{ taking } C' \text{ as origin.}$$

**Ex. 8.** If $(AA', BB') = (XC, AA') = (XC', BB') = -1$,
then $(AA', BB', CC')$ are in involution.

Project $X$ to infinity, and take the centre of the involution $(AA', BB')$ as origin.

**Ex. 9.** If $AA'$, $BB'$, $CC'$, $DD'$ be four segments in involution, and if
$$(LP, AA') = (LQ, BB') = (LR, CC') = (LS, DD');$$
show that $(PQRS)$ is independent of the position of $L$.

For $(PQ, RS) = (AB, CD) \times (A'B', CD)$.

**Ex. 10.** Deduce by Reciprocation a property of four pairs of rays in involution.

**Ex. 11.** If $(AA', BB')$ and $(AA', QQ')$ be harmonic, then
$$\frac{PA \cdot PA'}{QA \cdot QA'} \cdot BB' \cdot QQ' + \frac{PB^2}{QB} \cdot B'Q' + \frac{B'P^2}{QB'} \cdot Q'B = 0.$$
Take $A'$ at infinity.

**Ex. 12.** Deduce the relation
$$\frac{BB' \cdot QQ'}{QA \cdot QA'} + \frac{B'Q'}{QB} + \frac{Q'B}{QB'} = 0.$$
Take $P$ at infinity.

**7.** By considering sections of the involution pencil
$$V(AA', BB', CC', \ldots)$$
show that in Examples 1–4

**Ex. 1.** $\dfrac{\sin AVB \cdot \sin AVB'}{\sin A'VB \cdot \sin A'VB'} = \dfrac{\sin AVC \cdot \sin AVC'}{\sin A'VC \cdot \sin A'VC'} = \dfrac{\sin^2 AVE}{\sin^2 A'VE}$.

**Ex. 2.** $\sin AVB \cdot \sin B'VC \cdot \sin C'VA' = -\sin A'VB' \cdot \sin BVC' \cdot \sin CVA$.

**Ex. 3.** If $VR$ be either of the orthogonal rays, then
$$\tan RVA \cdot \tan RVA' \text{ is constant.}$$
Draw the transversal perpendicular to $VR$.

**Ex. 4.** If $VX$ be any line, then any pair $VP$, $VP'$ satisfies a relation of the form
$$\tan XVP \cdot \tan XVP' + l \cdot \tan XVP + l \cdot \tan XVP' + n = 0,$$
where $l$ and $n$ are constants.

**Ex. 5.** If $VA'$, $VB'$, $VC'$ be three bisectors of the angles $BVC$, $CVA$, $AVB$ (either three external, or one external and two internal), then
$$V(AA', BB', CC', \ldots)$$
is an involution.

**Ex. 6.** If $VX$ and $VY$ be fixed lines, and $VP$, $VP'$ be variable lines satisfying the condition
$$\sin XVP \div \sin YVP = -\sin XVP' \div \sin YVP',$$
then $VP$, $VP'$ generate an involution.

# CHAPTER XIX.

### INVOLUTION OF CONJUGATE POINTS AND LINES.

**1.** *The pairs of points on a line which are conjugate for a conic form an involution.*

Let $l$ be the line and $L$ its pole. Let $AA'$, $BB'$, $CC'$, ... be the pairs of conjugate points on $l$. Then the polar of $A$ which lies on $l$ passes through $L$. Also the polar of $A$ by hypothesis passes through $A'$. Hence $LA'$ is the polar of $A$. So $LA$ is the polar of $A'$, and so on. Hence $(AA'BB'CC'...)$ of poles $= L(A'AB'BC'C...)$ of polars $= (A'AB'BC'C...)$. Hence $(AA', BB', CC', ...)$ form an involution.

*The double points of the involution of conjugate points on a line are the meets of the line and the conic.*

For these meets are harmonic with every pair of conjugate points on the line.

This affords another proof that conjugate points on a line generate an involution.

**2.** *The pairs of lines through a point which are conjugate for a conic form an involution.*

Let $L$ be the point and $l$ its polar. Let $LA$, $LA'$; $LB$, $LB'$; $LC$, $LC'$; ... be the pairs of conjugate lines, the points $AA'$, $BB'$, $CC'$, ... being on $l$. Then the pole of $LA$ which passes through $L$ is on $l$. But the pole of $LA$ by hypothesis lies on $LA'$. Hence the pole of $LA$ is $A'$; so the pole of $LA'$ is $A$, and so on. Hence $L(AA'BB'CC'...)$ of polars $= (A'AB'BC'C...)$ of poles $= L(A'AB'BC'C...)$. Hence $L(AA', BB', CC', ...)$ form an involution.

*The double lines of the involution of conjugate lines through a point are the tangents from the point.*

For these tangents are harmonic with every pair of conjugate lines through the point.

This affords another proof that conjugate lines through a point generate an involution.

**Ex. 1.** *Through every point can be drawn a pair of lines which are conjugate for a given conic and also perpendicular.*

**Ex. 2.** *If two pairs of conjugate lines at a point are perpendicular, all pairs are perpendicular.*

**Ex. 3.** *Given that l is the polar of L, and given that ABC is a self-conjugate triangle, construct the tangents from L.*

3. An important case of conjugate lines is conjugate diameters, i. e. conjugate lines at the centre. The double lines of the involution of conjugate diameters are the tangents from the centre, i.e. the asymptotes.

**Ex. 1.** *Conjugate diameters of a hyperbola do not overlap, and conjugate diameters of an ellipse do overlap.*

**Ex. 2.** *Through the ends P and D of conjugate semi-diameters CP, CD of a conic are drawn parallels, meeting the conic again in Q and E; show that CQ, CE are also conjugate diameters.*

For if $CX$ bisect $PQ$ and $DE$, and $CY$ be the diameter conjugate to $CX$, then $CX$, $CY$ are the double lines of the involution $C(PQ, DE)$. Hence $C(XY, PD, QE)$ is an involution.

**Ex. 3.** *The joins of the ends of two pairs of conjugate diameters $PP'$, $DD'$ and $QQ'$, $EE'$ are parallel four by four.*

**Ex. 4.** *Two of the chords joining the ends of diameters parallel to a pair of tangents are parallel to the chord of contact.*

**Ex. 5.** *The conjugate diameters $CQ$, $CE$ cut the tangent at $P$ in $R$, $R'$; show that $RP \cdot PR' = CD^2$.*

For $P$ is the centre of the involution determined by the variable conjugate diameters $CQ$, $CE$ on the tangent at $P$. Also in the hyperbola the double points are on the asymptotes. Hence
$$RP \cdot PR' = -PT^2 = CD^2.$$
In the ellipse the diagonals of the quadrilateral of tangents at $P$, $P'$, $D$, $D'$ give a case of $CQ$, $CE$. Hence $RP \cdot PR' = CD^2$.

**Ex. 6.** *Parallel tangents of a conic cut the tangent at $P$ in $R$, $R'$; show that $RP \cdot PR' = CD^2$.*

**Ex. 7.** *The conjugate diameters $CQ$, $CE$ cut the tangents at the end of the diameter $PP'$ in $R$, $R'$; show that $PR \cdot P'R' = CD^2$.*

**4.** Defining the *principal axes* of a central conic as a pair of conjugate diameters which are at right angles, we can prove that—

*The principal axes of a conic are always real.*
For by XVIII. 4, one real pair of rays of an involution pencil is always orthogonal.

*A central conic (unless it be a circle) has only one pair of principal axes.*
For by XVIII. 4, if two pairs of rays of the involution pencil are orthogonal, then every pair is orthogonal, i.e. the conic is a circle.

**Ex.** *Given the centre of a conic and a self-conjugate triangle, construct the axes and asymptotes.*

Let $O$ be the centre and $ABC$ the triangle. Through $O$ draw $OA'$, $OB'$, $OC'$ parallel to $BC$, $CA$, $AB$. The asymptotes are the double lines, and the axes the orthogonal pair of the involution

$$O(AA', BB', CC').$$

**5.** *The feet of the normals which can be drawn from any point to a central conic are the meets of the given conic, and a certain rectangular hyperbola which has its asymptotes parallel to the axes of the given conic, and which passes through the centre of the given conic, and through the given point.*

Let $O$ be the given point. Take any diameter $CP$, and let the perpendicular $OY$ on $CP$ cut the conjugate diameter $CD$ in $Q$. Then, taking several positions of $P$, &c.,

$$C(Q_1 Q_2 \ldots) = C(D_1 D_2 \ldots) = C(P_1 P_2 \ldots)$$
$$= C(Y_1 Y_2 \ldots) = O(Y_1 Y_2 \ldots) = O(Q_1 Q_2 \ldots).$$

Hence the locus of $Q$ is a conic through $C$ and $O$.

This conic is a rectangular hyperbola with its asymptotes parallel to the axes, as we see by making $CP$ coincide with $CA$ and $CB$ in succession. Now let $R$ be the foot of a normal from $O$ to the given conic, then $R$ is on the above rectangular hyperbola; for, drawing $CP$ perpendicular to $OR$, $OY$ meets $CD$, i.e. $CR$, in $R$.

**Ex. 1.** *The same conic is the locus of points $Q$ such that the perpendicular from $Q$ on the polar of $Q$ passes through $O$.*

For $QO$, being perpendicular to the polar, is perpendicular to the diameter conjugate to $CQ$.

**Ex. 2.** *The normals at the four points where a conic is cut by a rectangular hyperbola which passes through the centre and has its asymptotes parallel to the axes, are concurrent at a point on the rectangular hyperbola.*

For let the normal at one of the meets $R$ cut the hyperbola again in $O$.

**Ex. 3.** *Eight lines can be drawn from a given point to meet a given central conic at a given angle.*

**Ex. 4.** *Deduce the corresponding theorems in the case of a parabola.*

Here the centre $\Omega$ is on the curve; hence one of the meets is the point $\Omega$. Rejecting this, we see that three normals and six obliques can be drawn from any point to a parabola.

**Ex. 5.** *If $OL$, $OM$, $ON$, $OR$ be concurrent normals to a conic, the tangents at $L$, $M$, $N$, $R$ touch a parabola which also touches the axes of the conic and the polar of $O$ for the conic.*

Reciprocate for the given conic.

**Ex. 6.** *$O$ is on the directrix of the parabola.*

Reciprocate for $O$.

**6.** A *common chord* of two conics is the line joining two common points of the conics.

*On a common chord of two conics the involution of conjugate points is the same for each conic, the double points being the common points.*

Conversely, *if two conics have on a line the same involution of conjugate points, this line is a common chord, the double points of the involution being the common points of the two conics.*

Two common chords of two conics which do not cut on the conics may be called *a pair of common chords*. We know that a pair of common chords meet in a vertex of a common self-conjugate triangle of the two conics. Conversely, *every point which has the same polar for two conics is the meet of a pair of common chords*.

Let $E$ be the point. Join $E$ to any common point $A$ of the two conics. Let $EA$ cut the polar of $E$ in $P$ and the conics again in $B$ and $B'$. Then $(EP, AB)$ is harmonic, and also $(EP, AB')$. Hence $B$ and $B'$ coincide, i.e. $EA$ passes through a second common point. So $EC$ passes through $D$.

Hence *two conics have only one common self-conjugate triangle*; for if $U'V'W'$ be a self-conjugate triangle, and $UVW$ the harmonic triangle belonging to the meets of the conics, then $U'$ coincides with $U$, $V$, or $W$, and so on. (See also XXV. 12.)

# Conjugate Points and Lines.

*If*, however, the *two conics touch at two points* the above proof breaks down, and *there is an infinite number of common self-conjugate triangles.*

Let the conics touch the lines $OP$ and $OQ$ at $P$ and $Q$. On $PQ$ take any two points $VW$ such that $(PQ, VW) = -1$. Then $OVW$ is clearly a common self-conjugate triangle.

Notice that *if two conics have three-point or four-point contact, the common self-conjugate triangle coincides with the point of contact.*

**Ex. 1.** *The common chords which pass through one of the vertices of the common self-conjugate triangle of two conics are a pair in the involution determined by the pairs of tangents from this point.*

$UV$, $UW$ being the double lines.

**Ex. 2.** Reciprocate Ex. 1.

**Ex. 3.** *The conic $\gamma$ touches the conic $\alpha$ at the two points $L$ and $M$, and touches the conic $\beta$ at the two points $N$ and $R$. Show that two common chords of $\alpha$ and $\beta$ meet at the intersection of $LM$ and $NR$.*

**7.** A *common apex* of two conics is the meet of two common tangents of the conics.

*At a common apex of two conics the involution of conjugate lines is the same for each conic, the double lines being the common tangents.*

Conversely, *if two conics have at a point the same involution of conjugate lines, the point is a common apex, the double lines of the involution being the common tangents of the two conics.*

*The join of a pair of common apexes of two conics has the same pole for both conics.*

Conversely, *if a line have the same pole for two conics, this line is the join of a pair of common apexes of the conics.*

These results follow by Reciprocation.

**8.** Since two conics have only one common self-conjugate triangle, it follows that *the harmonic triangle of the quadrangle of common points coincides with the harmonic triangle of the quadrilateral of common tangents.*

Let $UVW$ be the harmonic triangle of the quadrangle formed by the common points $a$, $b$, $c$, $d$, and let $AA'$, $BB'$, $CC'$ be the opposite vertices of the quadrilateral formed by the common tangents of the two conics. Then $AA'$ being a side

192                 *Involution of*                [CH.

of the common self-conjugate triangle, must coincide with
$UV$, $VW$, or $WU$, say with $VW$, as in the figure. So $BB'$
coincides with $WU$, and $CC'$ with $UV$.

*The polars of any common apex of two conics for the two
conics pass through the meet of two common chords of the conics.*

Take the common apex $B$. Now $B$ is on $WU$, the polar
of $V$. Hence the polar of $B$ for either conic passes through
$V$, the meet of the common chords $ad$, $bc$.

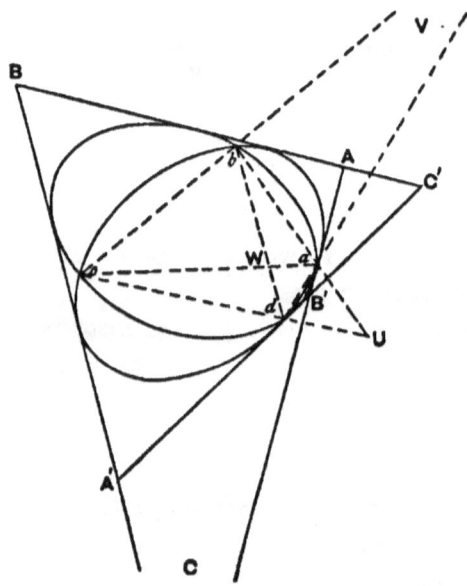

The common chords $ad$, $bc$ are said to *belong* to the common apex $B$. So to every common apex belong two common chords.

Similarly, *the poles of any common chord of two conics for the two conics lie on the join of two common apexes of the conics;* and these apexes are said to *belong* to the chord.

### Homothetic figures.

**9.** Given any figure of points $P$, $Q$, $R$, ..., and any point $O$ (called *the centre of similitude*), and any ratio $\lambda$ (called *the ratio of similitude*), we can generate another figure of points

$P'$, $Q'$, $R'$, ... thus—In $OP$ take the point $P'$, which is such that $OP' = \lambda \cdot OP$, and similarly construct $Q'$, $R'$, .... The figures $PQR$... and $P'Q'R'$... are called similar and similarly situated figures, or *homothetic figures*.

The following properties of homothetic figures follow from the definition by elementary geometry—

*Corresponding sides of the two figures (e.g. PQ and P'Q') are parallel and in the ratio $\lambda$ (i.e. $P'Q' = \lambda \cdot PQ$).*

*Corresponding angles of the two figures are equal*

$$(e.g. \angle PQR = \angle P'Q'R'.)$$

**Ex.** The triangles $ABC$, $A'B'C'$ are coaxal. $P$, $Q$, $R$ are three points on the axis. Show that if $AP$, $BQ$, $CR$ concur, so do $A'P$, $B'Q$, $C'R$.
Project the axis to infinity.

**10.** *If two conics are homothetic, the diameters conjugate to parallel diameters are themselves parallel.*

Consider the point corresponding to the centre of the first conic; it will be a point in the second conic, all chords through which are bisected at the point, i.e. it will be the centre of the second conic. Take any pair of conjugate diameters $PCP'$ and $DCD'$ of the first conic; and let $pcp'$ be the diameter of the second conic parallel to $PCP'$. Then, corresponding to $DCD'$ in the first conic, we shall have $dcd'$ in the second conic which bisects chords parallel to $pcp'$, i.e. $dcd'$ is the diameter conjugate to $pcp'$. Hence, to a pair of conjugate diameters of the first conic correspond a parallel pair of conjugate diameters of the second conic.

**11.** *Two conics will be homothetic, if two pairs of conjugate diameters of the one are parallel to two pairs of conjugate diameters of the other.*

For then every pair is parallel to some pair. Take any diameter $PCP'$ of the first conic, and through $P$ and $P'$ draw lines parallel to a pair of conjugate diameters; these lines meet in a point $Q$ on the conic. Let $pcp'$ be the diameter in the second conic parallel to $PCP'$, and through $p$ and $p'$ draw lines parallel to $PQ$ and $P'Q$. These will meet in a point $q$ on the second conic; for they are parallel to a

# 194 Involution of Conjugate Points and Lines.

pair of conjugate diameters of the first conic, and therefore parallel to a pair of conjugate diameters of the second conic. And clearly the points $Q$ and $q$ generate homothetic figures, the centre of similitude being the intersection of $Pp$ and $Cc$.

*Homothetic conics are conics which meet the line at infinity in the same points.*

If the conics are homothetic, their conjugate diameters are parallel. Hence the asymptotes, being the double lines of the involutions of conjugate diameters, are parallel, i.e. meet the line at infinity in the same points. And both conics pass through these points.

Conversely, if two conics pass through the same two points at infinity, they are homothetic. For since the conics pass through the same two points at infinity, the asymptotes of the two conics are parallel. Hence the conjugate diameters, being harmonic with the asymptotes, are parallel. Hence the conics are homothetic.

**Ex. 1.** *Through three given points, draw a conic homothetic to a given conic.*

To draw through $ABC$ a conic homothetic to $a$. Through the middle point of $AB$ draw a line parallel to the diameter bisecting chords of $a$ parallel to $AB$. This line passes through the centre $O$ of the required conic. Similarly $BC$ gives us another line through $O$. Hence the centre of the required conic and three points upon it are known.

**Ex. 2.** *Touching three given lines, draw a conic homothetic to a given conic.*

Draw tangents of the conic parallel to the sides of the given triangle. It will be found that we thus have four triangles homothetic to the given triangle. Taking any one of these triangles, and dividing the sides of the given triangle similarly, we get the points of contact of a homothetic conic.

# CHAPTER XX.

### INVOLUTION RANGE ON A CONIC.

**1.** The pairs of points $AA'$, $BB'$, $CC'$, ... on a conic are said to form an *involution range on a conic*, or briefly, to be in involution when the pencil $V(AA', BB', CC', ...)$ subtended by them at a point $V$ on the conic is in involution.

*If pairs of points $AA'$, $BB'$, $CC'$, ... be taken on a conic, such that $(AA'BC...) = (A'AB'C'...)$, then $(AA', BB', CC', ...)$ are in involution.*

For $V$ being any point on the conic, we have

$$V(AA'BC...) = (AA'BC...) = (A'AB'C'...) = V(A'AB'C'...).$$

Hence $V(AA', BB', CC', ...)$ is an involution pencil. Hence $(AA', BB', CC', ...)$ is an involution on the conic.

*An involution range on a conic has two double points, which form with any pair of points of the involution, two pairs of harmonic points on the conic.*

The double points $X$, $Y$ are the points in which the double lines of the involution pencil $V(AA', BB', CC',...)$ cut the conic.

**2.** *If the pairs of points $AA'$, $BB'$, $CC'$, ... on a conic be in involution, then the chords $AA'$, $BB'$, $CC'$, ... are concurrent; and conversely, if the chords $AA'$, $BB'$, $CC'$, ... of a conic be concurrent, then the pairs of points $AA'$, $BB'$, $CC'$, ... on the conic are in involution.*

If $(AA', BB', CC', ...)$ form an involution on the conic, we have $(AA'BB'CC'...) = (A'AB'BC'C...)$. Hence

$(AB; A'B')$, $(AC; A'C')$, and $(BC; B'C')$,

being cross meets, lie on the homographic axis of these two ranges on the conic. Hence the triangles $ABC$, $A'B'C'$ are coaxal and therefore copolar. Hence $AA'$, $BB'$, $CC'$ meet in a point, i.e. $CC'$ passes through the point $O$, where $AA'$ and $BB'$ meet. So all the lines $CC'$, $DD'$, ... pass through $O$.

Conversely, if $AA'$, $BB'$, $CC'$, ... be concurrent in $O$, then $(AA', BB', CC', ...)$ is an involution. For if not, let $C''$ be the mate of $C$ in the involution determined by $(AA', BB')$. Then by the first part $(AA', BB', CC'')$ meet in a point, i.e. $CC''$ passes through $O$. But $CC'$ passes through $O$; and $OC$ cannot cut the conic in three points. Hence $C''$ coincides with $C'$, i.e. $C'$ is the mate of $C$ in the involution $(AA', BB')$. So $D'$ is the mate of $D$, and so on.

Or thus, assuming the properties of poles and polars.

If $(AA', BB', CC', ...)$ be an involution on the conic, then $(AA'BB'CC'...)$ and $(A'AB'BC'C...)$ are homographic ranges on the conic; hence the meets $(A'B; AB')$, $(A'B'; AB)$, &c. lie on a fixed line, viz. the axis of homography. Hence $AA'$ and $BB'$ pass through the pole of this line; so for $CC'$, &c.

Again, if the chords $AA'$, $BB'$, $CC'$, ... of the conic meet in $O$, then the meets $(A'B; AB')$, $(A'B'; AB)$, &c. lie on the polar of $O$. Hence $(AA'BB'CC'...)$ and $(A'AB'BC'C...)$ are homographic ranges on the conic. Hence

$$(AA', BB', CC', ...)$$

are in involution.

The point $O$ where $AA'$, $BB'$, $CC'$, ... meet is called the *pole of the involution* $(AA', BB', CC', ...)$, and the line on which $(AB; A'B')$, &c. lie, i.e. the homographic axis of the two ranges $(AA'BB'...)$ and $(A'AB'B...)$, is called the *axis of the involution*. Note that *the axis of the involution is the polar for the conic of the pole of the involution*. For $(AB; A'B')$ and $(AB'; A'B)$, being cross meets, lie on the homographic axis of the ranges $(AA'BB'...)$ and $(A'AB'B...)$. Hence $O$ is the pole of the axis of involution. *The double points of the involution are the points* $X$, $Y$ *where the axis of involution cuts the conic.* For these are the common points of the homographic ranges $(AA'BB'...)$ and $(A'AB'B...)$. Hence, *the double points of an involution on a conic are real if the pole of the involution is outside the conic.*

**3.** *If two segments bounded by corresponding points (such as $AA'$, $BB'$) of an involution on a conic overlap, every two of such segments overlap, and the double points are imaginary, i.e. the pole of the involution is inside the conic. So in a non-overlapping involution on a conic, the double points are real and the pole is outside.*

For consider the pencil subtended at any point on the conic by the points in involution.

If $O$ is on the conic, the involution is nugatory.

**4.** Reciprocally, *a set of pairs of tangents to a conic are said to be in involution when they cut any tangent to the conic in pairs of points in involution.*

*Again, the meets of corresponding tangents lie on a line; and conversely, pairs of tangents from points on a line form an involution of tangents.*

*The double lines of the involution of tangents are the tangents at the meets of the above line with the conic.*

Notice that *if a set of pairs of tangents be in involution, the set of pairs of points of contact is in involution, and conversely.*

These propositions follow at once by Reciprocation.

**Ex. 1.** *A bundle of parallel lines cuts a conic in pairs of points in involution.*

**Ex. 2.** *A system of coaxal circles cuts a given circle in pairs of points in involution.*

**Ex. 3.** *Two chords $AA'$, $BB'$ of a conic cut in $U$, and $OT$ is the tangent at $O$; show that $O(AA', BB', TU)$ is an involution.*

**Ex. 4.** *Reciprocate Ex. 3.*

**Ex. 5.** *Three concurrent chords $AA'$, $BB'$, $CC'$ of a circle are drawn, show that*
$$\sin \tfrac{1}{2} AB \cdot \sin \tfrac{1}{2} B'C \cdot \sin \tfrac{1}{2} C'A' = -\sin \tfrac{1}{2} A'B' \cdot \sin \tfrac{1}{2} BC' \cdot \sin \tfrac{1}{2} CA,$$
*where $AB$ denotes the angle subtended by $AB$ at the centre.*

**Ex. 6.** *$A$, $B$, $C$ are points on a conic. $A'$, $B'$, $C'$ are points taken on the conic such that $(AA', BC) = (BB', CA) = (CC', AB) = -1$. Show that $(AA', BB', CC')$, $(AA', BC', B'C)$, $(BB', AC', A'C)$, and $(CC', AB', A'B)$ are involutions on the conic.*

Let the tangents at $A$, $B$, $C$ meet in $P$, $Q$, $R$. Then $AA'$ passes through $P$, $BB'$ through $Q$, and $CC'$ through $R$. But $PA$, $QB$, $RC$ meet in a point; hence $AA'$, $BB'$, $CC'$ meet in a point; i.e. $(AA', BB', CC')$ form an involution. Hence $(A'A, B'C') = (AA', BC) = -1$. Hence $(A'A, B'C') = (AA', CB)$. Hence $(AA', CB', BC')$ is an involution. And so on.

**Ex. 7.** *Through a given point $O$ draw the chord $XX'$ of a conic, such that $(AA', XX') = (BB', XX')$ where $A$, $A'$, $B$, $B'$ are any four points on the conic.*

Join $O$ to the meet of $AB'$ and $A'B$

**Ex. 8.** *$DE$ is a fixed diameter of a conic. $PQ$ is a variable chord of the conic. The tangent at $E$ meets $DP$ in $A$ and $DQ$ in $B$. If $A$, $B$ generate an involution, $PQ$ passes through a fixed point. If $EA \cdot EB$ be constant, the fixed point lies on $DE$. If $1/EA + 1/EB$ be constant, the fixed point lies on $EA$.*

One position of $PQ$ is in the first case $DE$, and in the second case, the tangent at $E$.

**Ex. 9.** *From a fixed point $O$ perpendiculars are drawn to the pairs of lines of a pencil in involution, meeting them in $AA'$, $BB'$, ...; show that the lines $AA'$, $BB'$, ... are concurrent.*

Consider the circle on $OV$ as diameter.

**Ex. 10.** *An involution of points on a conic subtends an involution pencil at every point on the axis of the involution.*

**Ex. 11.** *$AA'$, $BB'$, $CC'$, ... are concurrent chords of a conic; show that $(ABC...) = (A'B'C'...)$. Also reciprocate the proposition.*

**Ex. 12.** *Through fixed points $U$, $V$ are drawn the variable chords $RP$ and $RQ$ of a conic; show that $P$ and $Q$ generate homographic ranges on the conic, and that the common points lie on the line $UV$.*

**Ex. 13.** *Through a fixed point $O$ is drawn the variable chord $PP'$ of a conic. $A$ and $B$ are fixed points on the conic. $PB$, $P'A$ meet in $Q$, and $PA$, $P'B$ meet in $R$. Show that $Q$ and $R$ move on the same fixed conic.*

For $A(QR) = A(P'P) = B(PP') = B(QR)$.

**Ex. 14.** *Given two points $P$, $Q$ on the line of an involution, determine a segment of the involution which shall divide $PQ$ harmonically.*

Project the involution on to any conic through $PQ$, and join the pole of the involution to the pole of $PQ$.

**Ex. 15.** *Through a centre of similitude of two circles are drawn four lines meeting one circle in $ABCD$, $A'B'C'D'$, and the other circle in $abcd$, $a'b'c'd'$. Show that*
$(ABCD) = (A'B'C'D') = (abcd) = (a'b'c'd')$,
For $(ABCD) = (abcd)$ by similarity.

**Ex. 16.** *A range on a circle and its inverse are homographic.*

**Ex. 17.** *A range in involution, whether on a circle or a line, inverts into a range in involution, whether on a circle or a line.*

**Ex. 18.** *A variable circle passes through a fixed point, and cuts a given circle at a given angle; show that it determines on the circle two homographic ranges.*

Invert for the fixed point.

**Ex. 19.** *A variable circle cuts two given circles orthogonally; show that it determines on each circle a range in involution.*

Invert for a meet of the given circles.

**Ex. 20.** *A variable circle cuts a given circle and a given line orthogonally; show that it determines both on the circle and on the line a range in involution.*

**Ex. 21.** *The pole of the involution $(AA'BB'...)$ on a conic is the same as the homographic pole of the pencils subtended by $AA'BB'...$ and $A'AB'B...$ at any two points on the conic.*

For $AA'$ is one of the cross joins.

**Ex. 22.** *Given two pencils $V(ABC)$ and $V'(A'B'C')$, draw through $V$ and $V'$ a circle meeting the pencils in the points $abc$, $a'b'c'$, such that $(aa', bb', cc')$ is an involution on the circle.*

By Ex. 21 $aa'$ passes through a given point, and it is easy to see that its direction is given.

**Ex. 23.** *Two homographic ranges on the same circle or on equal circles can, in two ways, be placed so as to be in involution on the same circle.*

**Ex. 24.** *The pairs of tangents drawn to a parabola from points on a line are parallels to the rays of an involution pencil.*

**Ex. 25.** *On a fixed tangent of a conic are taken two fixed points $A$, $B$, and also two variable points $Q$, $R$, such that $(AB, QR) = -1$; find the locus of the meet of the other tangents from $Q$ and $R$.*

**Ex. 26.** *A variable tangent to a conic cuts two fixed lines in $A$, $A'$. Show that the points of contact $a$, $a'$ of the other tangents from $A$, $A'$ generate homographic ranges on the conic.*

Let $AA'$ touch in $a$. Then the ranges $a$ and $a$ are in involution, and the ranges $a$ and $a'$. Hence $(a...) = (a...) = (a'...)$.

**Ex. 27.** *The fixed tangent $OA$ of a conic meets a variable tangent in $X$, and the fixed tangent $OB$ meets the parallel tangent in $Y$. Show that $OX . OY$ is constant.*

Let the parallel tangent meet $OA$ in $X'$. Then $(X, X')$ generate an involution. Hence $(X) = (X') = (Y)$. And $O$ is the vanishing point of both ranges.

**Ex. 28.** *$AA'$ is a fixed diameter of a conic; on a fixed line through $A'$ is taken a variable point $P$, and the tangents from $P$ meet the tangent at $A$ in $Q$, $Q'$. Show that $AQ + AQ'$ is constant.*

$Q$, $Q'$ generate an involution, of which one double point (corresponding to $P$ being at $A'$) is at infinity. Hence the other double point $O$ bisects $QQ'$. Hence $AQ + AQ' = 2AO$.

**Ex. 29.** *If $P$ lie on a chord through $A$ instead of $A'$, then $1/AQ + 1/AQ'$ is constant.*

**5.** *Given two involution ranges on a conic or on a line, or two involution pencils at a point; find the pair of points or lines belonging to both involutions.*

The line joining the two poles $O_1 O_2$ of the involutions on the conic evidently cuts the conic in the required pair of points.

If the ranges are on a line, project the ranges on to a

conic by joining to a point on the conic, and project back on to the line the common points on the conic.

In the case of two involution pencils at a point, consider the involutions determined on any conic through the common vertex.

*If either of the pairs of double points (or lines) of the given involutions be imaginary, the common pair of points are real; and they are also real when both pairs of double points are real and do not overlap.*

For if the involution on the conic, of which $O_1$ is the pole, has imaginary double points, $O_1$ is inside the conic; hence $O_1 O_2$ cuts the conic whether $O_2$ is inside or outside the conic.

Also, if the double points are real and do not overlap, the points sought, being harmonic with both pairs, are the double points of a non-overlapping involution on the conic, and are therefore real.

The cases of involution on the same line or at the same point may be discussed as above.

**6.** *In a pencil in involution, one pair of rays is always orthogonal; and if two pairs of rays are orthogonal, then every pair is orthogonal.*

Let the rays of the involution pencil $V(AA'BB'CC'...)$ cut a circle through $V$ in the points $AA'$, $BB'$, $CC'$, .... Then $AA'$, $BB'$, $CC'$, ..., being chords joining pairs of points in involution on the circle, meet in a point $O$. Take $K$, the centre of the circle, and let $OK$ cut the circle again in $ZZ'$.

Then $VZ$, $VZ'$ is an orthogonal pair in the involution pencil. For, since $ZZ'$ passes through $K$, $ZVZ'$ is a right angle. And since $ZZ'$ passes through $O$, $ZZ'$ belong to the involution $(AA', BB', ...)$, i.e. $VZ$, $VZ'$ belong to the given involution pencil $V(AA', BB', ...)$.

Again, if two pairs of orthogonal rays exist, viz. $VX$, $VX'$ and $VY$, $VY'$, since $XX'$ and $YY'$ both pass through $K$, we see that $O$ coincides with $K$. Hence $AA'$, $BB'$, ... all pass through $K$. Hence all the angles $AVA'$, $BVB'$, ... are right angles.

**7. Chords of a conic which subtend a right angle at a fixed point on the conic meet in a point on the normal at the fixed point.**

Let the chords $QQ'$, $RR'$, ... of a conic subtend right angles at the point $P$ on the conic. Then $P(QQ', RR', ...)$ is an orthogonal involution pencil. Hence $(QQ', RR', ...)$ is an involution on the conic. Hence $QQ'$, $RR'$, ... all pass through a point $F$. Now suppose $PQ$ to coincide with the tangent at $P$; then $PQ'$ coincides with the normal, $Q$ coincides with $P$, and hence $QQ'$ coincides with the normal. Hence the normal is one such chord, and therefore $F$ lies on the normal at $P$.

The point $F$ is called the *Frégier point* of the point $P$.

**Ex. 1.** *Show that the theorem also follows by reciprocating for the point P.*

**Ex. 2.** *P and U are fixed points on a conic. Through U are drawn two lines meeting the conic in L, M, and the polar of the Frégier point of P in X, Y. Show that LM and XY subtend equal angles at P.*

Note—the polar of the Frégier point of $P$ is called the *Frégier line* of $P$.

**Ex. 3.** *In a parabola, PF is bisected by the axis.*
Take $PQ$ parallel to the axis.

**Ex. 4.** *In a parabola, the locus of F as P varies is an equal parabola.*

**Ex. 5.** *In a central conic, the angle PCF is bisected by the axes.*
Take $PQ$ parallel to the minor axis, then $F$ is on $CQ$.

**Ex. 6.** *In a central conic, the locus of F is a homothetic and concentric conic.*
For $CQ : CF :: CP : CF :: PG : GF$.
Now $PG = bb'/a$, and $Pg = ab'/b$. Hence, since $(PF, Gg)$ is harmonic, $PF$ can be found. Hence, $CQ : CF :: a^2 + b^2 : a^2 - b^2$.

**Ex. 7.** *If a triangle QPQ', right-angled at P, be inscribed in a rectangular hyperbola, the tangent at P is the perpendicular from P on QQ'.*
For, taking $PQ$, $PQ'$ parallel to the asymptotes, we see that the Frégier point of $P$ is at infinity.

**Ex. 8.** *If the chords PQ, PQ' of a conic be drawn equally inclined to the tangent at the fixed point P, then QQ' passes through a fixed point on the tangent at P.*

**8. To construct the double points of an involution range on a line or the double lines of an involution pencil.**

In the case of an involution range on a line, project the range on to any conic through a vertex on the conic; determine the double points of the involution on the conic; then the projections of these double points on the line are the double points of the involution on the line.

In the case of an involution pencil, draw any conic through the vertex, and join the vertex to the double points of the involution which the pencil determines on the conic. These joins are the double lines.

# CHAPTER XXI.

### INVOLUTION OF A QUADRANGLE.

**1.** *The three pairs of points in which any transversal cuts the opposite sides of a quadrangle are in involution.*

Let the transversal meet the sides of the quadrangle $ABCD$ in $aa'$, $bb'$, $cc'$. Then

$$A(DWBb') = C(DWBb').$$

Hence $(abc'b') = (cba'b')$.

Hence $(abc'b') = (a'b'cb)$.

Hence $(aa', bb', c'c)$ is an involution, i.e. $(aa', bb', cc')$ is an involution.

To determine the mate $c'$ of $c$ in the involution determined by

$$(aa', bb').$$

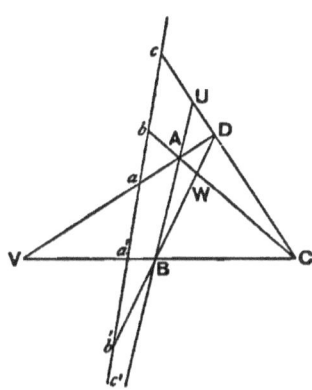

Take any point $V$. On $Va$ take any point $A$. Let $bA$ cut $Va'$ in $C$. Let $Cc$ cut $VA$ in $D$. Let $Db'$ cut $VC$ in $B$. Then $AB$ cuts $aa'$ in the required point $c'$.

**Ex. 1.** *Show that each diagonal of a quadrilateral is divided harmonically.*
Consider $CC'$ as the transversal of the quadrangle $ABA'B'$. Then $CC'$ are the double points.

**Ex. 2.** *If through any point parallels be drawn to the three pairs of opposite sides of a quadrangle, a pencil in involution is obtained.*

**Ex. 3.** *The same is true if the lines be drawn perpendicular to the sides. Hence show that if four circles be cut orthogonally by the same circle, the six radical axes form an involution.*

**Ex. 4.** *$U$, $V$, $W$ are the harmonic points of the quadrangle $ABCD$. If $U(PQ, BC) = -1 = V(PQ, CA)$, show that $W(PQ, AB) = -1$.*

**Ex. 5.** *If a meet of opposite sides of a quadrangle be joined to the middle point of the segment cut off from a given transversal by these opposite sides, the three lines so formed are concurrent.*

**Ex. 6.** *A transversal cuts the sides of a triangle in P, Q, R, and the lines joining the vertices to any point in P', Q', R'; show that PP', QQ', RR' are in involution.*

**Ex. 7.** *The three meets of any line with the sides of a triangle and the three projections of the vertices on this line form an involution.*

**Ex. 8.** *If from any point three lines be drawn to the vertices of a triangle, and three other lines parallel to the sides; these six lines form an involution.*

**Ex. 9.** *A transversal cuts the sides of a triangle ABC in P, Q, R, and PP', QQ', RR' form an involution on the transversal; show that AP', BQ', CR' are concurrent.*

**Ex. 10.** *Six points, A, B, C, A', B', C' are taken, and through any point O are drawn Oa, Oa', Ob, Ob', Oc, Oc' parallel to AA', BC, BB', CA, CC', AB. If the angles aOa', bOb', cOc' have the same bisectors, then AA', BB', CC' are concurrent.*

**Ex. 11.** Hesse's theorem. *If two opposite pairs of vertices of a quadrilateral are conjugate for a conic, then the third pair are conjugate for the same conic.*

Let $AA'$, $BB'$, $CC'$ be the opposite pairs of vertices. Take $P$ the pole of the side $ABC$. Let $ABC$ cut $PA'$ in $X$, $PB'$ in $Y$, $PC'$ in $Z$. Then $(AX, BY, CZ)$ are in involution (from quadrangle $PA'B'C'$). Also the polar of $A$ is $A'P$, if $AA'$ are conjugate; hence $AX$ are conjugate points. So $BY$ are conjugate, if $BB'$ are conjugate. Hence $CZ$ are conjugate. Hence $PZ$, i.e. $PC'$, is the polar of $C$; i.e. $CC'$ are conjugate.

### Involution of four-point conics.

**2. Desargues's theorem.**—*Any transversal cuts a conic and the opposite sides of any quadrangle inscribed in the conic in four pairs of points in involution.*

Let $ABCD$ be the inscribed quadrangle. Let the transversal cut the conic in $pp'$, $AC$ in $b$, $BD$ in $b'$, $CD$ in $c$, and $AB$ in $c'$. Then

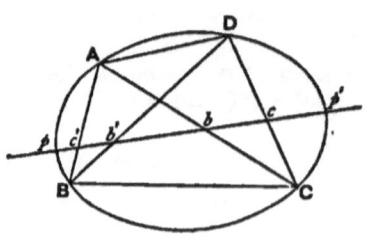

$$(pp'bc) = C(pp'AD)$$
$$= B(pp'AD) = (pp'c'b').$$

Hence $(pp'bc) = (p'pb'c')$. Hence $(pp', bb', cc')$ is an involution. Hence $cc'$ belong to the involution $(pp', bb')$. Similarly, $aa'$ belong to this involution. Hence $(pp', bb', cc', aa')$ is an involution.

**3.** *The system of conics which can be drawn through four given points are cut by any transversal in pairs of points in involution.*

For $pp'$ belong to the involution $(aa', bb', cc')$ determined by the opposite sides of the given quadrangle. And similarly for any other conic of the system.

Note that we have above given an independent proof that $(aa', bb', cc')$ is an involution. For through $ABCD$ and any point $p$ on the transversal we can draw a conic.

Note also that we should expect $aa'$, $bb'$, $cc'$ to belong to the involution $(pp', qq', ...)$ determined by the conics through the four points. For each pair of opposite sides of the quadrangle is a conic through the four points.

**Ex. 1.** *Any transversal cuts a conic in PQ and the successive sides of a four-sided inscribed figure in* 1, 2, 3, 4; *show that*
$$\frac{P1.P3}{Q1.Q3} = \frac{P2.P4}{Q2.Q4};$$
*and extend the theorem to any inscribed polygon of an even number of sides.*

**Ex. 2.** *On every line, there is a pair of points which are conjugate for every one of a system of conics through four given points.*

Viz. the double points of the involution.

**Ex. 3.** *Through the centres of a system of four-point conics can be drawn pairs of parallel conjugate diameters.*

Take the line in Ex. 2 at infinity.

**Ex. 4.** *Two conics can be drawn to pass through four given points and to touch a given line.*

Draw a conic $a$ through the four given points $A$, $B$, $C$, $D$, and through $e$, one of the double points of the involution of the quadrangle $ABCD$ on the given line $l$. Let $l$ cut $a$ again in $e'$. Then $e$, $e'$ are a pair in the involution of which $e$ is a double point. Hence $e'$ coincides with $e$. Hence $l$ touches $a$.

**Ex. 5.** *A fixed conic passes through one pair $AA'$ of an involution range, and $UU'$ are fixed points on the conic. $PP'$ is another pair of the involution. The conic meets UP again in $p$, and $U'P'$ again in $p'$. Show that $pp'$ passes through the mate of the meet of $UU'$ and $AA'$.*

**Ex. 6.** *The segment between the points of contact of a common tangent of two conics is divided harmonically by any opposite pair of common chords. Also the polars of a common apex of two conics form a harmonic pencil with a pair of common chords.*

For each point of contact, being a coincident pair of points in the involution, is a double point.

**Ex. 7.** *A conic passes through three out of four vertices of a quadrangle, and a line meets the six sides and the conic in an involution. Show that the conic also passes through the fourth.*

**Ex. 8.** *On the side $BC$ of the triangle $ABC$ inscribed in a circle (centre $O$) is taken a point $P$. The line through $P$ perpendicular to $OP$ meets $AB$ in $Q$, and* on $QP$ produced is taken a point $R$, such that $RP = PQ$. Show that $CR$, $AP$ meet on the circle.

$P$ is one of the double points on $RQ$.

**Ex. 9.** *$A$ is the middle point of a chord of a conic; $B$, $C$ are points on the chord equidistant from $A$; $BDE$ and $CFG$ are chords of the conic; show that $EF$ and $GD$ cut $BC$ in points equidistant from $A$.*

**Ex. 10.** *$A$ transversal parallel to a side of a quadrangle inscribed in a conic cuts the opposite side in $O$, and the conic and a pair of opposite sides in $AA'$, $BB'$; show that $OA \cdot OA' = OB \cdot OB'$.*

**Ex. 11.** *Three sides of a four-sided figure inscribed in a conic pass through three fixed points on a line; show that the fourth side passes through a fourth fixed point on the same line.*

**Ex. 12.** *Extend the theorem to any $2n$-sided figure.*

**Ex. 13.** *By taking the two vertices coincident which lie on the $2n$th side, deduce a simple solution of the problem—'Inscribe in a given conic a polygon of $2n-1$ sides, each side to pass through one of a set of $2n-1$ fixed collinear points.'*

Draw tangents from the $2n^{th}$ fixed point.

**Ex. 14.** *Show that the problem—'To inscribe in a given conic a polygon of $2n$ sides, each side to pass through one of a set of $2n$ fixed collinear points'—is either indeterminate or impossible.*

**Ex. 15.** *To deduce Carnot's theorem from Desargues's theorem.*

Let $BC$ cut $B_1 C_1$ in $L_1$ and $B_2 C_2$ in $L_2$. Then

$$AC_1 \cdot BL_1 \cdot CB_1 = AB_1 \cdot CL_1 \cdot BC_1 \text{ from } B_1 C_1 L_1,$$

and $AC_2 \cdot BL_2 \cdot CB_2 = AB_2 \cdot CL_2 \cdot BC_2$ from $B_2 C_2 L_2$.

Also $CL_1 \cdot CL_2 \cdot BA_1 \cdot BA_2 = BL_1 \cdot BL_2 \cdot CA_1 \cdot CA_2$,

since $L_1 L_2$, $A_1 A_2$, $CB$ are in involution. Now multiply up.

**Ex. 16.** *A conic $\beta$ is described through the points $A$, $B$, $C$, $O$, where $O$ is the pole of the triangle $ABC$ for the conic $a$. Show that $a$ and $\beta$ are so situated that triangles can be inscribed in $\beta$ which are self-conjugate for $a$.*

For let the polar of $A$ for $a$ cut $\beta$ in $PP'$, $AC$ in $b$, $AB$ in $c$, $OB$ in $B'$, $OC$ in $C'$; then $(PP', bB', cC')$ is an involution. Also $bB'$ are conjugate for $a$, and so are $cC'$; hence so are $PP'$. Hence $APP'$ is such a triangle.

**Ex. 17.** *If two conics $a$ and $\beta$ are so situated that triangles can be inscribed in $\beta$ which are self-conjugate for $a$, then the pole for $a$ of any triangle inscribed in $\beta$ lies on $\beta$.*

**Ex. 18.** *Reciprocate Ex. 16 and Ex. 17.*

**4.** If $A$ and $D$ become coincident, $AD$ becomes the tangent at $A$, $b$ and $c$ coincide, and $b'$ and $c'$ coincide. Hence, *if $ABC$ be a triangle inscribed in a conic, and if any transversal cut $BC$, $CA$, $AB$ in $a'$, $b$, $b'$, the tangent at $A$ in $a$, and the conic in $pp'$, then $pp'$ is a pair of points in the involution determined by $(aa', bb')$.*

**Ex. 1.** *A, B are the ends of a diameter of a conic, and C, D are fixed points on the conic; find a point P on the conic, such that PC, PD intercept on AB a segment bisected by the centre of the conics.*

The tangent at $P$ and $CD$ must meet $AB$ in points equidistant from the centre.

**Ex. 2.** *Through the fixed point B on a hyperbola are drawn the lines BP, BQ parallel to the asymptotes. Through the fixed point O on the hyperbola is drawn the variable chord OQPR cutting the curve again in R. Show that the ratio $PR:QR$ is constant.*

This is a particular case of the theorem—'$ABCO$ are fixed points on a conic. A line through $O$ meets $BC$ in $P$, $BA$ in $Q$, the conic in $R$, and $CA$ in $U$. Show that $(QPRU)$ is constant.'

Now $(QPRU) = (PQOT) = B(CAOT)$, $T$ being on the tangent at $B$.

**5.** If $A$ and $B$ coincide, and also $C$ and $D$, then $aa'bb'$ all lie on $AC$, at the point $E$, say; i.e. $E$ is a double point of the involution. Hence, *if any transversal cut a conic in $pp'$, the tangents at $A$ and $C$ in $cc'$, and $AC$ in $E$, then $E$ is a double point of the involution determined by $cc'$, $pp'$.*

**Ex. 1.** *Prove the following construction for the double points of the involution determined by $AA'$, $BB'$—Through $BB'$ describe any conic. Let the tangents from $A$ touch at $L, M$ and the tangents from $A'$ at $N, R$; then $LN$, $RM$ cut in one double point, and $LR$, $MN$ cut in the other.*

Consider first the quadrangle $LLNN$; then we see that $LN$ passes through a double point.

**Ex. 2.** *The tangents of a conic at P and Q meet in T. A transversal meets the conic in $AA'$, the tangents in $BB'$, and $PQ$ in $C$; show that*
$$CA \cdot CB' \cdot BA' = CA' \cdot BC \cdot B'A.$$

**Ex. 3.** *The tangents at the points $PQR$ on a conic meet in $P'Q'R'$, and the corresponding opposite sides of the triangles $PQR$, $P'Q'R'$ meet in $P''Q''R''$; show that* $(PP'', Q'R')$, $(QQ'', R'P')$, $(RR'', P'Q')$ *are harmonic ranges.*

**Ex. 4.** *The tangents of a conic at P and Q meet in T. A transversal parallel to PQ cuts the conic in $AA'$ and the tangents in $BB'$; show that $AB = A'B'$.*

For one double point is at infinity.

**Ex. 5.** *Any transversal cuts a hyperbola and its asymptotes in $AA'$, $BB'$; show that $AB = A'B'$.*

**Ex. 6.** *The tangents of a conic at P and Q meet in T. A line parallel to QT cuts PT in L, PQ in N, and the conic in M and R. Show that $LN^2 = LM \cdot LR$.*

**Ex. 7.** *Two parabolas with parallel axes touch at P. A transversal is drawn cutting the tangent at P in O, the diameter through P in E, and the curves in QQ', RR'. Show that $OE^2 = OQ \cdot OQ' = OR \cdot OR'$.*

**6.** If $A$, $B$ and $C$ coincide, then $a'$, $c'$ and $b$ coincide, and $a$, $b'$ and $c$ coincide. Hence, *if a system of conics be drawn having three-point contact at A, and passing through D, then any* *transversal cuts the conics in pairs of points in involution, one pair being the points on $AD$ and on the tangent at $A$.*

**Ex.** *The common tangent of a conic and its circle of curvature at P is divided harmonically by the tangent at P and the common chord.*

**7.** If $A$, $B$, $C$ and $D$ coincide, then $a$, $a'$, $b$, $b'$, $c$, $c'$ all coincide in the point $E$, where the tangent at $A$ cuts the transversal. Hence, *a system of conics having four-point contact at a point is cut by any transversal in pairs of points in involution, of which one double point is on the tangent at the point.*

**Ex. 1.** *The tangent at the point R to the circle of curvature at the vertex of a conic cuts the conic in P, Q, and the tangent at the vertex in T. Show that*
$$(TR, PQ) = -1.$$
For $R$ is the other double point.

**Ex. 2.** *If two conics have four-point contact at a point, the polars of any point on the tangent at this point coincide.*

**Ex. 3.** *If two conics touch, and if the polars of every point on the tangent at the point of contact coincide, the two conics have four-point contact at this point.*

For the opposite common chord coincides with the tangent.

**Ex. 4.** *Two equal parabolas which have the same axis have four-point contact at infinity.*

**8.** *If a transversal cut two pairs of opposite sides of the quadrangle $ABCD$ in $aa'$, $bb'$, and any two corresponding points $p$, $p'$ be taken in the involution $(aa', bb')$; then the six points $A$, $B$, $C$, $D$, $p$, $p'$ lie on a conic.*

For draw a conic through $ABCDp$; then the conic passes also through $p'$ by 'reductio ad absurdum.'

**Ex. 1.** *$ABCD$, $abcd$ are two quadrangles inscribed in a conic; $ab$, $cd$ meet $AD$, $BC$ in $E$, $F$, $G$, $H$; $ad$, $bc$ meet $AB$, $CD$ in $E'$, $F'$, $G'$, $H'$; show that $E$, $F$, $G$, $H$, $E'$, $F'$, $G'$, $H'$ are eight points on a conic.*

Let $F'daE'$ meet $AD$, $BC$ in $K$, $L$. Then $(ad, E'F', KL)$ are in involution. Hence $EFGHE'F'$ lie on a conic. And so on.

**Ex. 2.** *A line cuts two conics in $AB$, $A'B'$, and $E$, $F$ are the double points of the involution $AA'$, $BB'$ (or $AB'$, $A'B$); show that a conic through the meets of the given conics can be drawn through $E$, $F$.*

**Ex. 3.** *$AB$, $BC$, $CD$, $DA$ touch a conic. Through $U$ (the meet of $AC$, $BD$) is drawn any chord $PQ$ of the conic; show that the six points $A$, $B$, $C$, $D$, $P$, $Q$ lie on a conic.*

**Ex. 4.** *Four points $A$, $B$, $C$, $D$ are taken on a circle; $AB$ cuts another circle in $A'B'$, and $CD$ cuts this circle in $C'D'$; $BD$ cuts $A'D'$, $B'C'$ in $E$, $F$; and $AC$ cuts $A'D'$, $B'C'$ in $H$, $G$; show that $EFGH$ lie on a coaxal circle.*

**9.** *Every rectangular hyperbola which circumscribes a triangle passes through its orthocentre; and, conversely, every conic which circumscribes a triangle and passes through its orthocentre is a rectangular hyperbola.*

Let $D$ be the orthocentre of the triangle $ABC$. Let $O$ be the centre of a r. h. through $ABC$. Let the line at infinity cut the r. h. in $pp'$, and the sides of the quadrangle $ABCD$ in $aa'$, $bb'$, $cc'$. Join these points to $O$. Then $Op$ and $Op'$, being the asymptotes of the r. h., are orthogonal. Also $Oa$ and $Oa'$, being parallel to $AD$ and $BC$, are orthogonal; so $Ob$ and $Ob'$ are orthogonal, and $Oc$ and $Oc'$ are orthogonal. Hence $O(pp', aa', bb', cc')$ is an involution. Hence

$(pp', aa', bb', cc')$ is an involution.

Hence the conic $ABCpp'$ passes through $D$. Hence any r. h. through $ABC$ passes through the orthocentre of $ABC$.

Conversely, let $O$ be the centre of a conic through $ABCD$. Let the line at infinity cut this conic in $pp'$, and the sides of the quadrangle $ABCD$ in $aa'$, $bb'$, $cc'$. Then $(pp', aa', bb', cc')$ is an involution. But $aOa'$, $bOb'$, $cOc'$ are right angles. Hence $pOp'$ is a right angle. But $Op$, $Op'$ are the asymptotes of the conic. Hence the conic is a r. h.

**Ex. 1.** *Every conic through the meets of two r. h.s is a r. h.*

For let the meets be $ABCD$. Then if $D$ is not the orthocentre of $ABC$, let $D'$ be. Then the two r. h.s pass through $ABCDD'$, which is impossible. Hence $D$ is the orthocentre.

**Ex. 2.** *Every r. h. which passes through the middle points of the sides of a triangle passes through the circum-centre.*

# CHAPTER XXII.

### POLE-LOCUS AND CENTRE-LOCUS.

**1.** *The polars of a given point for a system of four-point conics are concurrent.*

Let $X$ be the given point. Let the polars of $X$ for two conics $a$, $\beta$ of the system meet in $X'$. Consider the involution $(pp', qq', rr', ...)$ determined by the conics $a$, $\beta$, $\gamma$, ... of the system on the line $XX'$. Since

$$(XX', pp') \text{ and } (XX', qq')$$

are harmonic, $XX'$ are the double points of the involution. Hence $(XX', rr')$, &c., are harmonic. Hence $XX'$ are conjugate points for every conic of the system. Hence the polars of $X$ for the system are concurrent in $X'$.

Clearly the polars of $X'$ for the system pass through $X$. Hence $X$, $X'$ are called *conjugate points for the system of four-point conics*.

**Ex. 1.** *Of a system of four-point conics, the diameters bisecting chords in a fixed direction are concurrent.*

**Ex. 2.** *The polars of a given point for the three pairs of opposite sides of a quadrangle are concurrent.*

For each pair is a conic of the system.

**Ex. 3.** *The polars of a given point for a system of conics touching two given lines at given points meet in a point on the chord of contact.*

For the chord of contact, considered as two coincident lines, is one of the four-point conics.

**2.** *Given a system of four-point conics and a line $l$, the locus of the poles of $l$ for conics of the system, is a conic, which coincides with the locus of points which are conjugate to points on $l$ for conics of the system.*

## Pole-locus and Centre-locus.

Let the poles of $l$ for conics $\alpha, \beta, \gamma, \ldots$ of the system be $L, M, N, \ldots$; and let $X', Y', \ldots$ be the conjugate points of the points $X, Y, \ldots$ on $l$ for the system. Then the polars of $X, Y, \ldots$ for $\alpha$ are $LX', LY', \ldots$. Hence
$$(XY\ldots) = L(X'Y'\ldots).$$
So $(XY\ldots) = M(X'Y'\ldots)$. Hence $L(X'Y'\ldots) = M(X'Y'\ldots)$. Hence $LMX'Y'\ldots$ lie on a conic. Hence all the points $X'Y'\ldots$ lie on a conic which passes through $L$ and $M$. Similarly the locus passes through $N\ldots$. Hence all the points $LMN\ldots$ and all the points $X'Y'\ldots$ lie on a single conic, called *the pole-locus of the line $l$ for the system of four-point conics*.

The pole-locus is also called *the eleven-point conic* because it passes through eleven points which can be constructed at once from the given line and the given quadrangle.

Three of these points are the harmonic points of the quadrangle. For $U$ is conjugate to the point in which $VW$ cuts $l$; and so on.

Six more of these points are the fourth harmonics of $a$ for $AD$, $b$ for $AC$, $c$ for $DC$, $a'$ for $BC$, $b'$ for $BD$, $c'$ for $BA$, taking the transversal of the figure of XXI. 1 as $l$. For the polar of $a$ for every conic of the system passes through the fourth harmonic of $a$ for $AD$, since $A$ and $D$ are on the conic.

The last two points are the double points of the involution determined by the conics on $l$. For these are clearly conjugate for each conic of the system.

**Ex. 1.** *If $l$ vary, all the eleven-point conics pass through three fixed points.*

**Ex. 2.** *If the quadrangle be a square, the pole-locus is a rectangular hyperbola.*

**Ex. 3.** *If $l$ pass through one of the harmonic points of the given quadrangle, the pole-locus breaks up into a pair of lines.*

Let $l$ pass through $W$. Then $UV$ contains four of the eleven points, viz. $UV$ and the fourth harmonics of $W$ for $AC$ and $BD$. Hence the locus cannot be a curved conic; hence it is two lines. It is easy to show that $UV$ contains five points, and that the other six ($W$ counting twice) lie on the fourth harmonic of $l$ for $WA$, $WD$.

**Ex. 4.** *If $l$ pass through $A$, then the pole-locus touches $l$ at $A$.*

For the conjugate points on $l$ coincide at $A$.

**Ex. 5.** *If l pass through A and C, the pole-locus is l and another line.*

**Ex. 6.** *The polars of any two points for conics of a four-point system form two homographic pencils.*
For $X'(LMN...) = Y'(LMN...)$.

**Ex. 7.** *The pencil of tangents at one of the four common points of a system of four-point conics is homographic with that at any other of the four points.*

**3.** Taking $l$ at infinity we deduce the following theorem—
*The locus of the centres of a system of conics circumscribing a given quadrangle is a conic which passes through the harmonic points of the quadrangle, through the middle points of the six sides of the quadrangle, and through the common conjugate points for the system on the line at infinity.*

The following is a direct proof of this proposition.

Let $ABCD$ be the given quadrangle, and $O$ the centre of one of the circumscribing conics. Join $O$ to the middle points $m, n, r, s$ of the sides $AB, BC, CD, DA$; and draw $Om', On', Or', Os'$ parallel to $AB, BC, CD, DA$.

Then since $Om$ bisects a chord parallel to $Om'$, $Om$ and $Om'$ are conjugate diameters. So $On, On'$, and $Or, Or'$, and $Os, Os'$ are conjugate diameters. Hence $O(mm', nn', rr', ss')$ is an involution. Hence $O(mnrs) = O(m'n'r's')$. But the rays of $O(m'n'r's')$ are in fixed directions. Hence $O(mnrs)$ is constant. Hence the locus of $O$ is a conic through the four points $m, n, r, s$.

Now define this locus by five of the centres, then the locus passes through the middle point of the side $AB$. Similarly the locus passes through the middle point of every side.

The locus also passes through the harmonic points of the quadrangles; for these are the centres of the three pairs of lines which can be drawn through the four points.

The locus also passes through the common conjugate points on the line at infinity; for these, being the double points of the involution in which the line at infinity cuts the conics, are the points of contact of the conics which can be drawn through the four points to touch the line at infinity, i. e. are the centres of the two parabolas which can be drawn through the four points.

Notice that the centre-locus by the former proof also passes through the conjugate point for the system of every point at infinity.

If the quadrangle is re-entrant, it is easy to see that the sides of the quadrangle cut the line at infinity in an overlapping involution. Hence the common conjugate points at infinity are imaginary, and the centre-locus is an ellipse. So if the quadrangle is not re-entrant, the centre-locus is a hyperbola.

Ex. 1. *Given four points on a conic, and a given point on the centre-locus as centre, construct the asymptotes.*

Ex. 2. *Five points ABCDE are taken. Show that the five conics which bisect the sides of the five quadrangles BCDE, ACDE, ABDE, ABCE and ABCD meet in a point.*

Ex. 3. *If a pair of opposite sides of the quadrangle be parallel, the centre-locus is a pair of lines.*

Ex. 4. *If a pair of sides, not opposite, be parallel, the centre-locus is a parabola.*

Ex. 5. *If two pairs of sides, not opposite, be parallel, the centre-locus is a line (and the line at infinity).*

Ex. 6. *A variable line cuts off from two given conics lengths which are bisected at the same point P. Show that the locus of P is the centre-locus belonging to the meets of the conics.*

Ex. 7. *The polars of any point on the centre-locus for conics of the system are parallel.*

Ex. 8. *The asymptotes of any conic of the system are parallel to a pair of conjugate diameters of the centre-locus.*

Let $O$ be the centre of that conic of the system which meets the line at infinity in $pp'$. Now the centre-locus meets the line at infinity in the double points $e, f$ of the involution $(pp', \ldots)$. Hence $(pp', ef) = -1$. Hence $Z(pp', ef) = -1$ where $Z$ is the centre of the centre-locus. But $Ze, Zf$ are the asymptotes of the centre-locus, Hence $Zp, Zp'$ are conjugate diameters of the centre-locus. And $Zp, Zp'$ are parallel to $Op, Op'$, which are the asymptotes of the conic whose centre is $O$.

Ex. 9. *If one of the four-point conics be a circle, the centre-locus is a rectangular hyperbola.*

For the common conjugate points at infinity, being conjugate for a circle, subtend a right angle at any finite point, i.e. the asymptotes of the centre-locus are perpendicular.

Ex. 10. *The axes of every conic circumscribed to a cyclic quadrangle are in the same directions.*

*Pole-locus and Centre-locus.*

**Ex. 11.** *The locus of the centres of rectangular hyperbolas circumscribing a given triangle is the nine-point circle.*

**Ex. 12.** *If two of the four-point conics be rectangular hyperbolas, the centre-locus is a circle.*

**Ex. 13.** *The nine-point circles of the four triangles formed by four points meet in a point.*

**Ex. 14.** *Given three points $A$, $B$, $C$ on a circle, and the ends $P$, $Q$ of a diameter; show that the centres of the rectangular hyperbolas $BCPQ$, $CAPQ$, $ABPQ$ lie on the nine-point circle of $ABC$.*

The centre of the r. h. $BCPQ$ is the middle point of $BC$, for the tangents at $B$ and $C$ are perpendicular to $PQ$.

**Ex. 15.** *The locus of the centres of all conics through the vertices of a triangle and its centroid is the maximum inscribed ellipse.*

**4. To find the centre of the centre-locus.**

Since $ms$ and $nr$ are parallel to $BD$, and since $mn$ and $sr$ are parallel to $AC$, hence $mnrs$ is a parallelogram. Also the centre of any conic circumscribing a parallelogram is the meet of the diagonals. Hence the required centre is the meet $Z$ of $mr$ and $sn$. Similarly $Z$ is on the join of the middle points of $AC$ and $BD$.

Note that $Z$ is the centre of mass of equal masses at $A$, $B$, $C$, $D$.

**Ex. 1.** *Several conics have three-point contact at $A$ and pass through $B$. Show that the centres of the conics lie on a conic whose centre $O$ is such that $3 \cdot AO = OB$.*

**Ex. 2.** *The six fourth harmonics of the ends of the six sides of a quadrangle for the meets with any transversal lie on a conic; and the lines joining opposite pairs of these points meet in a point.*

Project the transversal to infinity.

# CHAPTER XXIII.

### INVOLUTION OF A QUADRILATERAL.

**1.** *The three pairs of lines which join any point to the opposite vertices of a quadrilateral are in involution.*

Let $O$ be the point, and $AA'$, $BB'$, $CC'$ the opposite vertices of the quadrilateral. Let $OA$ cut $A'B'$ in $G$ and $A'B$ in $H$. Then $O(AA'BC)$
$= (HA'BC) = A(HA'BC)$
$= (GA'C'B') = O(AA'C'B')$.
Hence
$O(AA'BC) = O(A'AB'C')$.
Hence $O(AA', BB', CC')$ is an involution.

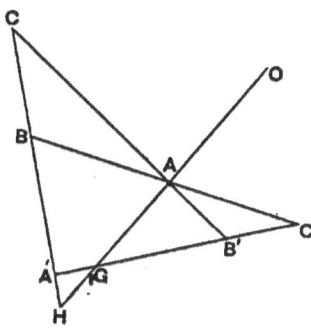

**Ex. 1.** Prove the theorem by considering the section of the quadrangle $OABC$ by $A'B'$.

**Ex. 2.** Deduce a construction for the mate $OC'$ of $OC$ in the involution determined by $O(AA', BB')$.

**Ex. 3.** Deduce the property of the harmonic points of a quadrangle.

**Ex. 4.** If any point be joined to the vertices of a triangle and to the meets of any line with the sides of the triangle, the pencil so formed is in involution.

**Ex. 5.** If any point $O$ be joined to the vertices $ABC$ of a triangle, and if $OA'$ $OB'$, $OC'$ be drawn parallel to $BC$, $CA$, $AB$, then $O(AA', BB', CC')$ is an involution.

**Ex. 6.** If any point $O$ be joined to the vertices $ABC$ of a triangle, and $A'B'C'$ be points on the sides of the triangle, such that $O(AA', BB', CC')$ is an involution; then $A'B'C'$ are collinear.

**Ex. 7.** The perpendiculars through $O$ to $OA$, $OB$, $OC$ meet $BC$, $CA$, $AB$ in collinear points.

**Ex. 8.** The six radical axes of four circles through the same point form an involution.

**Ex. 9.** *The orthogonal projections of the vertices of a quadrilateral on any line are in involution.*

**Ex. 10.** *If $AA'$, $BB'$, $CC'$ be the vertices themselves,*
$$\text{then } AB \cdot AB' \div AC \cdot AC' = A'B \cdot A'B' \div A'C \cdot A'C'.$$
For the ratios $AB' \div AC$, &c., are not altered by orthogonal projection.

**Ex. 11.** *Also if $P$, $Q$, $R$ bisect $AA'$, $BB'$, $CC'$,*
$$\text{then } AB \cdot AB' \div AC \cdot AC' = PQ \div PR.$$
For $PQR$ are collinear.

**Ex. 12.** *An infinite number of pairs of lines can be found which divide the diagonals of a quadrilateral harmonically.*

The pair of lines through any point $O$ are the double lines of the involution $O(AA', BB', CC')$.

### Involution of four-tangent Conics.

**2.** *The pair of tangents from any point to a conic and the pairs of lines joining this point to the opposite vertices of any quadrilateral circumscribing the conic are four pairs of lines in involution.*

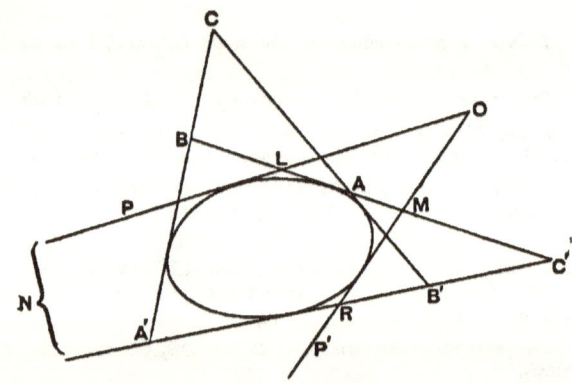

Let $AA'$, $BB'$, $CC'$ be the vertices of the quadrilateral. Let $OP$, $OP'$ be the tangents from the point $O$. Let the meets $(OP; AB)$, $(OP'; AB)$, $(OP; A'B')$, $(OP'; A'B')$ be called $L$, $M$, $N$, $R$.

Then $O(PP'AB) = (LMAB) = (NRB'A') = O(PP'B'A')$. Hence $O(PP'AB) = O(P'PA'B')$. Hence $O(PP', AA', BB')$ is an involution. Hence $OB$, $OB'$ belong to the involution determined by $O(PP', AA')$. Similarly $OC$, $OC'$ belong to this involution. Hence $O(PP', AA', BB', CC')$ is an involution.

**3.** *The system of conics which can be drawn to touch four given lines is such that the pairs of tangents from any point to conics of the system form an involution.*

For the tangents $OP$, $OP'$ to any conic of the system belong to the involution $O(AA, BB', CC')$, determined by the opposite vertices of the given quadrilateral of tangents.

Note that we have above given an independent proof that $O(AA', BB', CC')$ is an involution. For touching the four given lines and any other line $OP$ we can draw a conic.

Note also that we should expect $OA, OA'$ ; $OB, OB'$ ; $OC, OC'$ to belong to the involution $O(PP', QQ',...)$ of tangents. For each pair of opposite vertices may be considered to be a conic which touches the four lines ; and $OA, OA'$ are the tangents from $O$ to the conic $(A, A')$.

**4.** If two sides $BA$ and $AB'$ coincide, we get the theorem—*If a triangle $BA'B'$ be circumscribed to a conic, and if $A$ be the point of contact of $BB'$; then the tangents from $O$ are a pair in the involution $O(AA', BB')$.*

If the sides $CB$ and $C'B$ coincide and also the sides $CB'$ and $C'B'$, we get the theorem—*If a conic touch the lines $CB$, $CB'$ at $B$ and $B'$, then the tangents from $O$ are a pair in the involution $O(CC, BB')$ of which $OC$ is a double line.*

If the sides $BA$, $AB'$ and $B'A'$ coincide, we get the theorem—*If a system of conics have three-point contact with the line $BB'$ at $B'$ and touch a line through $B$, then the tangents from $O$ form an involution of which $OB$, $OB'$ are a pair.*

For three-point contact and three-tangent contact are equivalent.

If all four sides coincide, we get the theorem—*The tangents from $O$ to a system of conics having four-point contact at a point $B'$ form an involution of which $OB'$ is a double line.*

**Ex. 1.** *The pencil formed by the pairs of tangents from any point to two circles and the joins of the point to the centres of similitude is in involution.*

**Ex. 2.** *If the line joining the centres of similitude $SS'$ of two circles cut the circles in $AA'$, $BB'$ ; then $AA'$, $BB'$, $SS'$ are in involution.*

**Ex. 3.** *If $VP$, $VQ$ be the tangents from any point $V$ to a conic, and if 1, 2,*

218   *Involution of a Quadrilateral.*   [CH.

3, 4 be the successive vertices of a four-point figure circumscribed to the conic, show that
$$\frac{\sin PV_1 . \sin PV_3}{\sin QV_1 . \sin QV_3} = \frac{\sin PV_2 . \sin PV_4}{\sin QV_2 . \sin QV_4}.$$

**Ex. 4.** *Extend the theorem to any 2n-point circumscribed polygon.*

**Ex. 5.** *Through every point can be drawn a pair of lines which are conjugate for every conic of a four-tangent system.*

Viz. the double lines of the involution of tangents.

**Ex. 6.** *Through every point can be drawn a pair of lines to divide the diagonals of a given quadrilateral harmonically; and these meet any inscribed conic in harmonic points.*

For they are the common conjugate lines through the point.

**Ex. 7.** *Through every point can be drawn two conics of a four-tangent system; and the tangents to these conics at the point are the common conjugate lines at the point.*

Draw a conic α of the system to touch $OE$, one of the double lines of the involution of tangents at $O$. Then, since $OE$ is a double line, $OE$ is the other tangent from $O$ to α. Hence α passes through $O$.

**Ex. 8.** *The tangents at one of the intersections of two conics inscribed in the same quadrilateral are harmonic with the lines joining the point to any two opposite vertices of the quadrilateral.*

**Ex. 9.** $ABC$ *is a triangle and $O$ a given point. Through $O$, and parallel to the sides $BC$, $CA$, $AB$, are drawn the lines $OX$, $OY$, $OZ$; show that the double lines of the involution $O(XA, YB, ZC)$ are the tangents at $O$ to the two parabolas which can be inscribed in $ABC$ so as to pass through $O$.*

**Ex. 10.** $P$, $Q$, $R$ *are the points of contact of the lines $BC$, $CA$, $AB$ with a conic, and $OT$, $OT'$ are the tangents from any point $O$; show that $O(BC, PA, TT')$ and $O(RQ, AA, TT')$ are involutions.*

**Ex. 11.** *If $OP$, $OQ$ be a pair in the involution obtained by joining $O$ to the three pairs of opposite vertices of a quadrilateral, the lines $OP$, $OQ$ and the sides of the quadrilateral touch a conic.*

**Ex. 12.** *Three vertices of a four-point figure circumscribed to a conic lie on three fixed lines through a point; show that the fourth vertex lies on a fourth fixed line through the same point.*

**Ex. 13.** *Extend the theorem to any 2n-point figure.*

**Ex. 14.** *By taking the two sides coincident which pass through the 2nth vertex, deduce a simple solution of the problem—'Circumscribe to a given conic a polygon of $2n-1$ vertices, each vertex to lie on one of a set of $2n-1$ fixed concurrent lines.'*

**Ex. 15.** *Show that the problem—'To circumscribe to a given conic a polygon of $2n$ vertices, each vertex to lie on one of a set of $2n$ fixed concurrent lines'—is either indeterminate or impossible.*

**5.** *The three circles on the diagonals of any quadrilateral as diameters are coaxal.*

*The three middle points of the diagonals of a quadrilateral lie on a line* (called *the diameter of the quadrilateral*).

## Involution of a Quadrilateral.

*The directors of a system of conics touching the sides of a quadrilateral are coaxal, and three circles of the coaxal system are the three circles on the diagonals as diameters.*

*The centres of a system of conics touching the sides of a quadrilateral lie on a line which also contains the middle points of the diagonals of the quadrilateral.*

Let $AA'$, $BB'$, $CC'$ be the opposite vertices of the quadrilateral. Let the circles on $AA'$ and $BB'$ as diameters meet in $O$ and $O'$. Then in the involution pencil $O(AA', BB', CC')$, since $AOA'$ and $BOB'$ are right angles, $COC'$ is a right angle. Hence the circle on $CC'$ as diameter passes through $O$; and similarly through $O'$. Hence the circles on $AA'$, $BB'$, $CC'$ as diameters are coaxal. Hence their centres, viz. the middle points of $AA'$, $BB'$, $CC'$, are collinear.

Again, the tangents $OP$, $OP'$ from $O$ to any conic touching the sides of the quadrilateral belong to the involution $O(AA', BB', CC')$. Hence $POP'$ is a right angle. Hence the director of this conic passes through $O$; and similarly through $O'$. Hence this director, and similarly all the directors, belong to the above coaxal system. But the centre of a conic is the same as the centre of its director. Hence the centres of the conics lie on a line, viz. the line of centres of the coaxal system of circles.

The locus of centres is the diameter of the quadrilateral; for three circles of the system are the circles on $AA'$, $BB'$, $CC'$ as diameters.

*The radical axis of the coaxal system of directors is the directrix of the parabola of the system of conics.*

For the directrix is the limit of a director, and the radical axis is the limit of a coaxal, when each becomes a line.

*The limiting points of the coaxal system of directors are the centres of the rectangular hyperbolas of the system of conics.*

For when the coaxal becomes a point, the director becomes a point, and the conic becomes a rectangular hyperbola, the director being the centre of the r. h.

Note that the director of a conic which consists of two points is the circle on the segment joining the points as diameter, and the centre of the conic is the point half-way between the points.

**Ex. 1.** *The directors of all conics touching two given lines OP, OQ at P, Q are coaxal, the axis being the radical axis of the point O and the circle on PQ as diameter.*

**Ex. 2.** *The polar circle of a triangle circumscribing a conic is orthogonal to the director circle.*

Let $ABC$ be the triangle. Take any fourth tangent $A'B'C'$. Then the circle on $AA'$ as diameter passes through the foot $D$ of the perpendicular from $A$ on $BC$. Now $A$ and $D$ are inverse for the polar circle. Hence the polar circle is orthogonal to the circle on $AA'$, and similarly to the circles on $BB'$, $CC'$; and hence to the director, for this belongs to the same coaxal system.

**Ex. 3.** *The locus of the centre of a rectangular hyperbola which touches a given triangle is the polar circle of the triangle.*

For the polar circle cuts orthogonally the director circle which is the centre in a r. h.

**Ex. 4.** *If the nine-point circle of a triangle circumscribing a r. h. pass through the centre of the r. h.; show that the centre also lies on the circum-circle, and that the centre of the circum-circle lies on the r. h.*

The centre $O$ lies on the nine-point circle and on the polar circle and therefore on the circum-circle, as the three circles are coaxal. Let the asymptotes meet the circum-circle in $P$, $Q$. Then $ABC$, $OPQ$ are inscribed in the same conic, hence $PQ$ touches the r. h. Hence the point of contact is the centre of the circle.

**Ex. 5.** *The diameters of the five quadrilaterals which can be formed by five given lines are concurrent. Prove this, and deduce a construction for the centre of a conic, given five tangents.*

**Ex. 6.** *The axis of the parabola inscribed in a quadrilateral is parallel to the diameter of the quadrilateral.*

**Ex. 7.** *The diameter of a quadrilateral circumscribing a conic touches the centre-locus of the quadrangle formed by the points of contact.*

Otherwise the conic would have two centres.

**Ex. 8.** Steiner's theorem. *The orthocentre of a triangle circumscribing a parabola is on the directrix.*

For the involution subtended at the orthocentre by the quadrilateral formed by the sides of the triangle and the line at infinity is orthogonal.

**Ex. 9.** *The directrices of all parabolas touching a given triangle are concurrent.*

**Ex. 10.** Gaskin's theorem. *The circle circumscribing a triangle which is self-conjugate for a conic is orthogonal to the director circle of the conic.*

Take any tangent to the conic. Then from this tangent and the given self-conjugate triangle $UVW$, we can construct three other tangents such that $UVW$ is the harmonic triangle of the quadrilateral so formed. Let $AA'$, $BB'$, $CC'$ be the opposite vertices of this quadrilateral.

Then the circle about $UVW$ is clearly orthogonal to the circles on $AA'$, $BB'$, $CC'$ as diameters, for it cuts these diameters in inverse points.

Hence the circle about $UVW$ being orthogonal to three circles of a coaxal system is orthogonal to the director which belongs to the coaxal system.

**Ex. 11.** *The centre of a circle circumscribing a triangle self-conjugate for a parabola is on the directrix.*

**Ex. 12.** *The circle circumscribing a triangle self-conjugate for a rectangular hyperbola passes through the centre.*

**Ex. 13.** *Given five points on a conic, five self-conjugate triangles can be found, viz. the harmonic triangles of the inscribed quadrangles obtained by omitting one point; show that the ten radical axes of the circles circumscribing these triangles pass through the centre of the conic.*

**Ex. 14.** *Show that two, and only two, rectangular hyperbolas can be drawn to touch four given lines.*

Let the lines be $a$, $b$, $c$, $d$. Let the circle about the harmonic triangle of the quadrilateral meet the diameter of the quadrilateral in $L$ and $L'$. Then $L$ and $L'$ are the limiting points of the directors.

First take $L$, and let $a'$ be the reflexion of $a$ in $L$. Construct the conic touching $a$, $b$, $c$, $d$, $a'$. Then the centre of the conic, being the meet of the diameter of the quadrilateral and the line half-way between $a$ and $a'$, is $L$. Hence $L$ is the centre of the director. But the coaxal with centre at $L$ has zero radius. Hence the conic is a r. h.

So $L'$ gives another r. h. And there are only two; for the centre must be at $L$ or at $L'$.

**Ex. 15.** *Any transversal cuts the diagonals $AA'$, $BB'$, $CC'$ of a quadrilateral circumscribed to a conic in the points $P$, $Q$, $R$, and points $P'$, $Q'$, $R'$ are taken such that $(AA', PP')$, $(BB', QQ')$, $(CC', RR')$ are harmonic; show that $P'Q'R'$ and the pole of the transversal for the conic are collinear.*
Project $PQR$ to infinity.

**6.** *The locus of the poles of a given line for a system of four-tangent conics is a line.*

Let $P$ and $Q$ be the poles of the given line $LM$ for two of the conics; and let $LM$, $PQ$ cut in $U$. Then $UL$ and $UP$ are conjugate lines for two conics of the system, i.e. $UL$ and $UP$ are harmonic with two of the pairs of tangents from $U$. Hence $UL$ and $UP$ are the double 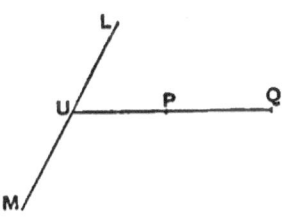 lines of the involution of tangents from $U$ to the system of conics. Hence $UL$ and $UP$ are harmonic with every pair of tangents from $U$, i.e. are conjugate for every conic of the system. Hence the pole of $LM$ for every conic of the system lies on $PQ$, i.e. $PQ$ is the locus of the poles of $LM$.

## Involution of a Quadrilateral.

Taking $LM$ at infinity, we again see that—

*The locus of the centres of a system of four-tangent conics is a line.*

**Ex.** *The three poles of a line for the three opposite pairs of vertices of a quadrilateral are collinear.*

**7.** By reciprocating the properties of the pole-locus (or directly) we can investigate the properties of *the polar-envelope of a point for a system of four-tangent conics.*

**Ex.** *From the fixed point $O$, tangents $OP$ and $OQ$ are drawn to one of a system of conics inscribed in the same quadrilateral. If $AA'$ be a pair of opposite vertices of the quadrilateral, and if $PP'$, $QQ'$ be such that $P(OP', AA')$, $Q(OQ', AA')$ are harmonic, then the envelope of the chords $PQ$, $PP'$, $QQ'$ is a single conic.*

# CHAPTER XXIV.

### CONSTRUCTIONS OF THE FIRST DEGREE.

**1.** *EXAMPLES of constructions in which the ruler only is to be used.*

**Ex. 1.** *Given the segment AC bisected in B; prove the following construction for a parallel to AC through P—Through B draw any line, cutting PA in E and PC in D; then if CE, DA meet in Q, PQ is the required line.*

For $B$ bisects $AC$, hence $PQ$ cuts $AC$ at infinity, since $P(AC, BQ)$ is harmonic.

**Ex. 2.** *Given two parallel segments AB and CD, prove the following construction for bisecting each—Let CB, AD meet in W, and AC, BD in V, then VW bisects both segments.*

For $U$ is at infinity.

**Ex. 3.** *Given a pair of parallel lines, draw through a given point a parallel to both.*

Use Ex. 2 and then Ex. 1.

**Ex. 4.** *Given a parallelogram, bisect a given segment.*

Let $AB$ be the segment. Through $A$ and $B$ draw parallels to the sides of the parallelogram meeting again in $C$ and $D$. Then $CD$ bisects $AB$.

**Ex. 5.** *Given two lines AB and CD which meet in an inaccessible point U, construct any number of points on the line joining U to a given point O.*

Through $O$ draw $LOM'$ and $MOL'$ meeting $AB$ in $LM$ and $CD$ in $L'M'$. Let $LL'$, $MM'$ meet in $W$. Then $U(AC, OW)$ is harmonic; hence the required line is the polar of $W$ for $AB$ and $CD$. To construct any other point on the line, draw any two lines $WNN'$ and $WRR'$ meeting $AB$ in $N$, $R$, and $CD$ in $N'$, $R'$. Then a point on the required line is the meet of $NR'$ and $N'R$.

**Ex. 6.** *Construct lines which shall pass through the meet of a given line with the line joining two given points, when this last line cannot be drawn.*

**Ex. 7.** *Given a segment AC bisected at B, join any point $P_1$ to ABC, on $P_1B$ take any point Q, join CQ cutting $AP_1$ in $L_1$, join AQ cutting $CP_1$ in $L_3$, join $L_1L_3$ cutting $BP_1$ in $L_2$, then $L_1L_2 = L_2L_3$, and $L_1L_3$ is parallel to AC. Again, let $AL_2$, $BL_3$ cut in $P_2$, and let $CP_2$ cut $L_1L_3$ in $L_4$, then $L_2L_3 = L_3L_4$. Again, let $AL_3$, $BL_4$ cut in $P_3$, and let $CP_3$ cut $L_1L_4$ in $L_5$, then $L_3L_4 = L_4L_5$. And so on.*

The first part comes from the quadrilateral $P_1L_1QL_3P_1$. The rest follows by Elementary Geometry.

This enables us *to divide a bisected segment into any number of equal parts*. To divide $AC$ into $n$ equal parts, construct the points $L_1 L_2 \ldots L_{n+1}$. Let $AL_1$ and $CL_{n+1}$ meet in $V$. With $V$ as vertex project $L_1 L_2 \ldots L_{n+1}$ on to $AC$.

**2.** *To construct a five-point conic.*

Let $A$, $B$, $C$, $D$, $E$ be the five given points on the conic. We shall construct the conic by finding the point in which any line $AG$ through $A$ meets the conic again. (See figure of XV. 1.) Let $AG$ and $CD$ meet in $L$, and $AB$ and $DE$ in $M$. Let $LM$ cut $BC$ in $N$. Then, by Pascal's theorem, $NE$ cuts $AG$ in the required point $F$ on $AG$. And since $AG$ is any line through $A$, we shall thus construct every point on the conic.

If any two of the points are coincident, the necessary modification of this construction is obvious, remembering that to be given two coincident points is to be given a point and the tangent at the point, and that the two coincident points lie on the tangent.

The case of three points being coincident is discussed in XXV. 17.

**Ex.** *Construct the polar of a given point for a five-point conic.*

**3.** As an example of coincident points, let us *construct a conic to touch two given lines at given points, and to pass through a given point*.

Suppose the conic is to touch $OP$ and $OQ$ at $P$ and $Q$, and to pass through $A$. Here $B$ and $C$ coincide with $P$, and the line $BC$ coincides with $OP$. So $D$ and $E$ coincide with $Q$, and $DE$ coincides with $OQ$. Hence the construction is—To find where any line $AG$ through $A$ cuts the conic again, let $AG$ and $PQ$ meet in $L$, and $AP$ and $OQ$ in $M$; let $LM$ cut $OP$ in $N$; then $NQ$ cuts $AG$ in the required point $F$.

**Ex. 1.** *Given four points and the tangent at one of them, construct the conic.*

**Ex. 2.** *Find a point $P$ at which the five points $A$, $B$, $C$, $D$, $E$, no three of which are collinear, subtend a pencil homographic with a given pencil.*

Take $DD'$ and $DE'$ so that $D(ABCD'E')$ shall be homographic with the given pencil. Draw a conic through $ABC$ to touch $DD'$ at $D$. Construct the point $F$ in which $DE'$ cuts this conic, and construct the point $P$ in which $FE$ cuts this conic. $P$ is the required point.

XXIV.] *Constructions of the First Degree.* 225

**4.** As an example of cases in which some of the given points are at infinity, let us *construct a conic, given one asymptote, the direction of the other asymptote, and two other points.*

Let $l$ be the given asymptote, and $m$ any line in the direction of the other asymptote, and $A$ and $B$ the two given points. We may take $C$ and $D$ to be the points at infinity on $l$, and $E$ to be the point at infinity on $m$. Then $M$ is the point at infinity on $AB$.

Hence the construction is—To find where any line $AG$ through $A$ cuts the conic again, let $AG$ and $l$ meet in $L$, and let a parallel through $L$ to $AB$ cut a parallel through $B$ to $l$ in $N$. Then a parallel through $N$ to $m$ cuts $AG$ in the required point $F$.

**Ex. 1.** *Given four points and the direction of an asymptote, construct the conic.*

**Ex. 2.** *Given three points on a conic and a tangent at one of them, and the direction of one asymptote; construct the conic.*

**Ex. 3.** *Given three points and the directions of both asymptotes, construct the conic.*

**Ex. 4.** *Given one point and both asymptotes, construct the conic.*

**Ex. 5.** *Given four points on a conic and the direction of one asymptote; construct the meet of the conic with a given line drawn parallel to the asymptote.*

**Ex. 6.** *Given three points on a conic and the directions of both asymptotes; find the meet of the conic with a given line parallel to one of the asymptotes.*

**Ex. 7.** *Given four points on a conic and the direction of one asymptote; find the direction of the other.*

**5.** As an example of drawing a parabola to satisfy given conditions, let us *construct a parabola, given three points and the direction of the axis.*

Let $ABC$ be the given points, and $l$ any line in the direction of the axis. We may consider $D$ and $E$ to coincide at the point at infinity upon $l$, so that the line $DE$ is the line at infinity. Then $M$ is the point at infinity on $AB$.

Hence the construction is—To find where any line $AG$ through $A$ cuts the conic again, let $AG$ cut a parallel through $C$ to $l$ in $L$; let a parallel through $L$ to $AB$ cut $BC$ in $N$; then a parallel through $N$ to $l$ cuts $AG$ in the required point $F$.

**Ex.** *Construct a parabola, given two points and the tangent at one of them, and the direction of the axis.*

**6.** *Given five points on a conic, to construct the tangent at one of them.*

Let $A, B, C, D, E$ be the five given points, and suppose $F$ to coincide with $A$; then $AF$ is the tangent at $A$. Hence the construction—Let $AB$ and $DE$ meet in $M$, and $BC$ and $AE$ in $N$, and let $MN$ cut $CD$ in $L$; then $LA$ is the tangent at $A$.

**Ex. 1.** *Given four points on a conic, and the tangent at one of them; construct the tangent at another of them.*

**Ex. 2.** *Given three points on a conic, and the tangents at two of them; construct the tangent at the third.*

**Ex. 3.** *Given both asymptotes of a hyperbola, and one point; construct the tangent at this point.*

**Ex. 4.** *Given three points on a parabola, and the direction of the axis; construct the tangent at one of the given points.*

**Ex. 5.** *Given two points on a parabola, the direction of the axis, and the tangent at one of the points; construct the tangent at the other point.*

**Ex. 6.** *Given four points on a conic, and the direction of one asymptote; construct that asymptote.*

**Ex. 7.** *Given three points on a conic, and the directions of both asymptotes; construct the asymptotes.*

**Ex. 8.** *Given two points on a conic, and one asymptote, and the direction of the other; construct the other asymptote.*

**7.** *Given five tangents of a conic, to construct the points of contact.*

Let $AB, BC, CE, EF, FA$ be the five given tangents. Then in the figure of XV. 4, if $D$ is the point of contact of $CE$, we may consider $CD, DE$ to be consecutive tangents of the conic. Hence the construction—Let $BE$ and $CF$ meet in $O$; then $AO$ cuts $CE$ in its point of contact. So the other points of contact can be constructed.

Hence given five tangents, we can at once construct five points; so that every construction which requires five points to be given, is available if we are given five tangents.

**Ex. 1.** *Given four tangents and the point of contact of one of them, construct the points of contact of the others.*

**Ex. 2.** *Given three tangents of a conic, and the points of contact of two of them; construct the point of contact of the third.*

**Ex. 3.** *Given both asymptotes of a hyperbola, and one tangent, construct the point of contact of the tangent.*

**Ex. 4.** *Given four tangents of a parabola, construct the points of contact, and the direction of the axis.*

**Ex. 5.** *Given two tangents of a parabola, and their points of contact, construct the direction of the axis.*

**8.** *Given five tangents of a conic, to construct the conic by tangents.*

Let $GB$, $BC$, $CD$, $DE$, $EH$ be the given tangents. Now every tangent cuts $GB$. Hence if we construct every other tangent from points on $GB$, we shall have constructed every tangent of the conic. On $GB$ take any point $A$. Let $AD$ and $BE$ meet in $O$. Let $CO$ meet $EH$ in $F$. Then, by Brianchon's theorem, $FA$ touches the conic, i.e. $AF$ is the other tangent from any point $A$ on $GB$.

**Ex. 1.** *Given four tangents of a conic, and the point of contact of one of them; construct the conic by tangents.*

**Ex. 2.** *Given four tangents of a parabola, construct the conic.*

**Ex. 3.** *Given three tangents of a conic, and the points of contact of two of them; construct the conic.*

**Ex. 4.** *Given the asymptotes of a conic, and one tangent; construct the conic.*

**Ex. 5.** *Given two tangents of a parabola, the point of contact of one of them, and the direction of the axis; construct the parabola.*

**Ex. 6.** *Given five tangents of a conic, construct the tangent parallel to one them.*

**Ex. 7.** *Given four tangents of a parabola, construct the tangent in a given direction.*

**Ex. 8.** *Construct the pole of a given line for a five-tangent conic.*

**Ex. 9.** *Ditto for a five-point conic.*

**9.** *Given three points on a conic and a pole and polar, to construct the conic.*

Let $A$, $B$, $C$ be the three given points, and $O$ the pole. Let $OA$ cut the polar in $a$, and take $A'$ such that $(Oa, AA')$ is harmonic. Similarly construct $\beta$ and $B'$. Through $ABCA'B'$ construct a conic. This will be the required conic; for since $(Oa, AA')$ and $(O\beta, BB')$ are harmonic, we see that $a\beta$ is the polar of $O$.

A reciprocal construction enables us to solve the problem—

## 228  *Constructions of the First Degree.*

*Given three tangents of a conic, and a pole and polar, to construct the conic.*

A simple case of each problem is—*Given three points (or three tangents) and the centre, to construct the conic.*

We obtain two more points (or tangents) by reflexion in the centre.

# CHAPTER XXV.

## CONSTRUCTIONS OF THE SECOND DEGREE.

**1.** *Construct the points in which a given line cuts a conic given by five points.*

Let $A, B, C, D, E$ be the five given points. Let the given line cut $DA, DB, DC$ in $a, b, c$, and cut $EA, EB, EC$ in $a', b', c'$, and cut the conic in $x, y$. Then

$$(xyabc) = D(xyABC) = E(xyABC) = (xya'b'c').$$

Hence $x, y$ are the common points of the two homographic ranges determined by $(abc)$ and $(a'b'c')$. Hence the two required points $x, y$ can be constructed by XVI. 6.

**2.** *Given five tangents to a conic, to construct the tangents from any point to the conic.*

Let three of the given tangents cut the other two in $ABC$ and $A'B'C'$. If a tangent from the given point $P$ cut these tangents in $X$ and $X'$, then $(ABCX) = (A'B'C'X')$; hence $P(ABCX) = P(A'B'C'X')$. But $PX$ and $PX'$ coincide; hence one of the tangents from $P$ is one of the common lines of the pencils $P(ABC)$ and $P(A'B'C')$. Hence the required tangents are the common lines of the homographic pencils determined by $P(ABC)$ and $P(A'B'C')$.

**3.** *Given five tangents to a conic, to construct the points in which any line cuts the conic.*

Construct first by XXIV. 7 the points of contact, and then proceed by § 1.

*Given five points on a conic, to construct the tangents from any point.*

Construct first by **XXIV. 6** the tangents at the points, and then proceed by § 2.

**4.** If instead of five points, we are given four points and the tangent at one, or three points and the tangents at two of them; or if, instead of five tangents, we are given four tangents and the point of contact of one, or three tangents and the points of contact of two of them, the necessary modifications of the above constructions are obvious.

**Ex. 1.** *Construct a line to cut four given lines in a given cross ratio and to pass through a given point.*
Let three of the lines cut the fourth in $BCD$. Take $A$ such that $(ABCD)$, is equal to the given cross ratio. Draw a conic to touch the three given lines and also to touch the fourth at $A$. Through the given point draw a tangent to this conic. This is the required line. There are therefore two solutions.

**Ex. 2.** *Give the reciprocal construction.*

**Ex. 3.** *Through a given point draw a line to cut three given lines in $A$, $B$, $C$, so that $AB : BC$ is a given ratio.*

**5.** *Given five points on a conic, to construct the centre, the axes, and the asymptotes.*

Let $A$, $B$, $C$, $D$, $E$ be the five given points. Through $A$ draw $AG$ parallel to $BC$, and construct the point $A'$ in which $AG$ cuts the conic again. Let $AC$ and $BA'$ cut in $H$, and $AB$ and $A'C$ cut in $K$. Then $HK$ bisects both $BC$ and $AA'$. Hence $HK$ is a diameter. Similarly construct another diameter. Then these diameters meet in the centre.

To construct the axes and asymptotes, we must first construct the involution of conjugate diameters. To do this—Through the centre $O$ draw $Oa$ parallel to $BC$, and let $Oa'$ be the diameter bisecting $AA'$ and $BC$. Then $Oa$, $Oa'$ are a pair of conjugate diameters. In the same way determine another pair $Ob$, $Ob'$. Then the rectangular pair of the involution determined by $O(aa', bb')$ are the axes; and the double lines of the same involution are the asymptotes.

If the diameters are parallel, the conic is a parabola; and the direction of the diameters is the direction of the axis of the parabola.

**Ex. 1.** *Given five points on a conic, construct a pair of conjugate diameters which shall make a given angle with one another.*

Let $CP$ and $CD$ be a pair of conjugate diameters. Take $CD'$ such that $\angle PCD'$ is equal to the given angle. Then the required lines are the common rays of the homographic pencils generated by $CD$ and $CD'$.

**Ex. 2.** *Through a given point $O$ draw a line meeting four given lines $a$, $a'$, $b$, $b'$ in points $A$, $A'$, $B$, $B'$, such that $OA \cdot OA' = OB \cdot OB'$.*

Through $O$ draw a parallel to either asymptote of the conic through the five points $ab$, $ab'$, $a'b$, $a'b'$, and $O$.

**Ex. 3.** *Through a given point $C$ draw a line meeting five given lines $a$, $a'$, $b$, $b'$, $c'$ in five points $A$, $A'$, $B$, $B'$, $C'$ such that $(AA', BB', CC')$ may be an involution.*

**6.** If we are *given five points* on a conic, the conic can be constructed by Pascal's theorem (see XXIV. 2). If we are *given five tangents* of the conic, the conic can be constructed by points (see XXIV. 7) or by tangents (see XXIV. 8).

*Given four points and one tangent, to construct the conic.*

Let $ABCD$ be the given points and $t$ the given tangent. Let $t$ cut the opposite sides of the quadrangle $ABCD$ in $aa'$, $bb'$, $cc'$. Take $e$, $f$, the double points of the involution $(aa', bb', cc')$. Then the two conics satisfying the required conditions are the conics through $ABCDe$ and through $ABCDf$. For let the conic through $ABCDe$ cut $t$ again in $e'$. Then $ee'$ belong to the involution $(aa', bb', cc')$, and $e$ is a double point of this involution; hence $e'$ coincides with $e$, i.e. $t$ touches the conic through $ABCDe$. So it touches the conic through $ABCDf$.

**7.** *Given four tangents and one point, to construct the conic.*

Let $OE$, $OF$ be the double lines of the involution subtended by the given quadrilateral at the given point $O$. Then it is proved, as above, that the required conics are those touching the given lines and also touching $OE$ or $OF$.

**Ex. 1.** *Show that when four points are given and one tangent, the solution is unique if the line pass through one of the harmonic points.*

The other conic degenerates into a pair of opposite sides.

**Ex. 2.** *Show that there is no curved solution if the line pass through two harmonic points.*

**Ex. 3.** *Reciprocate Ex. 1 and Ex. 2.*

**Ex. 4.** *Describe a parabola through four given points.*

**Ex. 5.** *Construct a parabola, given three tangents and one point.*

**8.** *Given three points and two tangents, to construct the conic.*
Let the three points be $A$, $B$, $C$, and the two tangents $TL$ and $TL'$. Let $AB$ cut $TL$ and $TL'$ in $c$ and $c'$, and let $AC$ cut $TL$ and $TL'$ in $b$ and $b'$. Take $z$, $z'$, the double points of the involution $(AB, cc')$, and $y$, $y'$ the double points of the involution $(AC, bb')$. Let any one, $yz$, of the four lines $yz$, $yz'$, $y'z$, $y'z'$ cut $TL$ and $TL'$ in $P$ and $P'$.

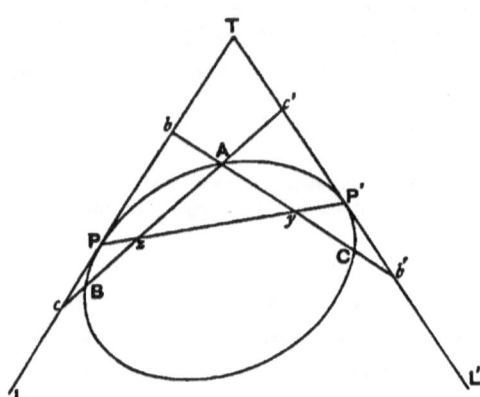

Then one conic satisfying the required conditions is the conic which passes through $A$ and touches $OL$ and $OL'$ at $P$ and $P'$. For let this conic cut $AB$ again in $B'$. Then $z$ is a double point of the involution $(AB, cc')$ and also of the involution $(AB', cc')$. Hence $B$ and $B'$ coincide, i.e. the conic passes through $B$. Similarly the conic passes through $C$.

So by taking any of the lines $yz'$, $y'z$, $y'z'$ instead of $yz$, we obtain another solution. Hence the problem has four solutions.

Note that since there are only four possible positions of the polar $PP'$ of $T$, we have proved that—*If the sides $BC$, $CA$, $AB$ of a triangle cut two lines $TL$ and $TL'$ in $aa'$, $bb'$, $cc'$, and if the double points $xx'$, $yy'$, $zz'$ of the involutions $(BC, aa')$, $(CA, bb')$, $(AB, cc')$ be taken, then the six points $xx'yy'zz'$ lie three by three on four lines.*

**9.** *Given two points and three tangents, to construct the conic.*
Let $A$ and $B$ be the given points, and $LM$, $MN$, $NL$ the given tangents. Take $MY$, $MY'$ the double lines of the involution $M(AB, LN)$, and take $NZ$, $NZ'$ the double points of the involution $N(AB, LM)$. Let $T$ be the meet of one of the lines $MY$, $MY'$ with one of the lines $NZ$,

## Constructions of the Second Degree. 233

*NZ'*. Describe a conic to touch *TA* and *TB* at *A* and *B* and to touch *MN*.

This is a conic satisfying the required conditions. For let *ML'* be the second tangent from *M* to this conic. Then *MY* is a double line of both the involutions *M(AB, NL)* and *M(AB, NL')*. Hence *ML'* coincides with *ML*, i.e. the conic touches *ML*. So the conic touches *NL*.

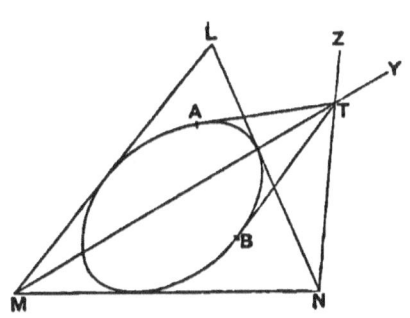

By taking one of the other four meets instead of the meet of *MY* and *NZ*, we obtain three other solutions.

**10.** *Given a triangle self-conjugate for a conic, and either two points on the conic, or one point on the conic and one tangent to the conic, or two tangents to the conic, to construct the conic.*

By V. 9, if we are given a self-conjugate triangle and one point, we are given three other points; and if we are given a self-conjugate triangle and one tangent, we are given three other tangents. In any of the above cases therefore the conic can now be constructed.

**11.** If we are given a focus, by XXVIII. 8 we are given two tangents. Hence the following problems belong to this chapter, but in each case a simpler solution can be given.

*Given a focus and three points, to construct the conic.*

Take the reciprocals of the given points for any circle with centre at the given focus, and draw a circle to touch these lines. The reciprocal of this circle is the required conic. Since four circles can be drawn, there are four solutions.

*Given a focus and two points and one tangent.*

Reciprocation gives four solutions, two of which are imaginary.

*Given a focus and one point and two tangents.*

Reciprocation gives two solutions.

*Given a focus and three tangents.*

Reciprocation gives one solution. In this case we can also solve the problem by determining the second focus by means of the theorem that two tangents to a conic are equally inclined to the focal radii to their meet.

**12.** *To construct a conic, given a self-conjugate triangle and a pole and polar.*

Let $ABC$ be the self-conjugate triangle, and let $L$ be the pole of $l$. Let $LA$ meet $BC$ in $A'$, and $l$ in $D$; let $LB$ meet $CA$ in $B'$, and $l$ in $E$; let $LC$ meet $AB$ in $C'$, and $l$ in $F$.

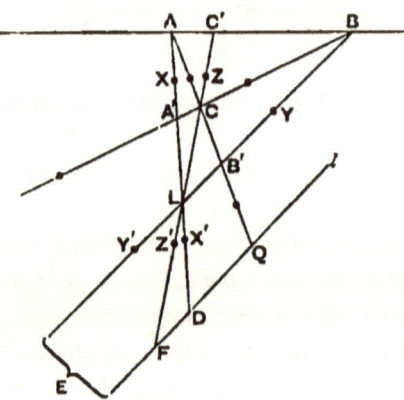

Now $A$ and $A'$ are conjugate points for the required conic, and so are $L$ and $D$. Hence the required conic must pass through $XX'$, the double points of the involution $(AA', LD)$. So the conic must pass through the double points $YY''$ of the involution $(BB', LE)$, and through the double points $ZZ'$ of the involution $(CC', LF)$.

Also the six points $XX'YY''ZZ'$ lie on a conic. For draw a conic through $XX'YY'Z$. Then since $LD$ are harmonic with $XX'$, and $LE$ with $YY'$, $l$ is the polar of $L$; also $(LF, ZZ') = -1$, and the conic passes through $Z$; hence the conic passes through $Z'$.

Again, the conic through $XX'YY'ZZ'$ satisfies the required conditions. We have proved that $l$ is the polar of $L$. Let $BC$ and $B'C'$ meet in $H$. Then the opposite vertices of the quadrilateral $BC, CB', B'C', C'B$ are $BB', CC',$ and $AH$. Now $BB'$ are conjugate for the conic, and so are $CC'$; hence so are $AH$. Hence the polar of $A$ passes through $H$; and also through $A'$. Hence $BC$ is the polar of $A$; so $CA$ is the polar of $B$, and $AB$ is the polar of $C$. Hence $ABC$ is a self-conjugate triangle for the conic.

This completes the theoretical solution of the problem; and we have shown that one, and only one, conic can be drawn satisfying the given conditions. Practically the above solution is worthless; for any pair of the points $XX'$, $YY'$, $ZZ'$ may be imaginary. The following is the practical construction when the conic is real.

We have already found two points upon $CL$. To find two points on $CA$. Let $AC$ cut $l$ in $Q$. Then $AC$ are conjugate points; and so are $B'Q$, for $Q$ is the pole of $LB$. Hence the two points upon $CA$ are the double points of the involution $(AC, B'Q)$. So two points can be found on $CB$. Hence six points on the conic are known, viz. those on $CA$, $CB$, and $CL$. Now if the conic is real, one of the points $ABC$ (say $C$) is inside the conic, and hence $CA$, $CB$, $CL$ all cut the conic in real points. Hence, by trying $ABC$ in succession, we get six real points on the conic.

If on trial we find that neither $A$ nor $B$ nor $C$ gives six real points, we conclude that the conic is imaginary.

We see again that *two conics cannot have two common self-conjugate triangles;* for since two such triangles more than determine a conic, the two conics would be coincident.

**Ex. 1.** *Given a pentagon $ABCDE$, construct a conic for which each vertex is the pole of the opposite side.*

Let $AB$ and $CD$ meet in $F$. The required conic is the one for which $ADF$ is self-conjugate, and $E$ is the pole of $BC$.

**Ex. 2.** *For this conic, the inscribed conic and the circumscribed conic are reciprocal.*

**Ex. 3.** *Given the centre of a conic and a self-conjugate triangle, construct the asymptotes.*

Draw $OX$, $OY$, $OZ$ parallel to $BC$, $CA$, $AB$; then the asymptotes are the double lines of $O(AX, BY, CZ)$.

**Ex. 4.** *Given a pole and polar and a self-conjugate triangle, construct the tangents from the pole.*

**Ex. 5.** *Given four points $A$, $B$, $C$, $D$ and a line $l$. With $A$ as pole of $l$ and with $BCD$ as a self-conjugate triangle, a conic is drawn; similarly the conics $(B, CDA)$, $(C, DAB)$, $(D, ABC)$ are drawn. Show that these four conics meet $l$ in the same two points.*

**13.** *Given five points on each of two conics, to construct the* conic which passes through the meets of these conics and also through a given point.

Through the given point $L$ draw any line $l$; and construct the points $pp'$, $qq'$ in which $l$ cuts the two conics. Then if $M$ be the other point in which the required conic cuts $l$, we know that $pp'$, $qq'$, $LM$ are pairs in an involution. Hence $M$ is known, i.e. a point on the conic is known on every line through $L$.

*Given five points on each of two conics, to construct the conic which passes through the four meets of these conics and also touches a given line.*

Construct the points in which the given line cuts the given conics, viz. $pp'$, $qq'$. Then the points of contact of the required conics are the double points $e$, $f$ of the involution determined by $pp'$, $qq'$. Then, taking either $e$ or $f$, we continue as above.

**Ex.** *Give the reciprocal constructions.*

**14.** *Given three points on a conic and an involution of conjugate points on a line, to construct the conic.*

If the given involution has real double points, draw a conic through the three given points and the two double points. This conic clearly satisfies the required conditions.

If the given involution is overlapping, proceed thus—Let $A$, $B$, $C$ be the given points, and $l$ the line on which the involution of conjugate points lies. Let $BC$ cut $l$ in $P$, and take $P'$, the mate of $P$, in the involution. Also take $P''$ such that $(BC, PP'') = -1$. Let $PA$ cut $P'P''$ in $a$, and take $A'$ such that $(AA', Pa) = -1$. So, using $CA$ and $QQ'$, $B'$ can be constructed.

Then the conic $ABCA'B'$ is the required conic. For since $(BC, PP'') = -1 = (AA', Pa)$, $P''a$ is the polar of $P$. Hence $PP'$ are conjugate points. So $QQ'$ are conjugate points. Hence the involution $(PP', QQ')$ (which is the given involution) is an involution of conjugate points for this conic.

If the given involution is overlapping, we have solved

xxv.] *Constructions of the Second Degree.* 237

the problem—*To draw a conic through five given points, two of which are imaginary.*

**15.** *Construct a conic to pass through four given points and to divide a given segment harmonically.*

Let $LM$ be the given segment. Let $E$, $F$ be the double points of the involution determined by the given quadrangle $ABCD$ on $LM$. Let the double points $P$, $Q$ of the involution $(LM, EF)$ be constructed. Then the conic through $ABCDP$ is the required conic. For let $LM$ cut this conic again in $Q'$. Then $PQ'$ belong to the involution of the quadrangle on $LM$. Hence $(PQ', EF) = -1$. Hence $Q'$ coincides with $Q$. And $(LM, PQ) = -1$. Hence the conic cuts $LM$ harmonically.

If the double points $E$, $F$ are imaginary, construct the involution of which $L$, $M$ are the double points, and let $P$, $Q$ be the common points of this involution and that of the quadrangle on $LM$. Then the required conic is $ABCDP$. For, as before, $LM$ cuts the conic again in $Q$, and
$$(LM, PQ) = -1.$$
Also, since $E$, $F$ are imaginary, this construction is real.

**Ex.** *Construct a conic which shall pass through four given points and through a pair (not given) of points of a given involution on a line.*

**16.** The following proposition will be used in the succeeding constructions—

*If a variable conic through four fixed points $A$, $B$, $C$, $D$ meet fixed lines through $A$ and $B$ in $P$ and $Q$, then $PQ$ passes through a fixed point upon $CD$.*

For consider the involution in which $CD$ cuts the conic and the four sides $AP$, $BQ$, $AB$, $PQ$ of the quadrangle $ABPQ$. Five of these points are fixed, viz. the meets with the fixed lines $AB$, $AP$, $BQ$, and the meets $C$, $D$ with the conic.

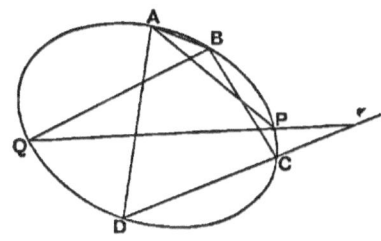

Hence the sixth meet is fixed, i.e. $PQ$ passes through a fixed point on $CD$.

The theorem may also be stated thus—

*A system of conics pass through ABCD. A fixed line through A cuts these conics in PP'..., and a fixed line through B cuts them in QQ'.... Then all the lines PQ, P'Q',... are concurrent in a point on CD.*

If $A$ and $B$ coincide, the theorem is—

*A system of conics touch at A and pass through CD. A fixed line through A cuts these conics in P, P',..., and another fixed line through A cuts them in Q, Q'.... Then all the lines PQ, P'Q',... are concurrent in a point on CD.*

If $A$, $B$ and $C$ coincide, the theorem is—

*A system of conics have three-point contact at A and pass through D. A fixed line through A cuts these conics in P, P',..., and another fixed line through A cuts them in Q, Q',.... Then all the lines PQ, P'Q',... are concurrent in a point on AD.*

**Ex. 1.** *Reciprocate all these theorems.*

**Ex. 2.** *Given three meets ABC of two five-point conics, prove the following construction for the fourth meet D—Take any two points L, M on either conic, and construct the points L', M' in which AL, BM cut the other conic. Join the meet of LM, L'M' to C. Then D is the meet of this line with either conic.*

**Ex. 3.** *Given two meets A, B of two five-point conics, prove the following construction for the other meets C and D—Take any two points L, M on either conic, and construct the points L', M' in which AL, BM cut the other conic. LM, L'M' meet in one point on CD. Similarly construct another point on CD. Now construct the points in which the joining line cuts either conic.*

**Ex. 4.** *Reciprocate the two preceding constructions.*

**Ex. 5.** *Prove the following construction for the directions of the axes of a conic given by five points—Draw a circle through three A, B, C of the given points; now construct the fourth meet P of the conic and the circle; then the directions of the axes bisect the angles between AB and CP.*

**17.** *Given five points on a conic, three of which are coincident, to construct the conic.*

Let $ABC$ be the three given coincident points, and $DE$ the other given points. Then to be given $ABC$ is equivalent to being given the point $A$, the tangent at $A$, and the circle of curvature at $A$. Let $AD$, $AE$ cut this given circle in $D'$, $E'$. Then $DE$, $D'E'$ meet on the common chord of the xxv.] *Constructions of the Second Degree.* 239

circle and the conic. Hence the point $P$ where this chord cuts the circle can be constructed. Now $P$ is on the conic. Hence we know four points $A$, $D$, $E$, $P$ on the conic and the tangent at one of them. Hence the conic can be constructed.

**Ex.** *Obtain, by using the reciprocal theorem, a solution of the problem— Given five tangents of a conic, three of which are coincident, construct the conic.*

Notice that the circle of curvature has three-tangent contact with the conic as well as three-point contact.

# CHAPTER XXVI.

## METHOD OF TRIAL AND ERROR.

**1.** GIVEN *two homographic ranges* $(ABC...)$ *and* $(abc...)$ *on different lines, and given two points* $V$ *and* $u$, *find two points* $XY$ *of the first range, such that the angles* $XVY$ *and* $xvy$ *may have given values,* $x$ *and* $y$ *being the points corresponding to* $X$ *and* $Y$ *in the homographic ranges.*

Try any point $P$ on $AB$ as a position of $X$. To do this, take $Q$ on $AB$, so that the angle $PVQ$ is equal to the given value of $XVY$. Take $p$ and $q$, the points corresponding to $P$ and $Q$, in the homographic ranges. Also take $r$ on $ab$, so that the angle $pvr$ may be equal to the given angle $xvy$. Then if $r$ coincides with $q$, the problem is solved.

If not, try several points $P_1, P_2 \ldots$. Then

$$(r_1 r_2 \ldots) = v(r_1 r_2 \ldots)$$
$$= v(p_1 p_2 \ldots) \quad \text{since the pencils are superposable}$$
$$= (p_1 p_2 \ldots) = (P_1 P_2 \ldots) \quad \text{since the ranges are homographic}$$
$$= V(P_1 P_2 \ldots) = V(Q_1 Q_2 \ldots) = (Q_1 Q_2 \ldots) = (q_1 q_2 \ldots).$$

Hence the ranges $(q_1 q_2 \ldots)$ and $(r_1 r_2 \ldots)$ are homographic. Now if $q$ and $r$ coincide, $q$ will be a position of $y$. Hence $y$ is either of the common points of the homographic ranges $(q_1 q_2 \ldots)$ and $(r_1 r_2 \ldots)$. Hence $Y$ and $X$ are known.

The problem has four solutions. Two are obtained above, and two more are obtained by taking the angles $PVQ$ and $pvq$ in relatively opposite directions.

Notice that we need only make three attempts; for the common points of two homographic ranges can be determined if three pairs of corresponding points are known.

## Method of Trial and Error.

The above process may be abbreviated by writing $(r)$ for the range $(r_1\ r_2\ ..\ )$, and so on.

The method is called by some writers *the method of False Positions*.

**Ex. 1.** *Find corresponding segments $XY$, $X'Y'$ of two given homographic ranges which shall be of given lengths.*

**Ex. 2.** *Given two homographic ranges on the same line, find a segment $XX'$ bounded by corresponding points, (i) which is bisected by a given point, or (ii) which divides a given segment harmonically.*

If $XX'$ satisfies either condition, $X$ and $X'$ generate ranges which are in involution and therefore homographic.

**Ex. 3.** *Find also $XX'$, given that (i) $AX : BX'$ is a given ratio, or (ii) that $XX'$ is of given length, or (iii) that $XX'$ divides a given segment in a given cross ratio.*

**Ex. 4.** *If $A$, $A'$ generate homographic ranges on two lines, show that through any given point two of the lines $AA'$ pass.*

**Ex. 5.** *Find corresponding points $X$, $X'$ of two homographic ranges on different lines, such that $XO$ and $X'O'$ meet at a given angle, $O$ and $O'$ being given points.*

The pencils at $O$ and $O'$ are homographic.

**Ex. 6.** *Given on the same line the homographic ranges $(ABC...) = (A'B'C'...)$, and the homographic ranges $(LMN...) = (L''M''N''...)$; find a point $X$ which has the same mate in both ranges.*

**Ex. 7.** *If $A$ and $A'$ generate homographic ranges on two lines, and $B$ and $B'$ generate homographic ranges on two other lines, find the positions of $A$, $B$, $A'$, $B'$ that both $AB$ and $A'B'$ may pass through a given point.*

2. *Between two given lines place a segment whose projections on two given lines shall be of given lengths.*

Let the projections lie on the lines $AB$ and $CD$. On $AB$ take a length $LM$ equal to the given projection on $AB$; through $L$ and $M$ erect perpendiculars to $AB$ to meet the given lines in $X$ and $Y$. Let the projection of $XY$ on $CD$ be $PQ$. If $PQ$ is of the required length, then the problem is solved.

If not, make $PQ'$ of the required length. Then the ranges generated by $Q'$ and $P$ are homographic, being superposable. Again, the ranges $P$ and $X$ are homographic, by considering a vertex at infinity. Similarly

$$\text{range } X = \text{range } L = \text{range } M = \text{range } Y = \text{range } Q.$$

Hence the ranges $Q'$ and $Q$ are homographic. Either of

the common points of these ranges gives a true position of $Q$.

**Ex. 1.** *On two given lines find points $A$ and $B$, such that $AB$ subtends given angles at two given points.*

**Ex. 2.** *Through a given point draw two lines, to cut off segments of given lengths from two given lines.*

**Ex. 3.** *Given two fixed points $O$ and $O'$ on two fixed lines, through a fixed point $V$ draw a line cutting the fixed lines in points $A$, $A'$, such that* (i) $OA \cdot O'A'$ *is constant, or* (ii) $OA : O'A'$ *is constant.*

**Ex. 4.** *Through a given point draw a line to include with two given lines a given area.*

**Ex. 5.** *Two sides of a triangle are given in position and the area is given; show that the base in two positions subtends a given angle at a given point.*

**Ex. 6.** *Given four points $A$, $B$, $C$, $D$ on the same line, find two points $X$, $Y$ on this line, such that $(AB, XY)$ and $(CD, XY)$ may have given values.*

**Ex. 7.** *Given two fixed points $A$ and $B$, find two points $P$, $Q$ on the line $AB$, such that $(AB, PQ)$ is given and also the length $PQ$.*

**Ex. 8.** *Given three rays $OA$, $OB$, $OC$, find three other rays $OX$, $OY$, $OZ$, such that the cross ratios $O(AB, XY)$, $O(BC, YZ)$, $O(CA, ZX)$ may have given values.*

**Ex. 9.** *Find the lines $OX$, $OX'$ such that $O(AA', XX')$ may be a given cross ratio and $XOX'$ a given angle, $OA$ and $OA'$ being given lines.*

**Ex. 10.** *Solve the equation $ax^2 + bx + c = 0$ by a geometrical construction.*

The roots are the values of $x$ at the common points of the homographic ranges determined by $axx' + bx + c = 0$.

**Ex. 11.** *Solve geometrically the equations*
$$y = lx + a, \quad z = my + b, \quad x = nz + c.$$
Obtain the common points of the homographic ranges $(x, x')$ determined by $y = lx + a$, $z = my + b$, $x' = nz + c$.

**Ex. 12.** *Solve geometrically the equations*
$$xy + lx + my + n = 0, \quad xy + px + qy + r = 0.$$

3. *Given two points $L$, $M$ on a conic, find a point $P$ on the conic, such that $PL$, $PM$ shall divide a given segment $UV$ in a given cross ratio.*

Take any position of $P$, and let $PL$, $PM$ meet $UV$ in $A$, $B$, and take $B'$ such that $(UV, AB')$ is equal to the given cross ratio. Then $(A) = L(A) = L(P) = M(P) = M(B) = (B)$. Also, since $(UV, AB')$ is constant, we have $(A) = (B')$. Hence $(B) = (B')$. Hence the required position of $B$ is either of the common points of the homographic ranges generated by $B$ and $B'$.

**Ex.** *Give two points L, M on a conic, find a point P on the conic such that the bisectors of the angle LPM may have given directions.*
Draw parallels to PL, PM through a fixed point.

**4.** *Inscribe in a given conic a polygon of a given number of sides, so that each side shall pass through a fixed point.*

Consider for brevity a four-sided figure. It will be found that the same solution applies to any polygon.

Suppose we have to inscribe in a conic a four-sided figure $ABCD$, so that $AB$ passes through the fixed point $U$, $BC$ through $V$, $CD$ through $W$, and $DA$ through $X$. On the conic take any point $A$. Let $AU$ cut the conic again in $B$. Let $BV$ cut the conic again in $C$. Let $CW$ cut the conic again in $D$. Let $DX$ cut the conic again in $A'$. So take several positions of $A$.

Then the range on the conic generated by $A$ is in involution with the range generated by $B$, since $AB$ passes through a fixed point $U$. Hence $(A) = (B)$. So

$$(B) = (C) = (D) = (A').$$

Hence the ranges $(A)$ and $(A')$ on the conic are homographic. A true position of $A$ is either of the common points of these homographic ranges.

Note that in the exceptional case of XXI. 3. Ex. 14, the common points lie on the line; and the above solution becomes nugatory.

**Ex. 1.** *Describe about a given conic a polygon such that each vertex shall lie on a given line.*
Inscribe in the conic a polygon whose sides pass through the poles of the given lines, and draw the tangents at its vertices.

**Ex. 2.** *Inscribe in a given conic a polygon of a given number of sides, such that each pair of consecutive vertices determine with two given points on the conic a given cross ratio.*

**Ex. 3.** *In the given figure ABCD inscribe the figure NPQR, so that RN, PQ meet in the fixed point U, and NP, RQ in the fixed point V.*

**Ex. 4.** *Construct a polygon, whose sides shall pass through given points and whose vertices shall lie on given lines.*

**Ex. 5.** *Construct a polygon, whose vertices shall lie on given lines and whose sides shall subtend given angles at given points.*

**Ex. 6.** *Construct a triangle ABC, such that A and B shall lie on given lines, and that the angle C shall be equal to a given angle, whilst the sides AB, BC, CA pass through fixed points.*

**Ex. 7.** *A ray of light starts from a given point, and is reflected successively from n given lines; find the initial direction that the final direction may make a given angle with the initial direction.*

**Ex. 8.** *Given two homographic ranges $(ABC\ldots) = (A'B'C'\ldots)$ on a conic, find the corresponding points $X$, $X'$, such that $XX'$ may pass through a given point.*

**Ex. 9.** *Given two points $AA'$ on a conic, find two points $XX'$ also on the conic, such that $(AA', XX')$ has a given value and $XX'$ passes through a given point.*

**Ex. 10.** *Through a given point $A$ is drawn a chord $PQ$ of a conic; $BC$ are fixed points on the conic; find the position of $PQ$ when $PB$ and $QC$ meet at a given angle.*

**Ex. 11.** *Through two given points describe a circle which shall cut a given arc of a circle in a given cross ratio.*

**Ex. 12.** *Through four given points draw a conic which shall cut off from a given line a length which is either given or subtends a given angle at a given point.*

# CHAPTER XXVII.

IMAGINARY POINTS AND LINES.

1. THE *Principle of Continuity* enables us to combine the elegance of geometrical methods with the generality of algebraical methods. For instance, if we wish to determine the points in which a line meets a circle, the neatest method is afforded by Pure Geometry. But in certain relative positions of the line and circle, the line does not cut the circle in visible points.

Here Algebraical Geometry comes to our help. For if we solve the same problem by Algebraical Geometry, we shall ultimately have to solve a quadratic equation; and this quadratic equation will have two solutions, real, coincident and imaginary. Hence we conclude that a line always meets a circle in two points, real, coincident or imaginary.

Another instance is afforded by XXIII. 5. Here we prove the proposition by using the points $O$ and $O'$ in which the circles on $AA'$ and $BB'$ as diameters meet. But these circles in certain cases do not meet in visible points. But we might have proved the same proposition by Algebraical Geometry, following the same method. Then it would have been immaterial whether the coordinates of the points $O$ and $O'$ had been real or imaginary, and the proof would have held good. Hence we conclude that we may use the imaginary points $O$ and $O'$ as if they were real.

In all solutions by Algebraical Geometry, points and lines will be determined by algebraical equations. Hence imaginary points and lines will occur in pairs. Hence we shall expect that in Pure Geometry, imaginary points and lines will occur in pairs.

**2.** The best way of defining the position of *a pair of imaginary points* is as the double points of a given overlapping involution; and the best way of defining the position of *a pair of imaginary lines* is as the double lines of a given overlapping involution.

Thus the points in which a line cuts a conic are the double points of the involution of conjugate points determined by the conic on the line; and these double points, i.e. the meets of the conic and line, are imaginary if the involution is an overlapping one.

So the tangents from any point to a conic are the double lines (real, coincident, or imaginary) of the involution of conjugate lines which the conic determines at the point.

Note that a pair of imaginary points is not the same as two imaginary points. For if $AA'$ are a pair of imaginary points and $BB'$ another pair of imaginary points, then $AB$ are two imaginary points but are not a pair.

**3.** *The middle point of the segment joining a pair of imaginary points is real.*

For it is the centre of the involution defining the imaginary points.

*A pair of imaginary points $AA'$ is determined when we know the centre $O$ and the square (a negative quanity) $OA^2$.*

For the involution defining the points is given by

$$OP \cdot OP' = OA^2.$$

*The fourth harmonic of a real point for a pair of imaginary points is real.*

For it is the corresponding point in the defining involution.

*The product of the distances of a pair of imaginary points from any real point on the same line is real and positive.*

Let $AA'$ be the pair, and $P$ any real point on the line

$AA'$. Take $O$ the middle point of the segment $AA'$. Then
$$PA \cdot PA' = (OA - OP)(OA' - OP)$$
$$= (OA - OP)(-OA - OP) = OP^2 - OA^2.$$
Now $OA^2$ is negative, or the involution would have real double points. Hence $PA \cdot PA'$ is real and positive.

**4.** *Two conics cut in four real points, or in two real and two imaginary points, or in four imaginary points.*

Since a conic is determined by five points, two conics cannot cut in more than four points, unless they are coincident.

Also we can draw two conics cutting in four points, e. g. two equal ellipses laid across one another.

Now if we were solving the problem by Algebraical Geometry, and were given that the problem could not have more than four solutions, and that it had four solutions in certain cases, we should be sure that the problem had in all cases four solutions, the apparent deficiencies, if any, being accounted for by coincident or imaginary points.

Hence it follows by the Principle of Continuity, that two conics always cut in four points, real, coincident, or imaginary.

Also imaginary points occur in pairs. Hence two or four of the points may be imaginary.

**5.** *If two conics cut in two real points, the line joining the other common points is real, even if the latter points are imaginary.*

For, by the principle of continuity, Desargues's theorem holds, even if two or four of the points on the conic are imaginary. Let any line cut the conics in $pp'$ and $qq'$ and the given real common chord in $a$. Then the real point $a'$, taken such that $(aa', pp', qq')$ is an involution, lies on the opposite common chord. Hence the opposite common chord is real, being the locus of the real point $a'$.

*If two conics cut in two real and two imaginary points, one pair of common chords is real and two imaginary.*

For if a second pair were real, the four common points would be real, being the meets of real lines.

**6.** *One vertex of the common self-conjugate triangle of two conics is always real.*

Take any line $l$; then the locus of the conjugate points of points on $l$ for both conics is a conic. Take any other line $m$; the locus of the conjugate points of points on $m$ for both conics is a second conic. These conics have one real point in common, viz. the conjugate point of the meet of $l$ and $m$. Hence they have another real point in common, say $U$.

Take the conjugate point $Q$ on $l$ of $U$ for both conics and the conjugate point $R$ on $m$ of $U$ for both conics. Then $QR$ is clearly the polar of $U$ for both conics; for the polar of $U$ for both conics passes through $Q$ and $R$. Hence $U$ is a real vertex of the common self-conjugate triangle of the two conics.

Similarly, the other two points, real or imaginary, in which the conics cut, are the other two vertices of the common self-conjugate triangle.

**7.** *The other two vertices of the common self-conjugate triangle of two conics are real if the conics cut in four real points or four imaginary points; but if the conics cut in two real and two imaginary points, the other two vertices are imaginary.*

If the four intersections are real, the proposition is obviously true.

If the four intersections are imaginary, one conic must be entirely inside or entirely outside the other. Hence the polar of the real vertex $U$ cuts the conics in either two non-overlapping segments $AA'$, $BB'$, or in one real segment and one imaginary, or in two imaginary segments. Now the other two vertices $VW$ are the points on the polar which are conjugate for both conics, i.e. are the common pair of the two involutions of conjugate points on the polar. And the double points $AA'$, $BB'$ of these involutions are either real and non-overlapping, or one pair (at least) is imaginary. Hence by XX. 5, $VW$ are real.

If two intersections are real and two imaginary, the meets of the given real common chord and of the opposite common chord (which is known to be real) gives a real position of $U$. But the opposite chord does not cut either conic; hence $U$ is outside both conics. Hence the polar of $U$, passing through the fourth harmonic of $U$ for the two real points, cuts the two conics in overlapping real segments. Hence $VW$, being the double points of the involution determined by these segments, are imaginary.

**8.** *One pair of common chords of two conics is always real.*

If all four intersections are real, it is clear that the six common chords are all real.

If all four intersections are imaginary, then $UVW$ are real. Take any point $P$ and its conjugate point $P'$ for the two conics. Then the common chords through $U$ are the double lines of the involution $U(VW, PP')$; for the polar of $P$ for these common chords passes through $P'$, and the polar of $V$ passes through $W$. Hence the common chords through $U$ are both real or both imaginary.

Also the common chords through two of the three points $UVW$ must be imaginary; for otherwise the four real common chords would intersect in four real common points of the conics. Let the chords through $V$ and $W$ be imaginary. Then taking $P$ inside the triangle $UVW$, we see that since $V(UW, PP')$ overlap, $P'$ must lie in the external angle $V$; so $P'$ must lie in the external angle $W$. Hence $P'$ lies in the internal angle $U$. Hence $U(VW, PP')$ does not overlap; hence the double lines of the involution are real, i.e. the common chords through $U$ are real.

If two intersections are real and two imaginary, we have already proved that two common chords are real.

**9.** *Two conics have four common tangents, of which either two or four may be imaginary.*

*If two conics have two real common tangents and two imaginary, the intersection of the real and also of the imaginary tangents is real; and the other four common apexes are imaginary.*

*One side of the common self-conjugate triangle of two conics is always real; the other two sides are real if the four common tangents are all real or all imaginary; otherwise the other two sides are imaginary.*

*One pair of common apexes of two conics is always real.*

These propositions can be proved similarly to the corresponding propositions respecting common points and common chords (or by Reciprocation).

**Ex.** *If two conics have three-point contact at a point, they have a fourth real common point, and a fourth real common tangent.*

# CHAPTER XXVIII.

### CIRCULAR POINTS AND CIRCULAR LINES.

1. THE *circular lines* through any point are the double lines of the orthogonal involution at the point.

*Every pair of circular lines cuts the line at infinity in the same two points* (called the *circular points*).

Take any two points $P$ and $Q$. Then to every ray in the orthogonal involution at $P$ there is a parallel ray in the orthogonal involution at $Q$, or briefly, the involutions are parallel. Hence the double lines are parallel. Hence the circular lines through $P$ and $Q$ meet the line at infinity in the same two points.

The notation $\infty$, $\infty'$ will be reserved for the circular points.

*Any two perpendicular lines are harmonic with the circular lines through their meet.*

For by definition the circular lines are the double lines of an involution of which the perpendicular lines are a pair.

*The points in which any two perpendicular lines meet the line at infinity are harmonic with the circular points.*

For the circular lines through the meet of the lines are harmonic with the given lines.

2. *The triangle whose vertices are any point $C$ and the circular points, is self-conjugate for any rectangular hyperbola whose centre is at $C$.*

For $C\infty$, $C\infty'$ being circular lines are harmonic with every orthogonal pair of lines through $C$, and are therefore harmonic with the asymptotes, i.e. with the tangents from $C$ to the r. h., and are therefore conjugate lines for the r. h.

252  *Circular Points and Circular Lines.*  [CH.

Also $C$ is the pole of $\infty\infty'$. Hence $C\infty\infty'$ is self-conjugate for the r. h.

**Ex. 1.** *All rectangular hyperbolas have a common pair of conjugate points.*
**Ex. 2.** *Every conic for which the circular points are conjugate is a r. h.*

**3.** *All circles pass through the circular points.*

Let $C$ be the centre of any circle. Then $C\infty$, $C\infty'$ are the asymptotes of the circle. For $C\infty$, $C\infty'$ are the double lines of the orthogonal involution at $C$, i.e. are the double lines of the involution of conjugate diameters of the circle. Now a conic passes through the points in which the line at infinity meets its asymptotes. Hence the circle passes through $\infty$ and $\infty'$.

Notice that we have proved that $C\infty$, $C\infty'$ touch at $\infty$, $\infty'$ any circle whose centre is at $C$.

**4.** *Every conic which passes through the circular points is a circle.*

Let $C$ be the centre of a conic through $\infty$, $\infty'$. Then since the lines joining the centre of a conic to the points where the conic meets the line at infinity are the asymptotes of the conic, we see that $C\infty$, $C\infty'$ are the asymptotes of the conic. Hence the involution of conjugate diameters of which the asymptotes are the double lines must be an orthogonal involution. Hence every pair of conjugate diameters is orthogonal. Hence the conic is a circle.

We now see the origin of the names circular points and circular lines. The circular points are the points through which all circles pass. A pair of circular lines is the limit of a circle when the radius is zero; the circle degenerating into a real point through which pass imaginary lines to the circular points. So that a pair of circular lines is both a circle and a pair of lines.

**5.** *Concentric circles have double contact, the line at infinity being the chord of contact.*

For all circles which have $C$ as centre, touch $C\infty$ at $\infty$ and $C\infty'$ at $\infty'$.

**Ex. 1.** *Every semicircle is divided harmonically by the circular points.*

**Ex. 2.** *The circle which circumscribes a triangle which is self-conjugate for a rectangular hyperbola passes through the centre.*

For five of the vertices of the two triangles consisting of the given triangle and $C\infty \infty'$ lie on the circle.

**Ex. 3.** Gaskin's theorem. *The circle about a triangle self-conjugate for a conic is orthogonal to the director.*

Let $V$ be a common point of the circle and the director. Let the polar of $V$ for the conic meet the circum-circle in $a, a'$; $V\infty, V\infty'$ in $\beta, \beta'$; and the tangent at $V$ to the circum-circle, and $\infty\infty'$ in $\gamma, \gamma'$. Then $Vaa'$ is a self-conjugate triangle. Hence $aa'$ are conjugate points for the conic. Again, $V\infty, V\infty'$ are conjugate lines, for the tangents from $V$ are orthogonal; hence $\beta\beta'$ are conjugate points. And $(aa', \beta\beta', \gamma\gamma')$ is an involution. Hence $\gamma\gamma'$ are conjugate points. Hence the polar of $\gamma'$ passes through $\gamma$. Now $\gamma'$ is at infinity, hence its polar $V\gamma$ passes through $C$; i.e. the tangent to the circum-circle at $V$ coincides with the radius of the director circle.

**Ex. 4.** *The axes of any one of the system of conics through four given points on a circle are in fixed directions.*

Take any point $V$ and join $V$ to the points at infinity $AA', BB', \ldots$ on the conics. Then $V(AA', BB', \ldots)$ is an involution pencil parallel to the asymptotes. But $V\infty, V\infty'$ is one pair, corresponding to the circle. Hence the double lines are at right angles, and therefore bisect the angles $AVA', BVB', \ldots$. Hence the axes are parallel to these double lines, and therefore are in fixed directions.

**Ex. 5.** *Two conics are placed with their axes parallel; show that their four meets are concyclic.*

**Ex. 6.** *Give a descriptive proof of the property of the director circle of a conic.*

Let $A$ and $B$ be any fixed points, and let $PA$ and $PB$ be any two lines through $A$ and $B$ which are conjugate for a conic. Draw the polar $b$ of $B$ cutting $PA$ in $Q$. Then $Q$ is the pole of $PB$. Hence
$$A(P) = A(Q) = (Q) = B(P).$$
Hence the locus of $P$ is a conic through $A$ and $B$.
Now let $R$ be any point on the director circle. Then $R\infty, R\infty'$ are conjugate for the conic, since the tangents from $R$, being perpendicular, are harmonic with $R\infty, R\infty'$. Hence the locus of $R$ is a conic through $\infty$ and $\infty'$, i.e. is a circle.

**6.** *If the pencil $V(ABC\ldots)$ be turned bodily through any angle about $V$ into the position $V(A'B'C'\ldots)$, then the common lines of the two homographic pencils $V(ABC\ldots)$ and $V(A'B'C'\ldots)$ are the circular lines through $V$.*

The pencils, being superposable, are homographic. Hence if they cut any circle through $V$ in $abc\ldots$ and $a'b'c'\ldots$, the two ranges $(abc\ldots)$ and $(a'b'c'\ldots)$ on the circle are homographic. One point on the homographic axis of these ranges is the meet of $ab'$ and $a'b$. But these lines are parallel. Hence this point is at infinity. So every point on the axis is at infinity. Hence the common points of the ranges $(abc\ldots)$ and $(a'b'c'\ldots)$ are the meets of the circle with the line at infinity, i.e. are $\infty\,\infty'$. Hence the common lines of the pencils $V(ABC\ldots)$ and $V(A'B'C'\ldots)$ are $V\infty$, $V\infty'$.

Hence—*The legs of a constant angle divide the segment joining the circular points in a constant cross ratio.*

Let the constant angles be $ALA'$, $BMB'$, $CNC'$, .... Through any point $V$ draw a circle and let parallels through $V$ to $LA$, $MB$, $NC$, ..., $LA'$, $MB'$, $NC'$... cut this circle in $a, b, c, \ldots, a', b', c'\ldots$. Then, as above, $\infty\,\infty'$ are the common points of the homographic ranges $(abc\ldots)$ and $(a'b'c'\ldots)$ on the circle. Hence

$$(\infty\,\infty', aa') = (\infty\,\infty', bb') = (\infty\,\infty', cc') = \ldots.$$

Hence $V(\infty\,\infty'\,aa')$ is constant. But the parallel lines $LA$ and $Va$ cut $\infty\,\infty'$ in the same point; so $LA'$ and $Va'$ cut $\infty\,\infty'$ in the same point. Hence $L(\infty\,\infty', AA')$ is constant. Hence $LA$ and $LA'$ divide the segment $\infty\,\infty'$ in a constant cross ratio.

**7.** *Coaxal circles are a system of four-point conics.*

For two circles meet in two points (real or imaginary)

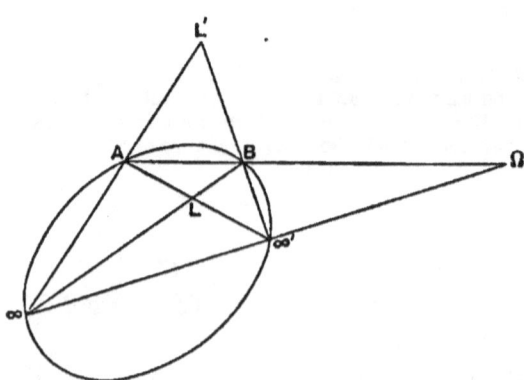

on the radical axis and also in the circular points. The adjoining ideal figure explains the relation of coaxal circles xxviii.] *Circular Points and Circular Lines.* 255

to the circular points. $A$ and $B$ are the common finite points on the radical axis, and $\Omega$ is the point at infinity on the radical axis.

$L$ and $L'$ are the limiting points. For since $LA$ and $LB$ are circular lines through $A$ and $B$, $L$ is a point-circle of the system. So for $L'$. Also $LL'\Omega$ is the common self-conjugate triangle of the coaxal system.

### Foci of a Conic.

**8.** *Every conic has four foci, which are inside the conic and lie two on each axis, those on either axis being equidistant from the centre.*

The tangents from a focus of a conic to the conic are the double lines of the involution of conjugate lines at the focus, i.e. are the double lines of an orthogonal involution, i.e. are circular lines, i.e. pass through $\infty$, $\infty'$. Hence a focus is an internal point, since the tangents from it are imaginary.

Also every intersection $S$ of the four tangents from $\infty$, $\infty'$ to the conic is a focus of the conic. For $S\infty$, $S\infty'$ being the tangents from $S$ and also circular lines, the involution of conjugate lines at $S$ is orthogonal, i.e. $S$ is a focus. Hence the foci of a conic are the other four meets of tangents to the conic from $\infty$ and $\infty'$.

Consider the adjoining ideal figure. Here $SS'FF'$ are the foci. Also $C$ is the centre; for the lines $SS'$, $FF'$, $\infty\infty'$ form a self-conjugate triangle, hence $C$ is the pole of $\infty\infty'$. Again, $SS'$ and $FF'$ are the axes. For

$C(\infty\infty', SF)$

is a harmonic pencil (from the quadrangle $SFS'F'$); hence $SS'$ and $FF'$ are orthogonal. Hence $SS'$

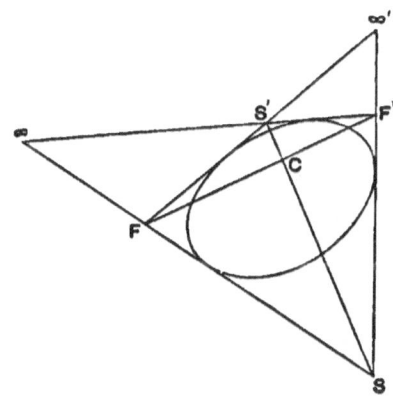

and $FF''$, being orthogonal conjugate lines at the centre, are the axes. Hence the foci lie two by two on the axes.

Again, $FF'$ cuts $\infty\infty'$ in a point $\Omega$, such that $(C\Omega, FF')$ is harmonic; hence $C$ bisects $FF'$. So $C$ bisects $SS'$.

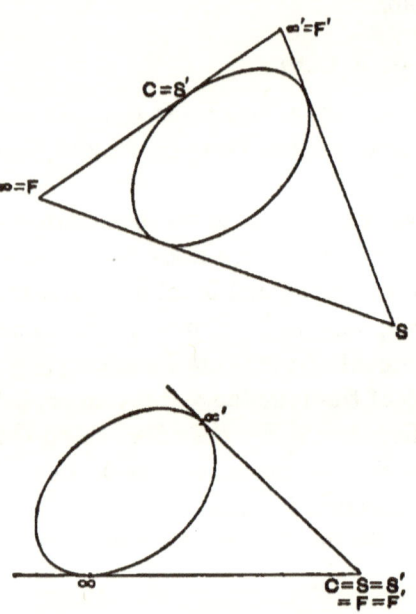

Hence the foci on each axis are equidistant from the centre.

It will be instructive to draw an ideal picture showing the relation of a parabola and of a circle to its foci.

In the case of a parabola $\infty\infty'$ touches the conic. Hence $F''$ coincides with $\infty'$ and $F$ with $\infty$. Also $C$ and $S'$ coincide at the point of contact of $\infty\infty'$.

In the case of a circle, $\infty$ and $\infty'$ are on the conic; and all the foci coincide with the centre $C$.

**Ex. 1.** *The sides of a triangle $ABC$ touch a conic a and meet a fourth tangent to a in $A'B'C'$; show that the double lines of the involution subtended by $(AA', BB', CC')$ at a focus are perpendicular.*

Being conjugate lines at a focus.

**Ex. 2.** *The circle described about a triangle which circumscribes a parabola, passes through the focus.*

For five of the vertices of the two triangles consisting of the given triangle and $S\infty\infty'$ lie on the circle.

**Ex. 3.** *A circle is drawn with centre on the directrix of a parabola to pass through the focus. At $R$, one of the meets of the parabola and the circle, are drawn the tangents to the circle and parabola, meeting the parabola and circle again in $P$ and $Q$. Show that $PQ$ is a common tangent to the two curves.*

Let $O$ be the centre on the directrix, and let the tangents from $O$ to the parabola meet the line at infinity in $\Omega$ and $\Omega'$. Then con-

XXVIII.] *Circular Points and Circular Lines.* 257

sidering the triangles $O\Omega\Omega'$ and $S\infty\infty'$, we see that the conics are related as in Ex. 14 of XIV. 2.

**9.** *The foci on one axis* (called the *focal axis*) *are real, and the foci on the other axis* (called the *non-focal axis*) *are imaginary.*

Take any point $P$, and through $P$ draw the orthogonal pair of the involution of conjugate lines at $P$, cutting one axis in $G$ and $H$ and the other axis in $g$ and $h$. Then $PG$ and $PH$ are harmonic with $P\infty$ and $P\infty'$ since $GPH$ is a right angle, and with the tangents from $P$ since $PG$ and $PH$ are conjugate. Hence $PG$ and $PH$ are the double lines of the involution $P(\infty\infty', SS', FF')$ to which the tangents belong.

Hence $P(SS', GH)$ and $P(FF', gh)$ are harmonic. And $C$ bisects $SS'$ and $FF'$. Hence $CS^2 = CG \cdot CH$ and

$$CF^2 = Cg \cdot Ch.$$

But on drawing the figure, we see that if $CG$ and $CH$ are of the same sign, $Cg$ and $Ch$ are of opposite signs. Hence, taking $CG \cdot CH$ positive, $CS^2$ is positive and $CF^2$ is negative. Hence $S$ and $S'$ are real and $F$ and $F'$ are imaginary.

**Ex. 1.** *Show that gh subtends a right angle at S and at S'.*

Now $Cg \cdot Ch = -CG \cdot CH$ by elementary geometry $= -CS^2 = CS \cdot CS'$. Hence $SS'gh$ lie on the circle whose diameter is $gh$.

**Ex. 2.** *Any line through G is conjugate to the perpendicular line through H; and the same is true of g and h.*

**Ex. 3.** *In a parabola, S bisects GH.*

**10.** *Confocal conics are a system of four-tangent conics.*

For if $S$ and $S'$ be the real foci, the conics all touch the lines $S\infty$, $S'\infty$, $S\infty'$, and $S'\infty'$.

Hence, *the tangents from any point to a system of confocals form an involution, to which belong the pairs* $(PS, PS')$, $(PF, PF')$ *and* $(P\infty, P\infty')$, $P$ *being the given point.*

*Through every point can be drawn a pair of lines which are conjugate for every one of a system of confocals.*

s

Viz. the double lines $PG$, $PH$ of the above involution.

$PG$ and $PH$ are perpendicular.

For they are harmonic with $P\infty$, $P\infty'$.

*The pairs of tangents from any point to a system of confocals and the focal radii to the point have a common pair of bisectors.*

For the double lines $PG$ and $PH$ of the involution are perpendicular.

**Ex. 1.** *In a parabola, $PG$ and $PH$ are the bisectors of the angles between $PS$ and a parallel through $P$ to the axis.*

**Ex. 2.** *From a given point $O$, lines are drawn to touch one of a system of confocal conics in $P$ and $Q$; show that $PQ$ and the normals at $P$ and $Q$ touch a fixed parabola which touches the axes of the confocals.*

Viz. the polar-envelope of the point $O$ for the system of four-tangent conics. The normal $PG$ at $P$ touches the polar-envelope, because it is conjugate to $OP$ for every conic of the system. Also $\infty\infty'$ and the axes touch, since they are the harmonic lines of the quadrilateral.

**Ex. 3.** *The directrix of the parabola is $CO$, $C$ being the common centre.*

For the tangents at $O$ to the two confocals through $O$ are two positions of $PQ$.

**Ex. 4.** *The circle about $OPQ$ passes through a second fixed point.*

Let the normals at $P$ and $Q$ meet in $R$. Then the circle about $OPQ$ is the circle about $PQR$, which passes through the focus of the parabola.

**Ex. 5.** *The locus of the orthocentre of $PQR$ is a line.*

Viz. the directrix of the parabola.

**Ex. 6.** *The conic through $OPQ$ and the foci passes through a fourth fixed point.*

Let the perpendiculars at $S$, $S'$ to $OS$, $OS'$ meet in $U$. Then
$$S(UO, PQ) = S'(UO, PQ) = -1.$$

**11.** *The locus of the poles of a given line for a system of confocals is the normal at the point of contact of the given line with a confocal.*

For let the given line $l$ touch a confocal at $P$, and let $PG$ be the normal, and $PH (= l)$ the tangent to this confocal. Then $PG$ and $PH$ are perpendicular. Hence $P(GH, \infty\infty')$ is harmonic. But $PH$ is one of the double lines of the involution of tangents from $P$ to the confocals, being the pair of coincident tangents from $P$ to the confocal which

$PH$ touches. And $P\infty$, $P\infty'$ is a pair of this involution. Hence $PG$ is the other double line. Hence $PG$ and $PH$, being harmonic with every pair of tangents, are conjugate for every confocal. Hence the locus of the poles of $l$ is $PG$.

#### Reciprocation of circular points and lines.

**12.** Circular lines are the double lines of the orthogonal involution at a point $P$. Hence, *the reciprocal of a pair of circular lines* is a pair of points on a line $p$ which are the double points of the involution on the line which subtends an orthogonal involution at the origin $O$ of reciprocation, in other words, are the meets of $p$ with the circular lines through the origin of reciprocation.

Circular points are the points on the line at infinity which are the double points of the involution on the line at infinity which subtends an orthogonal involution at $O$. Hence *the reciprocals of the circular points* are the double lines of the orthogonal involution at $O$, i.e. are the circular lines through the origin of reciprocation.

*The reciprocal of a circle for the point $O$ is a conic with focus at $O$.*

For since the circle passes through the circular points, the reciprocal touches the circular lines through $O$, i.e. $O$ is a focus of the reciprocal.

*To reciprocate confocal conics into coaxal circles.*

Confocal conics are conics inscribed in the quadrilateral $S\infty$, $S'\infty$, $S\infty'$, $S'\infty'$. Reciprocate for $S$. Then since $S\infty$, $S\infty'$ touch the given conics, the circular points lie on the reciprocal conics, i.e. the reciprocal conics are circles. Also the given conics have two other common tangents; hence the reciprocal conics have two other common points, i.e. are coaxal circles.

*To reciprocate coaxal circles into confocal conics.*

Coaxal circles are conics circumscribed to the quadrangle $AB\infty\infty'$. (See figure of § 7.) Reciprocate for $L$. Then the given conics pass through four fixed points, two on each circular line through the origin of reciprocation. Hence the reciprocal conics touch four fixed lines, two through each of the circular points; i. e. the tangents to all the reciprocal conics from $\infty$, $\infty'$ are the same, i. e. the reciprocal conics are confocal.

# CHAPTER XXIX.

PROJECTION, REAL AND IMAGINARY.

1. *To project a given conic into a circle and at the same time a given line to infinity.*

Take $K$, the pole of the given line $l$ which is to be projected to infinity. Through $K$ draw two pairs of conjugate lines cutting $l$ in $AA'$, $BB'$.

On $AA'$ and $BB'$ as diameters describe circles cutting in $V$ and $V'$. About $AA'$ rotate $V$ out of the plane of the paper. With $V$ as vertex project the given figure on to any plane parallel to the plane $VAA'$.

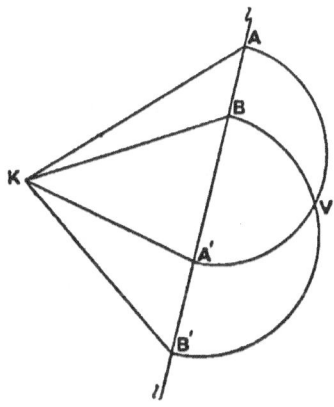

Then $KA$ will be projected into a line parallel to $VA$, and $KA'$ into a line parallel to $VA'$. Hence $AKA'$ will be projected into a right angle. So $BKB'$ will be projected into a right angle. Again, since $KA$ and $KA'$ are conjugate for the given conic, their projections will be conjugate for the conic which is the projection of the given conic. So $KB$ and $KB'$ will be conjugate in the figure obtained by projection. Again, $K$ is the pole for the given conic of the given line $l$ which is projected to infinity. Hence in the second figure, $K$ is the pole of the line at infinity, i.e. is the centre of the conic.

Hence in the second figure $KA$, $KA'$ and $KB$, $KB'$ are two pairs of orthogonal conjugate lines at the centre, i.e. the second conic has two pairs of orthogonal conjugate diameters. Hence the second conic is a circle.

2. The above construction fails when the line to be projected to infinity is the line at infinity itself. The problem then becomes—

*To project a given conic into a circle, so that the centre of the conic may be projected into the centre of the circle.*

If the conic is an ellipse, this can be done at once by Orthogonal Projection. If the conic is a hyperbola, we must use an imaginary Orthogonal Projection. If the conic is a parabola, the projection is impossible.

**Ex.** *Project a system of homothetic conics into circles.*

3. *To project a given conic into a circle and a given point into its centre.*

Take $K$ to be the given point and $l$ its polar.

*To project a given conic, so that one given point may be projected into the centre and another given point into a focus.*

To project $L$ into the centre and $K$ into a focus, take $l$ in the above construction to be the polar of $L$ instead of the polar of $K$, using $K$ and $l$ as before. Then $L$ is projected into the pole of the line at infinity, i. e. into the centre, and $K$ is projected into a point at which two pairs of conjugate lines are orthogonal, i. e. into a focus.

*To project a given conic, so that two given points may be projected into its foci.*

To project $K$, $K'$ into the foci. Take $L$ and $L'$, the double points of the involution $(PP', KK')$, $P$ and $P'$ being the points in which $KK'$ cuts the conic. Now project $K$ into a focus and $L$ into the centre. Then $(KK', LL')$ is harmonic; also $L'$ is at infinity, for since $(PP', LL')$ is harmonic, $L'$ is on the polar of $L$. Hence $KK'$ is bisected at $L$, i. e. $K'$ is the other focus.

**Ex. 1.** *Project a given conic in a given plane into a circle in another given plane.*

Take the line $AA'$ parallel to the intersection of the two planes, and take $V$ in the plane through $AA'$ parallel to the second plane.

**Ex. 2.** *Project a given conic into a parabola, and a given point into its focus, and a given point on the conic into the vertex of the parabola.*

Suppose we want to project $S$ into the focus, and $P$ into the vertex of a parabola. Let $SP$ cut the conic again in $P'$. Take the tangent at $P'$ as vanishing line.

**Ex. 3.** *Project a given conic into a rectangular hyperbola, and a given point into a focus.*

Let two conjugate lines at $S$ cut the conic in $P$ and $P'$. Take $PP'$ as vanishing line.

4. In the fundamental construction of § 1, if the point $K$ be outside the conic, the pencil of conjugate lines at $K$ is not overlapping; hence the segments $AA'$, $BB'$ do not overlap; hence the points $V$ and $V'$ are imaginary. In this case we say that the vertex of projection is imaginary, and that we can by an *imaginary projection* still project the conic into a circle and $l$ to infinity. Also by the Principle of Continuity proofs which require an imaginary projection are valid; in fact we need not pause to inquire whether the projection is real or whether it is imaginary.

*Prove Pascal's theorem by projection.*

See figure of XV. 1. Project $MN$ to infinity and the conic into a circle. Then in a circle we have $AB$ parallel to $DE$, and $BC$ parallel to $EF$. It follows by elementary geometry that $AF$ is parallel to $CD$. Hence in the original figure $L$ is on $MN$.

**Ex. 1.** *Prove by Projection that the harmonic triangle* (i) *of an inscribed quadrangle,* (ii) *of a circumscribed quadrilateral are self-conjugate for the conic.*

Project in each case into a parallelogram. Notice that a parallelogram inscribed in a circle must be a rectangle.

**Ex. 2.** $A, B, C, D$ *are four points on a conic. Show that the harmonic triangle of the quadrilateral $AB, BC, CD, DA$ is generally not self-conjugate.*

**Ex. 3.** *Show that the harmonic triangles of a quadrangle inscribed in a conic and of the quadrilateral of tangents at the vertices of the quadrangle are coincident.*

**Ex. 4.** $A, B, C, D, A', B', C', D'$ *are eight points on a conic. $AB, CD, A'B', C'D'$ are concurrent, and so are $BC, DA, B'C', D'A'$; show that $CA, DB, C'A', D'B'$ meet in a point, and that a conic can be drawn touching $A'A, B'B, C'C, D'D$ at $A, B, C, D$.*

**Ex. 5.** *The chords $PP'$, $QQ'$, $RR'$, $SS'$ of a conic meet in $O$. Show that the two conics $OPQRS$ and $OP'Q'R'S'$ touch at $O$.*

Project the conic into a circle and $O$ into its centre. Then the two conics are the reflexions of one another in $O$. Hence the tangents at $O$ coincide.

**Ex. 6.** *If two homologous triangles be inscribed in (or circumscribed to) a conic, the c. of h. is the pole of the a. of h.*
Project the polar of the c. of h. to infinity and the conic into a circle. Then in the new figure each triangle is the reflexion of the other in the centre. Hence the sides are parallel. Hence the a. of h. is at infinity; i. e. the a. of h. is the polar of the c. of h. Hence the same is true in the original figure.

**Ex. 7.** *Two homologous triangles are inscribed in (or circumscribed to) a conic; show that any transversal through the centre of homology cuts the sides in pairs of points in involution.*

**Ex. 8.** *Reciprocate Ex. 7.*

**Ex. 9.** *$A$ is a fixed point; $P$ is any point on its polar for a given conic; the tangents from $P$ meet a given line in $Q$, $R$. Show that the meets of $AR$, $PQ$ and of $AQ$, $PR$ lie on a fixed line.*
Project the conic into a circle and $A$ into its centre.

**Ex. 10.** *The lines joining the vertices of a triangle $ABC$ inscribed in a conic to a point $O$ meet the conic again in $a, b, c$; and $Ab$, $Bc$, $Ca$ meet the polar of $O$ in $R$, $P$, $Q$. Show that the lines joining any point on the conic to $P$, $Q$, $R$ meet $BC$, $CA$, $AB$ in collinear points.*

**Ex. 11.** *The lines $AB$ and $AC$ touch a conic at $B$ and $C$. The lines $PQ$ and $PR$ touch the conic at $Q$ and $R$. Show by Projection that the six points $A$, $B$, $C$, $P$, $Q$, $R$ lie on a conic. Through $A$ is drawn a line cutting the conic in $L$ and $M$ and cutting $QR$ in $N$, and a point $U$ is taken such that $(LM, NU) = -1$. Show that $U$ lies on the conic $ABCPQR$.*

**Ex. 12.** *If from three collinear points $X$, $Y$, $Z$ pairs of tangents be drawn to a conic, and if $ABC$ be the triangle formed by one tangent from each pair, and $DEF$ the points in which the remaining three tangents meet any seventh tangent, the lines $AD$, $BE$, $CF$ meet at a point on $XYZ$.*
Reciprocating, we have to prove the theorem—'If $AOA'$, $BOA'$, $COC'$ be chords of a conic, and $P$ any point on the conic, then the meets of $AB$, $PC'$, of $BC$, $PA'$, and of $CA$, $PB'$ lie on a line through $O$.' Project to infinity the line joining $O$ to the meet of $AB$, $PC'$, and at the same time the conic into a circle. The theorem becomes—'If $AA'$, $BB'$, $CC'$ be parallel chords of a circle and $P$ a point on the circle such that $PC'$ is parallel to $AB$, then $PB'$ is parallel to $CA$ and $PA'$ to $BC$.' This theorem follows by elementary geometry.

**Ex. 13.** *$ABC$ is a triangle inscribed in a conic of which $O$ is the centre. $OA'$, $OB'$, $OC'$ bisect $BC$, $CA$, $AB$. Through $P$, any point on the conic, are drawn lines parallel to $OA'$, $OB'$, $OC'$ meeting $BC$, $CA$, $AB$ in $X$, $Y$, $Z$; show that $X$, $Y$, $Z$ are collinear.*
By an Orthogonal Projection, real or imaginary, project the given conic into a circle with $O$ as centre. Then in the circle, $OA'$ is perpendicular to $BC$, $OB'$ to $CA$, and $OC'$ to $AB$. Hence the theorem becomes—'The feet of the perpendiculars drawn from any point situated on a circle upon the sides of a triangle inscribed in the circle are collinear.'

**Ex. 14.** *Reciprocate Ex. 13.*

**Ex. 15.** *Through a fixed point $O$ is drawn a chord $PP'$ of a conic; show that the locus of the middle point of $PP'$ is a homothetic conic through $O$ and through the points of contact of tangents from $O$.*

**5.** *To project any two given imaginary points into the circular points.*

Let the two imaginary points $E$, $F$ be given as the double points of the overlapping involution $(AA', BB')$. Take any point $K$ in the given plane and proceed as in § 1 to project the angles $AKA'$ and $BKB'$ into right angles and $AA'$ to infinity. Then $KE$ and $KF$ are the double points of the orthogonal involution $K(AA', BB')$, and $E$ and $F$ are at infinity; hence $E$ and $F$ are the circular points.

If $E$ and $F$ are real points, we can project them into the circular points by an imaginary projection; and proofs in which imaginary projection is employed are valid by the Principle of Continuity.

*To project any two imaginary lines into a pair of circular lines.*

Let the given lines $KE$, $KF$ be defined as the double lines of the involution $K(AA', BB')$. Draw any transversal $AA'BB'$. Then proceed as in § 1 to project the angles $AKA'$ and $BKB'$ into right angles. Then $KE$ and $KF$, being the double lines of an orthogonal involution, are circular lines.

*To project any conic into a rectangular hyperbola.*

Project any two conjugate points into the circular points.

*To project a system of angles which cut a given line in two homographic ranges, into equal angles.*

Project the common points into the circular points.

**Ex. 1.** *Deduce the construction for drawing a conic to touch three lines and to pass through two points from the construction for drawing a circle to touch three lines.*

**Ex. 2.** *The pole-locus of four given points $A$, $B$, $C$, $D$ and a given line $l$, touches the sixteen conics which can be drawn through the common conjugate points on $l$ to touch the sides of one of the triangles $ABC$, $ACD$, $ADB$, $BCD$.*

Project these conjugate points into the circular points; then $l$ goes to infinity. Also $AD$, $BC$ meet the line at infinity in points harmonic with the circular points; hence $AD$, $BC$ are perpendicular. Similarly $BD$, $AC$ are perpendicular, and also $CD$, $AB$. Also the pole-locus becomes the nine-point circle of each of the four triangles; and this is known to touch any circle which touches the sides of any one of the four triangles.

**6. *To project any two conics into circles.***

Project any two common points into the circular points, or project one conic into a circle and a common chord to infinity.

There are six solutions, as there are six common chords. But the projection is only real if we take a real common chord which meets the conics in imaginary points, for the line at infinity satisfies these conditions.

*To project a system of four-point conics into a system of coaxal circles.*

Proceed as above.

**Ex. 1.** *Points P, Q, R are taken on BC, CA, AB, and conics are described through AQRLM, BRPLM, CPQLM, where L, M are any two points. Show that these conics meet in a point.*
Project LM into the circular points.

**Ex. 2.** *Given two tangents and two points on a conic, the locus of the meet of the tangents at these points is two lines.*

**Ex. 3.** *Two conics pass through ABCD. AEF, BGH cut the conics in EG, FH; show that CD, EG, FH are concurrent.*

**Ex. 4.** *A variable conic passing through four fixed points A, B, C, D meets a fixed conic through AB in PQ; show that PQ passes through a fixed point.*

**Ex. 5.** *A, B, C, D are four fixed points on a fixed conic. BC, DA meet in F, and AB, CD meet in G. A variable conic through ACFG cuts the fixed conic again in PQ. Show that PQ passes through the pole of BD for the fixed conic.*

**Ex. 6.** *If a conic pass through two given points and touch a given conic at a given point, its chord of intersection with the given conic passes through a fixed point.*

**Ex. 7.** *On each side (UW) of the common self-conjugate triangle of two conics lie two common apexes (BB') and the two poles (PP' and QQ') of two common chords (bc and ad) of the conics. Also (PP', BB') and (QQ', BB') are harmonic.*

See figure of XIX. 8. $B, B'$ lie on $UW$ because $BB'$ and $UW$ are both sides of the self-conjugate triangle. $P, P'$ lie on $UW$ because $bc$ passes through $V$; so $Q, Q'$ lie on $UW$. Now project $bc$ into the circular points. Then $P$ and $P'$ are the centres of the circles, and $B$ and $B'$ are the centres of similitude. Hence $(PP', BB') = -1$. So by projecting $ad$ into the circular points, we prove that $(QQ', BB') = -1$.

**Ex. 8.** *Reciprocate Ex. 7.*

**Ex. 9.** *Of two circles, the poles of the radical axis and the centres of similitude form a harmonic range.*

**Ex. 10.** *If tangents be drawn from any point on any common chord of two conics, touching one conic in AB and the other in CD; show that the lines AC, AD, BC, BD meet two by two in the common apexes corresponding to the common chord.*

**Ex. 11.** *If through any common apex of two conics a line be drawn cutting the conics in the points AB and CD, at which the tangents are ab and cd; show that the points ac, ad, bc, bd lie two by two on the corresponding common chords.*

**Ex. 12.** *If the joins of any point on any common chord of two conics to the poles of this chord cut the conics in AB and CD; show that the lines AC, AD, BC, BD meet two by two in the common apexes corresponding to the common chords.*

**Ex. 13.** *If three conics have two points in common, the opposite common chords of the conics taken in pairs, are concurrent.*

**Ex. 14.** *If three conics have two points in common, the three pairs of common apexes corresponding to the chord lie three by three on four lines.*

**Ex. 15.** *Reciprocate Ex. 13 and Ex. 14.*

**Ex. 16.** *Two conics $a$ and $\beta$ meet at B, C, and touch at A. DEG touches $a$ at E and $\beta$ at G. DFH touches $a$ at F and $\beta$ at H. Show that EF, BC, GH meet (at K say) on the tangent at A, and that the poles of BC for $a$ and $\beta$ lie on DA and divide it harmonically. Show also that*

$$A(KD, BC) = D(AK, EF) = K(FH, AC) = -1.$$

**Ex. 17.** *The envelope of a line which meets two given conics in pairs of harmonic points is a conic which touches the eight tangents to the conics at their meets.*

Let the conics meet in $ABCD$. Project $AB$ into the circular points. Then by Ex. 2 of III. 6, the envelope of the line is a conic which touches the four tangents at $C$ and $D$. So by projecting $CD$ into the circular points, we prove that the envelope touches the tangents at $A$ and $B$.

**Ex. 18.** *Prove Ex. 17 by one projection.*

**Ex. 19.** *If the given conics be two parabolas with axes parallel, the envelope is a parabola with axis parallel to these axes.*

**Ex. 20.** *The locus of a point the tangents from which to two given conics are pairs of a harmonic pencil is a conic on which lie the eight points in which the given conics touch their common tangents.*

**Ex. 21.** *Two equal circles touch. Show that the locus of a point, the pairs of tangents from which to the circles are harmonic, is a pair of lines.*

For if the circles touch at $A$ and the common tangents touch them at $BC$, $DE$, the lines $BAE$, $CAD$ contain the eight points, four being at $A$.

**Ex. 22.** *If $SA$, $SA'$, $S'A$, $S'A'$ be the common tangents of two circles, $S$ and $S'$ being the centres of similitude, and if the angles at $A$ and $A'$ be right, show that the above locus breaks up into a pair of lines.*

For the four polars of the other two common apexes bisect the angles between $SS'$ and $AA'$.

**Ex. 23.** *The tangents to a system of four-point conics at their meets form four homographic pencils.*

**Ex. 24.** *Reciprocate Ex. 23.*

**Ex. 25.** *If two conics be so situated that two of their meets AB subtend at another meet C an angle which divides harmonically the tangents at C, the same is true for AB at D, for CD at A, and for CD at B.*

Apply Ex. 23 to the four conics consisting of the two given conics and the pair of lines $AC$, $BD$ and the pair $AD$, $BC$..

**Ex. 26.** *In such conics, the envelope of the lines which divide the two conics harmonically degenerates into two points.*

**Ex. 27.** *Reciprocate Ex. 25 and Ex. 26.*

**Ex. 28.** *Four parabolas are drawn with their axes in the same direction to touch the four triangles formed by four points; show that they have a common tangent.*

A particular case of the more general theorem—'Four conics are drawn to touch two given lines and to touch, &c.'

Reciprocate, and project the given points into the circular points.

**Ex. 29.** *A polygon is inscribed in one of a system of four-point conics, and each side but one touches a conic of the system; show that the remaining side also touches a conic of the system.*

For the theorem is true for coaxal circles by Poncelet's theorem.

**Ex. 30.** *Reciprocate Ex. 29; and deduce a property of confocal conics.*

**7.** *To project any two conics into confocal conics.*

Let the opposite vertices of the quadrilateral circumscribed to both conics be $AA'$, $BB'$, $CC'$. Project $AA'$ into the circular points; then the conics have the foci $BB'$, $CC'$ in common, i.e. are confocal.

*To project a system of conics inscribed in the same quadrilateral into confocal conics.*

Project a pair of opposite vertices of the circumscribing quadrilateral into the circular points.

**Ex. 1.** *A variable conic touches four fixed lines; from the fixed points $B$, $C$ taken on two of these lines the other tangents are drawn; find the locus of their meet.*

Project $BC$ into the circular points.

**Ex. 2.** *The line $PQ$ touches a conic. Find the locus of the meet of tangents of the conic which divide $PQ$ (i) harmonically, (ii) in a constant cross ratio.*

**Ex. 3.** *If a series of conics be inscribed in the same quadrilateral of which $AA'$ is a pair of opposite vertices, and from a fixed point $O$, tangents $OP$, $OQ$ be drawn to one of the conics, the conic drawn through $OPQAA'$ will pass through a fourth fixed point.*

Project $AA'$ into the circular points, and see Ex. 4. of XXVIII. 10.

**Ex. 4.** *Reciprocate Ex. 3.*

**Ex. 5.** *If two conics be inscribed in the same quadrilateral, the two tangents at any of their meets cut any diagonal of the quadrilateral harmonically.*

**Ex. 6.** *Given the cross ratio of a pencil, three of whose rays pass through fixed points and whose vertex moves along a fixed line, the envelope of the fourth ray is a conic touching the three sides of the triangle formed by the given points.*

XXIX.] *Projection, Real and Imaginary.* 269

**Ex. 7.** *The locus of the point where the intercept of a variable tangent of a central conic between two fixed tangents is divided in a given ratio is a hyperbola whose asymptotes are parallel to the fixed tangents.*

This is a particular case of the theorem—'If a tangent of a conic meet two fixed tangents $AB$, $AC$ in $P$, $Q$ and a fixed line $l$ in $U$, and if $R$ be taken such that $(PQ, RU)$ is constant; then the locus of $R$ is a conic through the meets $B$, $C$ of $l$ with the fixed tangents.' To prove this project $BC$ into $\infty \infty'$. Then we have to prove that—'If through the focus $S$ of a conic, a line $SR$ be drawn making a given angle with a variable tangent $QR$, then the locus of $R$ is a circle.' This can be proved by Geometrical Conics.

**8.** *To project any two conics into homothetic conics.*

Project any common chord to infinity. The new conics will pass through the same two points at infinity, and hence are homothetic. (See XIX. 11, end.)

*To project any two conics which have double contact into homothetic and concentric conics.*

Project the chord of contact to infinity. The pole of the chord of contact projects into the common centre.

**Ex.** *The point $V$ on a conic is connected with two fixed points $L$ and $M$. Show that chords of the conic which are divided harmonically by $VL$ and $VM$ pass through a fixed point $O$. Also as $V$ varies, the locus of $O$ is a conic touching the given conic at two points on the join of the fixed points $L$ and $M$.*

**9.** *To project any two conics having double contact into concentric circles.*

Project the two points of contact into the circular points. Then the conics will both pass through the circular points, i.e. will both be circles. Also they will both have the same pole of the line at infinity, i.e. they will be concentric.

**Ex. 1.** *Conics having the same focus and corresponding directrix can be projected into concentric circles.*

For the focus $S$ has the same polar, and the tangents from $S$ are the same. Hence the conics have double contact.

**Ex. 2.** *Through the fixed point $O$ is drawn a chord $OAB$ of a conic, and on $OAB$ is taken the point $P$ such that $(OABP)$ is constant. Find the locus of $P$.*

**10.** *The lines which join pairs of corresponding points of two homographic ranges on a conic, touch a conic having double contact with the given conic at the common points of the ranges.*

Let $(ABC...)$ and $(A'B'C'...)$ be the two homographic ranges, and $E$, $F$ their common points. Project the conic into a circle and the homographic axis $EF$ to infinity. Then $E$, $F$ are projected into the circular points.

Now in the second figure, $AB'$ and $A'B$ meet on the homographic axis. Hence $AB'$ and $A'B$ are parallel. So $AC'$ and $A'C$ are parallel, and so on. Hence the arcs $AA'$, $BB'$, $CC'$, ... are all equal. Hence the envelope of $AA'$ is a concentric circle, i.e. a circle having double contact with the circle which is the projection of the given conic at the circular points $E$, $F$. Hence in the original figure the envelope of $AA'$ is a conic having double contact with the given conic at the double points of the two homographic ranges.

**Ex. 1.** *Two conics have double contact, and a tangent to one conic meets the other conic in $A$ and $A'$. Show that $A$ and $A'$ generate homographic ranges, and find the common points of these ranges.*

**Ex. 2.** *If $(ABC..)$ and $(A'B'C'...)$ be two homographic ranges on a conic, show that the locus of the poles of $AA'$, $BB'$, ... is a conic having double contact with the given conic.*

**Ex. 3.** *The points of contact of the tangents $AA'$, $BB'$. $CC'$, ... form a range on the envelope homographic with the ranges $ABC...$ and $A'B'C'...$.*

**Ex. 4.** *Show that the tangents at $ABC...$ and $A'B'C'...$ cut the homographic axis in homographic ranges.*
For equal angles cut the line at infinity in homographic ranges.

**Ex. 5.** *If $O$ be the pole of the homographic axis of the two homographic ranges on a conic, then $O(ABC...) = O(A'B'C'...)$.*

**Ex. 6.** *If all but one of the sides of a polygon pass through fixed points and all the vertices lie on a conic, then the envelope of the remaining side is a conic having double contact with the given conic.*
For the last side determines homographic ranges on the conic.

**Ex. 7.** *If all but one of the vertices of a polygon move on fixed lines and all the sides touch a conic, the locus of the remaining vertex is a conic having double contact with the given conic.*

**Ex. 8.** *Two sides of a triangle inscribed in a conic pass through fixed points; show that the envelope of the third is a conic touching the given conic at the meets of the given conic with the join of the given points.*

**Ex. 9.** *A triangle $PQR$ is inscribed in a conic; $PQ$, $PR$ are in given direction; show that $QR$ envelopes a conic.*

**Ex. 10.** *The envelope of chords of a conic which subtend a given angle at a given point on the conic is a conic having double contact with the given conic.*

**Ex. 11.** *$A$, $B$ are two fixed points on a conic, and $P$, $Q$ two variable points on the conic such that $(AB, PQ)$ is constant; show that $PQ$ envelopes a conic which touches the given conic at $A$ and $B$.*

**Ex. 12.** *Show also that the locus of the meet of $AQ$ and $BP$, and the locus of the meet of $AP$ and $BQ$, are both conics having double contact with the given conic at $A$ and $B$.*
For $A(ABQ...) = B(ABP...)$ and $A(ABP...) = B(ABQ...)$.

**Ex. 13.** *Inscribe in a given conic a polygon of any given number of sides, each side of which shall touch some fixed conic having double contact with the given conic.*

**Ex. 14.** *If tangents be drawn from points on a conic to a conic having double contact with it, the points of contact generate homographic ranges on the conic.*

**Ex. 15.** *A conic is drawn through the common points $E$, $F$ of two homographic ranges $A$, $B$, $C$, ... and $A'$, $B'$, $C'$, ... on the same line. A pair of tangents moves so as to pass through a pair of points of these ranges. Show that the points of contact generate homographic ranges on the conic, whose common points are $E$ and $F$.*

**Ex. 16.** *Also if $P$ be any point, and $PA$ cut the conic in $a\alpha$, and $A'a$ cut the conic in $a'$; show that $aa'$ generate homographic ranges on the conic.*

**Ex. 17.** *Through a point $P$ is drawn a chord cutting a conic in $a\alpha$, and a point $a'$ is taken on the conic such that the angle $a\alpha a'$ is constant; show that $aa'$ generate homographic ranges.*

Here $AB...\ A'B'...$ is at infinity.

**Ex. 18.** *Reciprocate examples* 15, 16 *and* 17.

**Ex. 19.** *If two conics $\alpha$ and $\beta$ have double contact at the points $L$ and $M$; and through $LM$ be described any conic $\gamma$, then the opposite two common chords of $\alpha\gamma$ and $\beta\gamma$ meet on $LM$.*

**Ex. 20.** *Any angle whose legs pass through $L$ and $M$ respectively, intercepts chords on $\alpha$ and $\beta$ which meet on $LM$.*

A particular case of Ex. 19.

**Ex. 21.** *If two hyperbolas have the same asymptotes, any two lines parallel to the asymptotes intercept parallel chords of the hyperbola.*

**Ex. 22.** *Any two lines parallel to the asymptotes of a hyperbola intercept parallel chords on the hyperbola and its asymptotes.*

**Ex. 23.** *Reciprocate Ex.* 22.

**Ex. 24.** *If tangents at the two points $P$, $Q$ on one of two conics having double contact at $L$ and $M$ meet the other in $AB$ and $CD$, show that two of the chords $AC$, $AD$, $BC$, $BD$ meet $PQ$ on $LM$, and the other two meet $PQ$ in points $UV$ such that a conic can be drawn touching these chords at $U$ and $V$ and touching the conics at $L$ and $M$.*

**Ex. 25.** *Reciprocate Ex.* 24.

**Ex. 26.** *If a tangent to a conic meet a homothetic and concentric conic in $P$ and $P'$, show that $CP$ and $CP'$ generate homographic pencils whose common lines are the common asymptotes, $C$ being the common centre.*

# CHAPTER XXX.

### GENERALISATION BY PROJECTION.

**1.** In the previous chapter we have investigated theorems by projecting the given figure into the simplest possible figure. In this chapter we shall deal with the converse process, viz. of deriving from a given theorem the most general theorem which can be deduced by a projection, real and imaginary. This process is called *Generalising by Projection*.

In our present advanced state of knowledge of Pure Geometry, Generalisation by Projection is not a very valuable instrument of research. In fact the student will often find that it is more easy to prove the generalised theorem than the given theorem.

Many things are as general already as they can be. For instance, if we generalise by projection a point, a line, a conic, a harmonic range, a range having a given cross ratio, two conics having double contact, and so on, we obtain the same thing.

**2.** The properties of any figure have an intimate relation with the circular points $\infty, \infty'$. Hence the generalised figure will have an intimate relation with the projections of the circular points. But in the second figure there will also be a pair of circular points. Hence, to avoid confusion, we shall call the projections of the circular points $\varpi$ and $\varpi'$.

**3.** Since any two points can be projected into the circular points, *the circular points* generalise into any two points $\varpi$ and $\varpi'$, real or imaginary.

## Generalisation by Projection.

Since a pair of circular lines pass through the circular points, *a pair of circular lines* generalise into a pair of lines, one through $\varpi$ and one through $\varpi'$.

Since all circles pass through the circular points, *a circle* generalises into a conic which passes through $\varpi$ and $\varpi'$, where $\varpi$ and $\varpi'$ are any two points.

Since concentric circles touch one another at the circular points, *concentric circles* generalise into conics touching one another at $\varpi$ and at $\varpi'$.

Since the line at infinity touches a parabola, *a parabola* generalises into a conic touching the line $\varpi\varpi'$.

Notice that we cannot generalise the distinction between *a hyperbola* and *an ellipse*; for by an imaginary projection a pair of real points may be projected into a pair of imaginary points and vice versâ.

Since a rectangular hyperbola is a conic for which the circular points are conjugate, *a rectangular hyperbola* generalises into a conic for which $\varpi$, $\varpi'$ are a pair of conjugate points.

Since the centre of a conic is the pole of the line at infinity, *the centre of a conic* generalises into the pole of the line $\varpi\varpi'$.

Hence *a circle on AB as diameter* generalises into a conic passing through $AB\varpi\varpi'$, and such that the pole of the line $\varpi\varpi'$ is on $AB$.

Since parallel lines meet on the line at infinity, *parallel lines* generalise into lines which meet at a point on the line $\varpi\varpi'$.

Note that throughout this chapter, $\varpi$ and $\varpi'$ are any two points, real or imaginary.

4. If $B$ bisects the segment $AC$, then the range $(AC, B\Omega)$ is harmonic; hence '*B bisects AC*' generalises into 'If $AC$ meet $\varpi\varpi'$ in $I$, then $B$ is such that $(AC, BI)$ is harmonic, $\varpi$ and $\varpi'$ being any two points.'

*Generalise by Projection the theorem*—'*Given two concentric circles, any chord of one which touches the other is bisected at the point of contact.*'

T

The result is—'Given two conics touching one another at any two points $\varpi$ and $\varpi'$, if any chord $PP'$ of one, touch the other at $Q$ and meet $\varpi\varpi'$ in $I$, then $(PP', QI)$ is harmonic.'

Or, without mentioning $\varpi$ and $\varpi'$,—'Given two conics having double contact, if any chord $PP'$ of one, touch the other at $Q$ and meet the chord of contact in $I$, then $(PP', QI)$ is harmonic.'

The student should convince himself by trial that the second theorem can be projected into the first, and that the second theorem is the most general theorem which can be projected into the first.

Generalise by Projection the following theorems—

**Ex. 1.** *Given three concentric circles, any tangent to one is cut by the other two in four points whose cross ratio is constant.*

**Ex. 2.** *The middle points of parallel chords of a circle lie on a line which passes through the centre of the circle.*

**Ex. 3.** *If the directions of two sides of a triangle inscribed in a circle are given, then the envelope of the third is a concentric circle.*

**Ex. 4.** *Given four points on a conic, the locus of the centre is the conic through the middle points of the six sides of the quadrangle formed by the four given points.*

5. If $AVA'$ is a right angle, then $VA$ and $VA'$ divide the segment joining the circular points harmonically; hence a right angle $AVA'$ generalises into an angle $AVA'$, such that $VA$ and $VA'$ divide the segment joining any two points $\varpi$, $\varpi'$ harmonically.

*Generalise by Projection the theorem*—'*The perpendiculars to the sides of a triangle at the middle points of the sides meet at the centre of the circum-circle.*'

The result is—'If the sides $BC$, $CA$, $AB$ of a triangle meet the segment joining any two points $\varpi$ and $\varpi'$ in $L$, $M$, $N$; and if $X$, $Y$, $Z$ be taken such that $(\varpi\varpi', XL)$, $(\varpi\varpi', YM)$, $(\varpi\varpi', ZN)$ are harmonic; and if $D$, $E$, $F$ be taken such that $(BC, DL)$, $(CA, EM)$, $(AB, FN)$ are harmonic; then $DX$, $EY$, $FZ$ meet at the pole of $\varpi\varpi'$ for the conic which passes through $ABC\varpi\varpi'$.'

Generalise by Projection the following theorems—

**Ex. 1.** *A tangent of a circle is perpendicular to the radius to the point of contact.*

*Generalisation by Projection.*

**Ex. 2.** *The feet of the perpendiculars from any point on a circle on the sides of an inscribed triangle are collinear.*

**Ex. 3.** *The locus of the meet of perpendicular tangents of a conic is a concentric circle.*

**Ex. 4.** *The circle about any triangle self-conjugate for a conic is orthogonal to its director circle.*

**Ex. 5.** *The chords of a conic which subtend a right angle at a fixed point on the conic pass through a fixed point on the normal at the point.*

**Ex. 6.** *If a triangle PQR, right-angled at P, be inscribed in a rectangular hyperbola, the tangent at P is the perpendicular from P on QR.*

**6.** Since all circles pass through the circular points, *a system of circles* generalises into a system of conics passing through the same two points ($\varpi$ and $\varpi'$).

Since coaxal circles pass through the same four points of which two are the circular points, *coaxal circles* generalise into a system of conics which pass through the same four points (of which two are $\varpi$ and $\varpi'$).

Since the limiting points of a system of coaxal circles are the two vertices of the common self-conjugate triangle which lie on the line joining the poles of $\infty \infty'$, *the limiting points* generalise into the two vertices of the common self-conjugate triangle of a system of four-point conics which lie on the line joining the poles of any common chord ($\varpi \varpi'$), i.e. they generalise into any two vertices of the common self-conjugate triangle.

Since the centres of similitude of two circles are the two intersections of common tangents which lie on the line joining the poles of $\infty \infty'$ for the circles, *the centres of similitude of two circles* generalise into the two intersections of common tangents of two conics (through $\varpi$ and $\varpi'$) which lie on the line joining the poles of any common chord ($\varpi \varpi'$) for the conics, i.e. they generalise into any pair of opposite common apexes of two conics.

**7.** *Generalise by Projection the theorem*—'*Any common tangent of two circles subtends a right angle at either limiting point.*

The result is—'If $\varpi$ and $\varpi'$ be any two common points of two conics, and if $L$ and $L'$ be the two vertices of the common self-conjugate triangle which are collinear with the poles of $\varpi\varpi'$, then any common tangent of the conics subtends at $L$ (and at $L'$) an angle whose rays divide the segment $\varpi\varpi'$ harmonically.'

In other words,—'Any common tangent of two conics subtends at any vertex of the common self-conjugate triangle an angle which divides harmonically every common chord which does not pass through this vertex.'

Generalise by Projection the theorems—

**Ex. 1.** *Any transversal meets a system of coaxal circles in pairs of points in involution.*

**Ex. 2.** *The circle of similitude of two circles is coaxal with them.*

**8.** Since a focus of a conic is one of the four meets of the tangents from the circular points to the conic, *a focus of a conic* generalises into one of the meets of the tangents from any two points ($\varpi$ and $\varpi'$) to the conic.

*The two foci of a conic* generalise into a pair of opposite vertices of the quadrilateral of tangents from any two points ($\varpi$ and $\varpi'$).

Since the line joining the circular points touches a parabola, *the focus of a parabola* generalises into the meet of tangents from any two points ($\varpi$ and $\varpi'$) lying on any tangent of a conic.

Since confocal conics touch the same four tangents from the circular points (viz. $S\infty$, $S'\infty$, $S\infty'$, $S'\infty'$), *confocal conics* generalise into conics inscribed in the same quadrilateral (of which $\varpi$ and $\varpi'$ are a pair of opposite vertices).

Since conics which have the same focus $S$ and the same corresponding directrix $l$ touch $S\infty$, $S\infty'$, where $l$ meets these lines, *conics which have the same focus $S$ and the same corresponding directrix $l$* generalise into conics having double contact, the common tangents passing through $S$ (and through $\varpi$ and $\varpi'$), and touching the conics at points on $l$.

*A conic having $S$ as focus* generalises into a conic touching any two lines ($S\varpi$ and $S\varpi'$) through $S$.

**9.** Generalise by Projection the theorem—'*The circle which circumscribes a triangle whose sides touch a parabola passes through the focus of the parabola.*'

The result is—'The conic which passes through the points $A$, $B$, $C$, $\varpi$, $\varpi'$, where $\varpi$ and $\varpi'$ are any two points, and $A$, $B$, $C$ are the vertices of a triangle whose sides touch a conic which touches the line $\varpi\varpi'$, passes through the meet of tangents to the latter conic from $\varpi$ and $\varpi'$.'

In other words—'The conic, which passes through five out of the six vertices of two triangles which circumscribe a given conic, passes through the sixth also'.

**Ex. 1.** *Given two points on a conic, find the locus of the pole of their join, given also either (i) two tangents, or (ii) a tangent and a point.*

Generalise by Projection the following theorems—

**Ex. 2.** *Any line through a focus of a conic is perpendicular to the line joining its pole to the focus.*

**Ex. 3.** *Given a focus and two tangents of a conic, the locus of the other focus is a line.*

**Ex. 4.** *The locus of the centre of a circle which touches two given circles is a conic having the centres of the circles as foci.*

**Ex. 5.** *The locus of the centre of a circle which passes through a fixed point and touches a fixed line is a parabola of which the point is the focus.*

**Ex. 6.** *Confocal conics cut at right angles.*

**Ex. 7.** *The envelope of the polar of a given point for a system of confocals is a parabola touching the axes of the confocals and having the given point on its directrix.*

**10.** Since the rays of an angle of given size divide the segment joining the circular points in a given cross ratio, *a constant angle* generalises into an angle whose rays divide the segment joining any two points ($\varpi$ and $\varpi'$) in a constant cross ratio.

Generalise by Projection the theorem—'*The envelope of a chord of a conic which subtends a constant angle at a focus $S$ is another conic having $S$ as focus; and the two conics have the same directrix corresponding to $S$.*'

The result is—'The envelope of a chord of a conic which subtends at $S$, one of the meets of a tangent from any point $\varpi$ with a tangent from any point $\varpi'$, an angle whose rays divide $\varpi\varpi'$ in a constant cross ratio, is another conic, touching $S\varpi$ and $S\varpi'$; and the two conics have the same polar of $S$.'

In other words—'If $SQ$ and $SR$ be the tangents from any point $S$ to a conic, the envelope of a chord $PP'$ of the conic such that $S(QRPP')$ is constant, is a conic having double contact with the given conic at the points of contact of $SQ$ and $SR$.'

**Ex. 1.** *Generalise*—'*a regular polygon.*'

A regular polygon may be defined as a polygon which can be inscribed in a circle so that each side subtends the same angle at the centre of the circle.

Generalise by Projection the following theorems—

**Ex. 2.** *The envelope of a chord of a circle which subtends a given angle at any point of the circle is a concentric circle.*

**Ex. 3.** *If from a fixed point $O$, $OP$ be drawn to a given circle, and $TP$ be drawn making the angle $TPO$ constant, the envelope of $TP$ is a conic with $O$ as focus.*

**Ex. 4.** *If from a focus of a conic a line be drawn making a given angle with a tangent, the locus of the point of intersection is a circle.*

**Ex. 5.** *The locus of the intersection of tangents to a parabola which meet at a given angle is a hyperbola having the same focus and corresponding directrix.*

**11.** *Generalise*—'*The bisectors of an angle.*'

If $AD$, $AE$ are the bisectors of the angle $BAC$, then $A\,(BC, DE)$ is harmonic, and also $A\,(\infty\,\infty\,', DE)$ since $EAD$ is a right angle. Hence the bisectors of the angle $BAC$ generalise into the double lines of the involution

$$A\,(BC,\ \varpi\varpi'),$$

where $\varpi$ and $\varpi'$ are any two points.

**Ex.** *Generalise by Projection* — '*The pairs of tangents from any point to a system of confocals have the same bisectors.*'

**12.** *Generalise*—'*a segment divided in a given ratio.*'

Let $AB$ be divided at $C$ in a given ratio. Then $AC:CB$ is constant; hence $(AB, C\Omega)$ is constant, where $\Omega$ is the point at infinity upon $AB$. Hence *a segment $AB$ divided in a given ratio at $C$ generalises into a segment $AB$ divided at $C$ so that $(AB, CI)$ is constant, $I$ being the meet of $AB$ and the segment joining any two points ($\varpi$ and $\varpi'$).*

**Ex. 1.** *Generalise the equation* $AB + BC + CA = 0$ *connecting three collinear points.*

The given equation may be written
$$-(AC, B\Omega) + 1 - (AB, C\Omega) = 0.$$
This generalises into $-(AC, BI) + 1 - (AB, CI) = 0$, i.e. into
$$AB \cdot CI + AI \cdot BC + AC \cdot IB = 0.$$
Hence the generalised theorem is—'If $A, B, C, D$ be any four collinear points, then
$$AB \cdot CD + AC \cdot DB + AD \cdot BC = 0.'$$

**Ex. 2.** *If $ABCD$ be collinear, show that the ratio $AB \div CD$ generalises into* $-(BC, AE) \div (DA, CE)$.

**Ex. 3.** *If $AB$ and $CD$ be parallel and if $AC, BD$ meet in $M$, show that the ratio $AB \div CD$ generalises into $(AC, ME)$, $E$ being the meet of $AB$ and $CD$.*

**13.** *Two fixed points $A$ and $B$ on a conic are joined to a variable point $P$ on the conic, and the intercept $QR$ cut off from a given line $l$ by $PA$ and $PB$ is divided at $M$ in a given ratio; show that the envelope of $PM$ is a conic touching parallels to $l$ through $A$ and $B$.*

Let $\Omega$ be the point at infinity on $l$. Then $(QR, M\Omega)$ is a given cross ratio. Hence $P(AB, M\Omega)$ is given. Project $A$ and $B$ into the circular points and let $I$ be the projection of $\Omega$. Then $P(\infty \infty', MI)$ is given, i.e. $IPM$ is a given angle.

Hence the theorem becomes—'A fixed point $I$ is joined to a variable point $P$ on a circle, and $PM$ is drawn making a given angle with $IP$: show that the envelope of $PM$ is a conic touching $I\infty$ and $I\infty'$, i.e. is a conic having $I$ as focus.' And this is true (see VIII. 17). Hence the original theorem is true.

## Generalisation by Reciprocation.

**14.** If we first generalise a given theorem by projection and then reciprocate the generalised theorem, we obtain another general theorem. This process is called *Generalising by Projection and Reciprocation*, or briefly *Generalising by Reciprocation*.

*Generalise by Reciprocation the theorem*—'All normals to a circle pass through the centre of the circle.'

Generalising by Projection we get—'If $t$ be the tangent at any point $P$ of a conic which passes through any two points $\varpi$, $\varpi'$, and if the line $n$ be taken such that $t$ and $n$ are harmonic with $P\varpi$ and $P\varpi'$, then $n$ passes through the pole of $\varpi\varpi'$ for the conic.'

Reciprocating this theorem we get—'If on the tangent at any point $T$ of a conic, a point $N$ be taken such that the segment $TN$ is divided harmonically by the tangents from the fixed point $O$, then $N$ lies on the polar of $O$ for the conic.' This is the required theorem.

**Ex.** *Generalise by Projection and Reciprocation the theorem—'The envelope of a chord of a circle which subtends a constant angle at the centre is a concentric circle.'*

# CHAPTER XXXI.

## HOMOLOGY.

1. Two *figures* in the same plane are said to be *in homology* which possess the following properties. To every point in one figure corresponds a point in the other figure, and to every line in one figure corresponds a line in the other figure. Every two corresponding points are collinear with a fixed point called *the centre of homology*, and every two corresponding lines are concurrent with a fixed line called *the axis of homology*. The line joining any two points of one figure corresponds to the line joining the two corresponding points of the other figure. The point of intersection of any two lines of one figure corresponds to the point of intersection of the two corresponding lines of the other figure.

The two figures are said to be homologous, and each is called the *homologue* of the other. The figures may be said to be in plane perspective; and the centre of homology is then called the centre of perspective, and the axis of homology is called the axis of perspective.

2. Homologous figures exist for—

*If we take two figures in different planes, each of which is the projection of the other, and if we rotate one of the figures about the meet of the two planes until the planes coincide, then the figures will be homologous.*

For let $ABC$ be the projections of $A'B'C'$ from the vertex $V$. Then $AA'$, $BB'$, $CC'$ meet in $V$. That is, the triangles $ABC$, $A'B'C'$ (in different planes) are copolar. Hence they are coaxal; i.e. $BC$, $B'C'$ meet in $a$, and $CA$, $C'A'$ meet in $\beta$, and $AB$, $A'B'$ meet in $\gamma$ on the meet of the two planes. Similarly every two lines which are the projections, each of the other, meet on the intersection of the two planes.

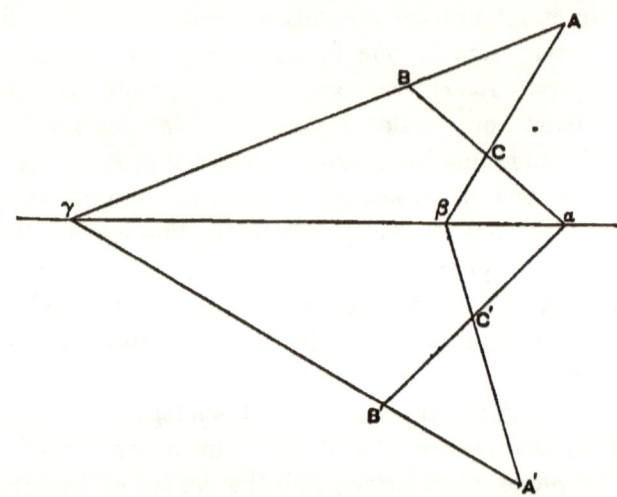

Now rotate one figure about the line $\alpha\beta\gamma$ until the two figures are in the same plane. Then the two triangles are still coaxal (for $BC$, $B'C'$ still meet at $a$, and so for the rest). Hence the two triangles are also copolar; i.e. $AA'$, $BB'$, $CC'$ meet in a point. Call this point $O$. Then $O$ may be defined as the meet of $AA'$ and $BB'$, and we have proved that every other line such as $CC'$ passes through $O$.

Now the two figures are in the same plane. Also to every point in one figure corresponds a point in the other figure, viz. the point which was its projection; and to every line corresponds a line, viz. its former projection. Also, corresponding points are concurrent with a fixed point $O$, and corresponding lines are collinear with a fixed line $\alpha\beta\gamma$. Also, the join of two points corresponds to the join of the corresponding points; for in the former figure the one is the projection of the other. For the same reason, the meet of two lines corresponds to the meet of the corresponding lines.

Hence the two figures are homologous.

## Homology.

**3.** *If two figures are homologous, and we turn one of them about the axis of homology, the figures will be the projections, each of the other.*

For suppose the three lines $BC$, $CA$, $AB$ in one figure to be homologous to $B'C'$, $C'A'$, $A'B'$ in the other figure. Let $BC$, $B'C'$ meet in $a$, let $CA$, $C'A'$ meet in $\beta$, and let $AB$, $A'B'$ meet in $\gamma$. Rotate one of the figures about the axis of homology $a\beta\gamma$, so that the figures may be in different planes. The figures will now be each the projection of the other.

For the triangles $ABC$, $A'B'C'$ (in different planes) are coaxal; hence they are copolar. Hence $AA'$, $BB'$, $CC'$ meet in a point $V$. This point $V$ may be defined as the meet of $AA'$ and $BB'$; and we have proved that in the displaced position the join $CC'$ of any two homologous points passes through a fixed point $V$. Hence the homologous figures in the displaced position are projections, each of the other.

*A homologue of a conic is a conic.*

For after rotating one figure about the axis of homology, the figures are each the projection of the other; and the projection of a conic is a conic.

*A homologue of a figure has all the properties of a projection of the figure.*

For it can be placed so as to be a projection of the figure.

Hence *a range and the homologous range are homographic; also a pencil and the homologous pencil are homographic.*

**4.** *If one of two figures in perspective (i.e. either homologous or each the projection of the other), be rotated about the axis of perspective, the figures will be in perspective in every position; and the locus of the centre of perspective is a circle.*

For take any two corresponding triangles $ABC$ and $A'B'C'$. Then in every position these triangles will remain coaxal; hence in any position they will be copolar, i.e. $CC'$ will pass through the fixed point $V$ determined as the meet of $AA'$ and $BB'$. Hence the figures will be in perspective in any position obtained by rotating one figure about the axis of perspective.

To find the locus of $V$. Take any position of $V$, and

through $V$ draw a plane $P'LP$ at right angles to the planes of the figures, cutting them in $LP'$ and $LP$.

Let a parallel to $LP'$ through $V$ cut $LP$ in $J$, and a parallel to $LP$ through $V$ cut $LP'$ in $I'$. Let the point at infinity on $LP$ be called $I$, and the point at infinity on $LP'$ be called $J'$.

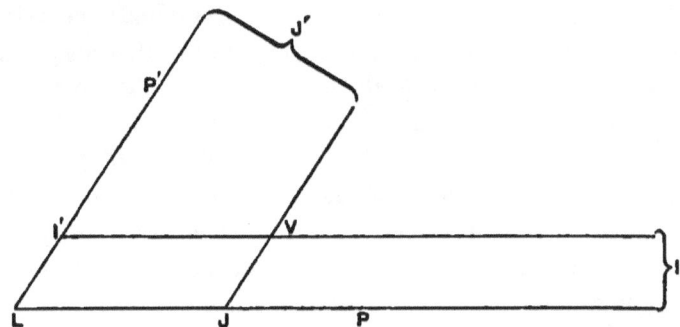

Then, since $I'V$ and $LJ$ are parallel, we see that $I'V$ passes through $I$, i.e. $I'$ is the projection of $I$ for this position of $V$; and so $J$ is the projection of $J'$.

Now rotate the moving plane about the axis of perspective into any other position. The new position of the centre of perspective (or vertex of projection) is got by joining any two pairs $AA'$, $BB'$ of corresponding points. Hence in the new position $II'$ and $JJ'$ will cut in $V$. Also $LJ$ is still parallel to $I'V$, for $I$ is at infinity; so $JV$ is parallel to $LI'$. Also if $LJ$ is the trace on the fixed plane, then $LJ$ is constant in magnitude and position. Also $LI'$ is constant in magnitude, although it changes its position by rotation about the axis of perspective. It follows that $LJVI'$ is a parallelogram, in which $J$ is fixed, and $JV$ is given in magnitude. Hence the locus of $V$ is a circle in a plane perpendicular to the planes of the figures, with centre $J$ and radius $LI'$.

To form a clear conception of figures in homology, imagine that they are the projections, each of the other, the vertex of projection very nearly coinciding with the centre of homology, and the planes of the figures very nearly coinciding with one another.

**5.** *Coaxal figures are copolar, and copolar figures are coaxal; that is to say, if two figures, (in the same plane or not,) correspond, point to point, line to line, meet of two lines to meet of corresponding lines, and join of two points to join of corresponding points; then, if corresponding lines cut on a fixed line, the joins of corresponding points will pass through a fixed point, and if the joins of corresponding points pass through a fixed point, corresponding lines will cut on a fixed line.*

*Coaxal figures are copolar.* Take two fixed points $A$, $B$ in one figure, and let $A'$, $B'$ be the corresponding points in the other figure. Take any variable point $P$ in one figure, and let $P'$ be the corresponding point in the other figure. Then, by definition, $AP$, $A'P'$ are corresponding lines, for they join corresponding points; hence $AP$ and $A'P'$ meet on the axis. Similarly $BP$, $B'P'$ meet on the axis; and $AB$, $A'B'$ meet on the axis. Hence the triangles $ABP$, $A'B'P'$ are coaxal, and therefore copolar. Hence $AA'$, $BB'$, $PP'$ meet in a point, i.e. $PP'$ passes through a fixed point, viz. the meet of $AA'$ and $BB'$.

*Copolar figures are coaxal.* Take two fixed lines, viz. $AP$ and $AQ$, and a variable line $PQ$ in one figure, and let $A'P'$, $A'Q'$, $P'Q'$ be the corresponding lines in the other figure. Then the points $A$, $P$, $Q$ correspond to $A'$, $P'$, $Q'$. Hence the triangles $APQ$, $A'P'Q'$ are copolar, and therefore coaxal. Hence $PQ$, $P'Q'$ meet on a fixed line, viz. the join of the meets of $AP$, $A'P'$ and of $AQ$, $A'Q'$. Hence the figures are coaxal.

**Ex. 1.** *If one of two figures in homology be turned through two right angles about the axis of homology, the figures will again be in homology.*

**Ex. 2.** *If one of two figures in homology be turned through two right angles about an axis which passes through the centre of homology and is perpendicular to the plane of the figures, the figures will again be in homology.*

**Ex. 3.** *Given two homologous figures $ABC...$, $A'B'C'...$; let $A''B''C''...$ be a projection of $ABC...$ on any plane through the axis of homology; then will $A''B''C''...$ be also a projection of $A'B'C'...$, and the vertices of projection and the centre of homology will be collinear.*

For $VO$ is one of the lines $A'A''$, &c.

This construction enables us to place any two homologous figures in projection with the same figure.

**Ex. 4.** *Show that the two complete quadrangles determined by $ABCD$ and $A'B'C'D'$ will be homologous provided the five points of intersection of $AB$ with $A'B'$, of $BC$ with $B'C'$, of $CA$ with $C'A'$, of $AD$ with $A'D'$, and of $BD$ with $B'D'$ are collinear.*

**Ex. 5.** *Show that the two complete quadrilaterals whose vertices are $ABCDEF$ and $A'B'C'D'E'F'$ will be homologous if $AA'$, $BB'$, $CC'$, $DD'$, $EE'$ meet in a point.*

**Ex. 6.** *The sides $PQ$, $QR$, $RP$ of a variable triangle pass through fixed points $CAB$ in a line. $Q$ moves on a fixed line. Show that $P$ and $R$ describe homologous curves.*

For $PR$ and $P'R'$ pass through $B$, and $RR'$, $PP'$ meet on $QQ'$, $P'Q'R'$ being a second position of $PQR$.

**Ex. 7.** *If the axis of homology be at infinity, show* (i) *that corresponding lines are parallel,* (ii) *that corresponding sides of the figures are proportional,* (iii) *that corresponding angles of the figures are equal.*

Such figures are called *homothetic figures*, and the centre of homology in this case is called *the centre of similitude*, and the constant ratio of corresponding sides is called *the ratio of similitude*.

**Ex. 8.** *If, with any vertex of projection, we project homologous figures on to any plane, we obtain homologous figures; and if the plane of projection be taken parallel to the plane containing the vertex of projection and the axis of homology, we obtain homothetic figures.*

Hence homologous figures might have been defined as the projections of homothetic figures.

**Ex. 9.** *If the centre of homology be at infinity, show that the joins of corresponding points are all parallel; and that if one figure be rotated about the axis of homology, the vertex of projection will always be at infinity.*

This may be called *parallel homology*.

**Ex. 10.** *In parallel homology, show that to a point at infinity corresponds a point at infinity, and that the line at infinity corresponds to itself.*

**Ex. 11.** *In parallel homology, show that a parallelogram corresponds to a parallelogram.*

**Ex. 12.** *In parallel homology, show that, when rotated about the axis of homology into different planes, the figures have the same orthogonal projection; and that the ratios of two areas is the same as that of the corresponding areas.*

**6.** The abbreviation c. of h. will be used for centre of homology, and a. of h. for axis of homology.

*Given the c. of h. and the a. of h. and a pair of corresponding points, construct the homologue of a given point.*

Let $O$ be the c. of h., and let $A'$ be the given homologue of $A$. To find the homologue of $X$; let $AX$ cut the a. of h. in $L$, then $LA'$ cuts $OX$ in the required point $X'$.

*With the same data, construct the homologue of a given line.*

Draw through $O$ any transversal cutting the given line in $X$; construct the homologue $X'$ of $X$, then the join of $X'$ to the point $M$, where the given line cuts the a. of h., is the homologue of the given line.

*Given the c. of h. and the a. of h. and a pair of corresponding lines, construct the homologue (i) of a given point, (ii) of a given line.*

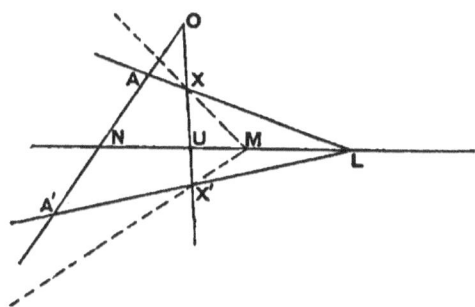

Let any transversal through $O$ cut the given lines in $A$ and $A'$. Then $A$, $A'$ are corresponding points, and we may proceed as above.

*Given the c. of h. and the a. of h. and a pair of corresponding points, one of which is at infinity, construct the homologue of a given point.*

$LX'$ is parallel to $AA'$, if $A'$ is at infinity.

**7.** *The homologue of the c. of h. is the c. of h.; the homologue of any point on the a. of h. is the point itself; if the homologue of any other point be itself, then the homologue of every point is itself.*

For let us construct the homologue of $O$. We draw $AO$ cutting the a. of h. in $N$; we draw $NA'$ cutting $OO$ in the required point. Now $OO$ is indeterminate, but $NA'$ cuts every line through $O$ in $O$, and hence cuts $OO$ in $O$. Hence the homologue of $O$ is $O$.

Next, let us construct the homologue of any point $L$ on the a. of h. We draw $AL$ cutting the a. of h. in $L$; we draw $A'L$ cutting $OL$ in the required point. Hence the homologue of $L$ is $L$.

Lastly, suppose a point (which is not at the c. of h. nor on the a. of h.) to coincide with its homologue. Take these as the points $A$, $A'$ in the above construction. To construct the homologue of $X$, we draw $AX$ cutting the a. of h. in $L$; then $A'L$ cuts $OX$ in the required point $X'$. Hence $X'$ coincides with $X$, for $A'L$ coincides with $AL$, i.e. with $AX$.

**Ex. 1.** *Show that the only lines which coincide with their homologues are the a. of h. and lines through the c. of h.*

**Ex. 2.** *Given the homologues $A'$, $B'$, $C'$ of three points $A$, $B$, $C$; construct the homologue of a given point $D$.*

The triangles give the centre and axis of homology.

**8.** The homologue of a point at infinity of one figure is called *a vanishing point* of the homologous figure.

The homologue of the line at infinity considered as belonging to one of the figures is called *the vanishing line* of the homologous figure.

*All the vanishing points of either figure lie on the vanishing line of that figure.*

For a vanishing point is the homologue of a point on the line at infinity of the other figure, and hence lies on the homologue of the line at infinity.

*Each vanishing line is parallel to the a. of h.*

For corresponding lines meet on the a. of h. Hence a vanishing line meets the a. of h. at a point on the line at infinity, i.e. a vanishing line is parallel to the a. of h.

**Ex. 1.** *If any transversal through $O$ cut the axis in $N$, and the vanishing lines in $I$ and $J'$, then $OI = J'N$, and $OI \cdot OJ' = J'N \cdot IN$.*

For $(NI, O\Omega) = (N\Omega', OJ')$.

**Ex. 2.** *The product of the perpendiculars from any two homologous points, each on the vanishing line of its figure, is constant.*

For $(PQ, I\Omega) = (P'Q', \Omega'J')$.

**Ex. 3.** *Given a parallelogram $ABCD$, prove the following construction for drawing through a given point $E$ a parallel to a given line $l$—Let $AB$, $CD$, $AC$, $BC$, $AD$ cut $l$ in $K$, $L$, $M$, $N$, $R$. Through $M$ draw any line cutting $EK$, $EL$ in $A'$, $C'$. Let $RA'$ and $NC'$ cut in $F$. Then $EF$ is parallel to $l$.*

For $EF$ is the vanishing line.

**9.** *Given the c. of h., the a. of h., and a pair of corresponding points; construct the vanishing lines.*

Let $AA'$ be the pair of corresponding points. Let us first construct the homologue of the line at infinity, considered to belong to the same figure as $A$. In the construction of

§ 6, $X$ and $M$ are both at infinity. Hence the construction is—Through the c. of h. $O$, draw any line $OX$ ($X$ being at infinity). Through $A$ draw $AL$ parallel to $OX$, cutting the a. of h. in $L$; then $LA'$ cuts $OX$ in $X'$; and the required line is $X'M$, i.e. a parallel through $X'$ to the a. of h.

Similarly we construct the vanishing line of the other figure.

*Given the c. of h., the a. of h., and one vanishing line, to construct the homologue of a given point.*

Let any transversal through the c. of h. cut the vanishing line in $A$. Then the homologue of the point $A$ is the point at infinity $A'$ on $OA$.

Two cases arise. (i) The given point $X$ belongs to the same figure as the finite point $A$. Let $AX$ cut the a. of h. in $L$; draw through $L$ a parallel to $OA$ to cut $OX$ in $X'$. Then $X'$ is the homologue of $X$. (ii) The given point $X'$ belongs to the same figure as the point at infinity $A'$. Through $X'$ draw a parallel to $OA$ cutting the a. of h. in $L$. Then $AL$ cuts $OX'$ in the required point $X$.

**Ex.** *Given the c. of h. and the a. of h. and one vanishing line, construct the other vanishing line.*

10. *The angle between two lines in one figure is equal to the angle subtended at the c. of h. by the vanishing points of the homologous lines.*

Let $AP$ and $AQ$ be the given lines, $P$ and $Q$ being at infinity. Then $P'$ and $Q'$ are the vanishing points of the homologous lines $A'P'$ and $A'Q'$. Also $OP'$ is parallel to $AP$, and $OQ'$ to $AQ$. Hence the angles $P'OQ'$ and $PAQ$ are equal.

11. *Construct the homologue of a given conic, so that the homologue of a given point $S$ shall be a focus.*

Take any line as a. of h., and any parallel line as vanishing line; and let two conjugate lines at $S$ meet the vanishing line in $P$ and $Q$, and let two other conjugate lines at $S$ meet it in $U$ and $V$. On $PQ$ and $UV$ as diameters describe circles, and take either of the intersections of these circles as c. of h.

Then since the vanishing points $P$ and $Q$ of the lines $SP$ and $SQ$ subtend a right angle at the c. of h., the homologues $S'P'$, $S'Q'$ will be at right angles. So $S'U'$, $S'V'$ will be at right angles. Hence at $S'$ we shall have two pairs of conjugate lines at right angles. Hence $S'$ is a focus of the homologous conic.

**12.** *The homologue of a conic, taking a focus as c. of h. and the corresponding directrix as vanishing line and any parallel as a. of h., is a circle, of which the focus is the centre.*

Let $S$ be the given focus and $XM$ the corresponding directrix. With $S$ as c. of h. and $XM$ as vanishing line, and any parallel line as a. of h., describe a homologue of the given conic. The homologue of $S$ is $S$, and of $XM$ is the line at infinity; hence in the homologous conic, $S$ is the pole of the line at infinity, i.e. $S$ is the centre of the homologous conic.

Let $SP$, $SP'$ be a pair of conjugate diameters of the homologous conic. The homologue of $SP$ is $SP$, the homologue of $SP'$ is $SP'$; and the homologues of conjugate lines are conjugate lines. Hence in the given figure, $SP$ and $SP'$ are conjugate lines; and $S$ is the focus, hence $SP$ and $SP'$ are perpendicular. Hence every pair $SP$, $SP'$ of conjugate diameters of the homologous conic is orthogonal. Hence the homologous conic is a circle. And we have already proved that the focus is the centre of the circle.

Note that the homologue of an angle at $S$ is an equal (in fact, coincident) angle at $S$.

This case of homology is the limit of Focal Projection when the two figures are in the same plane.

**Ex. 1.** *Any homologue of a conic, taking a focus $S$ as c. of h., is a conic with $S$ as focus; and the homologue is a circle only if the vanishing line is the corresponding directrix.*

**Ex. 2.** *Any homologue of a conic, taking the polar of a given point $P$ as vanishing line, is a conic with $P$ as centre; and the homologue is a circle only if $P$ is a focus of the given conic.*

**13.** *If two curves be in homology, the c. of h. must be a meet of common tangents, and the a. of h. must be a join of common points.*

For let $OT$ be a tangent from the c. of h. $O$ to one of the curves. Let $OPQ$ be a chord of the curve very near $OT$. Then $OPQ$ meets the homologous curve in the homologous points $P'$, $Q'$. Now let $P$ and $Q$ coincide in $T$; then $P'$ and $Q'$ also coincide, in $T'$ the homologue of $T$. Hence $OT$ touches the homologous curve.

Again, let $L$ be one of the points where one of the curves cuts the a. of h. Then $L$, being on the a. of h., is its own homologue. Hence the homologous curve passes through $L$.

Hence if two curves are in homology, the c. of h. must be looked for among the meets of common tangents; and the a. of h. must be looked for among the joins of common points.

**14.** *Any two circles are homologous in four real ways.*

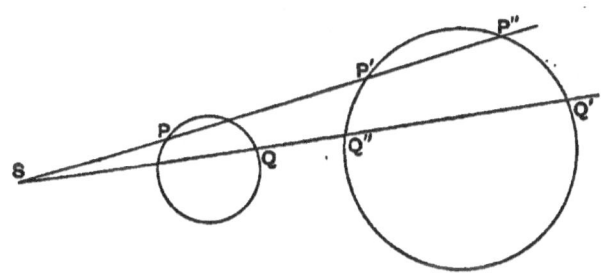

Let $S$ be either of the centres of similitude of the two circles. Take any point $P$ on one circle, and let $SP$ cut the other circle in $P'$ and $P''$. Then one of these points, and only one (viz. $P'$ in the figure), possesses the property that $SP : SP'$ is the ratio of the radii. We may call $P$, $P'$ similar points, and $P$, $P''$ non-similar points.

*If we take either centre of similitude as centre of homology and the straight line at infinity as axis of homology, then the circles are homologous, each point being homologous to its similar point.*

For take any two pairs of similar points, viz. $P$, $P'$ and $Q$, $Q'$. Then $SP:SP'::SQ:SQ'$; hence $PQ$ is parallel to $P'Q'$, i.e. every chord joining two points on one circle is parallel to the chord joining the similar points on the other circle. Hence the two circles are homologous, the straight line at infinity being the axis of homology, and similar points being homologous points.

*If we take either centre of similitude as centre of homology and the radical axis as the axis of homology, then the circles are homologous, each point being homologous to its non-similar point.*

For take any two pairs of non-similar points, viz. $P$, $P''$ and $Q$, $Q''$. Then $SP:SP'::SQ:SQ'$, and
$$SP'.SP'' = SQ'.SQ''.$$
Hence $SP.SP'' = SQ.SQ''$. Hence $PP''QQ''$ are concyclic; hence, if $PQ$, $P''Q''$ meet in $X$, we have
$$XP.XQ = XP''.XQ'',$$
i.e. $X$ has the same power for both circles, i.e. $X$ is on the radical axis of the circles. Hence we have proved that the chord joining any two points on one circle and the chord joining the non-similar points on the other circle meet on the radical axis of the circles, which is therefore the axis of homology.

Hence, since with either centre of similitude we may take the straight line at infinity or the radical axis, the circles are in homology in four real ways.

**15.** *Two conics which have double contact are homologous in two ways, the c. of h. being the common pole and the a. of h. the common polar in both cases.*

Let $O$ be the common pole and $MN$ the common polar, the points $M$ and $N$ being on both conics. Let any line through $O$ cut one conic $a$ in $A$, $D$ and the other conic $\beta$ in $B$, $C$. Then $a$ is determined by the five points $AMMNN$, the points $MM$ being on $OM$ and the points $NN$ being on $ON$. Now form the homologue of $a$, taking $O$ as c. of h., $MN$ as

a. of h., and $B$ as the homologue of $A$. The homologue of a conic is a conic. The homologues of the points $AMMNN$ are the points $BMMNN$.
Hence the homologue of $a$ is the conic through $BMMNN$, i.e. is the conic $\beta$.

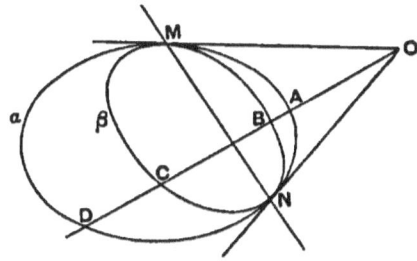

Again, with the same c. of h. and a. of h., but with $C$ as the homologue of $A$, form the homologue of $a$. The homologue is now the conic through $CMMNN$, i.e. is the conic $\beta$.

Now in the first case $C$ and $D$ are homologous, and in the second case $B$ and $D$ are homologous. So there are not four ways, but two ways, in which the conics are homologous.

In the first way, every point $P$ on $a$ is homologous with the point $P'$ in which $OP$ cuts $\beta$ on the same side of $MN$ as $P$; and in the second way, every point $P$ on $a$ is homologous with the point $P''$ in which $OP$ cuts $\beta$ on the opposite side of $MN$ to $P$.

**Ex. 1.** *Prove the theorem directly by showing that the figures are coaxal.*

**Ex. 2.** *In the above figure, show that $(OD, AB)$ is a constant cross ratio as the chord $OABD$ moves round $O$.*

For taking another chord $OA'B'D'$, then $AA'$, $BB'$, $DD'$ meet on $MN$.

**Ex. 3.** *Through $O$, the common pole of two conics having double contact, are drawn four chords cutting one conic in $ABCD$ and the other in $A'B'C'D'$; show that $(ABCD) = (A'B'C'D')$, if all the points lie on the same side of the common polar.*

**Ex. 4.** *A conic is its own homologue, any point and its polar being c. of h. and a. of h.*

**Ex. 5.** *Give a direct proof of Ex. 4 by means of the quadrangle construction for the polar of a given point.*

**Ex. 6.** *If a conic be its own homologue, show that if the c. of h. be given, the a. of h. must be the polar of the c. of h.*

**Ex. 7.** *Through any point $O$ are drawn the four chords $OAA'$, $OBB'$, $OCC'$, $ODD'$ of a conic; show that the conics $OABCD$ and $OA'B'C'D'$ touch at $O$.*

The given conic is its own homologue, $O$ being the c. of h.; also $O$ is its own homologue. Hence the five points $O, A, B, C, D$ are homologous to the five points $O, A', B', C', D'$; hence the conics through them are homologous.

Take a chord of these conics through $O$, viz. $OPP'$. Then when $P$ coincides with $O$, so does $P'$. Hence $OPP'$ ultimately touches both conics at $O$, i.e. the conics touch at $O$.

**Ex. 8.** *Tangents from $P$ to a conic meet any line $l$ in $L$, $M$, and the other tangents from $L$, $M$ meet in $P'$; show that $P$, $P'$ generate homologous figures, $l$ being the a. of h.*

**16.** *Any two conics are in homology.*

Take any meet $O$ of common tangents $TT'$, $UU'$ as c. of h.

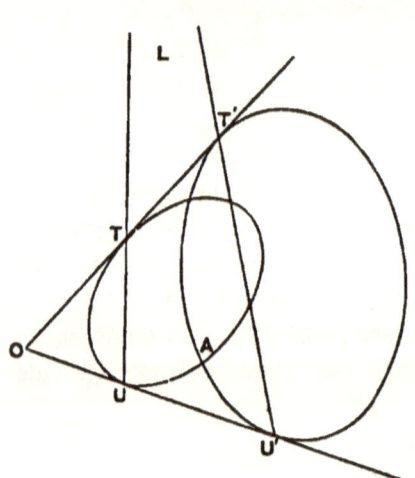

Let $TU$ and $T'U'$, the polars of $O$, cut in $L$. Let $A$ be one of the four common points of the two conics. Take $LA$ as a. of h. Also take $UU'$ as a pair of corresponding points.

The homologue of the conic $TUA$ can now be found. Suppose the conic $TUA$ to be given by the five points $TTUUA$, where $TT$ are the coincident points in which $OT$ touches the conic, and $UU$ are the coincident points in which $OU$ touches the conic. The homologue of $A$ is $A$, for $A$ is on the a. of h. The homologues of $UU$ are $U'U'$ by hypothesis. Again, since $TT'$ passes through $O$, and $TU$, $T'U'$ meet on the a. of h., hence $T'$ is the homologue of $T$, i.e. $T'T'$ are the homologues of $TT$. Hence the homologues of $TTUUA$ are $T'T'U'U'A$. Hence the homologue of the conic $TUA$ is a conic passing through $A$ and touching $OT'$ at $T'$ and touching $OU'$ at $U'$; i.e. the homologue of one given conic is the other given conic.

Hence *two conics are homologous in twelve ways.*

For we may take as c. of h. $O$ any one of the six meets of common tangents of the two conics. We may then take as $A$ any one of the four common points of the two conics. But this will only give us two possible axes of h. For the point where $LA$, the a. of h., meets either conic again will be another common point. Hence there are only two positions of $LA$, when the position of $O$ has been chosen.

**Ex. 1.** *Show by using the reciprocal solution to the above that we may take any common chord of the conics as a. of h.*

**Ex. 2.** *Two conics in different planes may be placed in projection by two rotations, viz.* (i) *about the meet of the planes until the planes coincide, and* (ii) *about any common chord of the conics (when placed in one plane).*

**Ex. 3.** *Show that two homothetic conics have two centres of similitude.*
Viz. the common apexes belonging to the line at infinity.

**Ex. 4.** *Show by using the circular points that any two conics which have a common focus are in homology, the common focus being the c. of h.*

**Ex. 5.** *Any conic is homologous with any circle whose centre is at a focus of the conic, the focus being the c. of h.*

**17.** *If two conics touch, they are homologous, taking the point of contact as c. of h.*

This follows as a limiting case of the general theorem; *or thus directly.* Through $O$, the point of contact, draw any chord cutting one conic in $P$ and the other in $P'$. Take $O$ as c. of h., and the common chord $AB$, which does not pass through $O$, as a. of h. Also take $P$, $P'$ as homologous points. Consider the homologue of the conic determined by $OOPAB$. It is a conic through $OOP'AB$, i.e. it is the other conic.

**Ex. 1.** *Find the envelope of a chord of a conic subtending a given angle at a given point on the conic.*
Draw any circle touching the given conic at the given point and passing through any other point on the conic. This circle is homologous to the given conic; and the homologous chord clearly envelopes a concentric circle. Hence the given chord envelopes the homologue of a concentric circle, i.e. a conic having imaginary double contact with the given conic.

**Ex. 2.** *Obtain by homology the theorem:—The envelope of a chord of a conic which subtends a right angle at a given point on the conic is a point on the normal at the given point.*

**18.** *If the join $XX'$ of any two homologous points cut the a. of h. in $U$, then $(OXUX')$ is constant, $O$ being the c. of h.*

For take any fixed pair of homologous points $AA'$. Then $AX$, $A'X'$ meet on the a. of h., say at $L$. Hence if $AA'$ cut the a. of h. in $N$, we have
$$(OXUX') = (OANA') = \text{constant}.$$

This proof fails if $AA'XX'$ all lie on the same line through $O$. In this case take any pair of homologous points $BB'$ which do not lie on $AA'XX'$, and let $OBB'$ cut the a. of h. in $R$.

Then $(OXUX') = (OBRB') = (OANA') =$ constant.

Conversely, *if a point $X'$ be taken such that $(OXUX')$ is constant, $O$ being a fixed point and $U$ the meet of $OX$ with a fixed line, then the figures generated by $X$ and $X'$ will be homologous, $O$ being the c. of h. and the fixed line the a. of h.*

For if $AA'$, $XX'$ be two pairs of points thus obtained, since $(OANA') = (OXUX')$, it follows that $AX$, $NU$, $A'X'$ meet in a point. Hence the join of any two points meets the join of the corresponding points on a fixed line. Hence the figures are homologous.

$(OU, XX')$ is called *the parameter* of the homology.

**Ex. 1.** *If two homologous lines $LX$ and $LX'$ cut the a. of h. in $L$, show that $L(OXMX')$ is constant, $M$ being any other point on the a. of h.; and conversely, if $LA'$ be determined as the corresponding line to $LA$ by this definition, show that the figures generated by $LA$ and $LA'$ are homologous.*

**Ex. 2.** *If $(OU, XX') = -1$, show that the figure made up of a figure and its homologue is its own homologue.*

This is called *harmonic homology*.

Notice that harmonic homology bears the same relation to ordinary homology as an involution range bears to two homographic ranges on the same line. In fact the figure $(ABC...A'B'C'...)$ is homologous to the figure $(A'B'C'...ABC...)$, if the two figures $(ABC...)$ and $(A'B'C'...)$ are in harmonic homology.

**Ex 3.** *In harmonic homology, if the c. of h. be at infinity in a direction perpendicular to the a. of h., then each figure is the reflexion of the other in the a. of h.*

**Ex. 4.** *In harmonic homology, if the a. of h. be at infinity, then each figure is the reflexion of the other in the c. of h.*

**Ex. 5.** *If a conic be its own homologue, show that the homology is harmonic, and that the homologue of the line at infinity is halfway between the c. of h. and its polar. Also show that a conic is an ellipse, parabola, or hyperbola, according as the line halfway between any point and its polar cuts the conic in imaginary, coincident, or real points.*

**Ex. 6.** $AA'$, $BB'$, $CC'$ *are the three pairs of opposite vertices of a quadrilateral. Through any point $D$ on $CC'$ are drawn $DA$ meeting $BA'C'$ in $E'$, and $DA'$ meeting $AB'C$ in $E$. Show that $EE'$, $AA'$, $BB'$ are concurrent, and also $B'E'$, $BE$ and $CC'$.*

By harmonic homology, taking the meet of $AA'$, $BB'$ as c. of h. and $CC'$ as a. of h.

**Ex. 7.** *Show that two figures in homology reciprocate into two figures in homology, and that the parameters of homology are numerically equal.*

**Ex. 8.** *The parameter of homology of two homothetic figures is the reciprocal of the ratio of similitude.*

**Ex. 9.** *The parameter of homology of two figures in parallel homology is the constant ratio of the ordinates.*

**Ex. 10.** *Keeping the same c. of h., show that the two parameters of homology of two circles are equal but of opposite signs.*

**Ex. 11.** *Keeping the radical axis as a. of h., show that the two parameters of homology of two circles are equal but of opposite signs.*

**Ex. 12.** *The poles of the radical axis of two circles divide the join of the two centres of similitude harmonically.*

For the poles $X$ and $X'$ are homologous if the radical axis be the a. of h., whichever centre of similitude we take as the c. of h. Hence $(SN, XX') = -(S'N, XX')$.

**Ex. 13.** *If the radical axis of two circles be taken as the a. of h., and if the vanishing lines and the radical axis cut the line of centres in $IJ'N$; show that*
$$SI : IN :: r : r', \text{ and } SJ' : J'N :: r' : r.$$

**Ex. 14.** *OX is the perpendicular to the line $l$ from $O$, and $A$ is any point on $OX$. From a variable point $P$ the perpendicular $PM$ is drawn to $l$, and $MA$, $PO$ meet in $P'$. Show that $P$ and $P'$ generate homologous figures.*

**Ex. 15.** *If $A, B, C$ be fixed points and $P, P'$ variable points such that $B(APP'C) = A(BPP'C) = $ constant; show that $P$ and $P'$ generate homologous figures, of which $C$ is the c. of h. and $AB$ is the a. of h.*

**19.** *If $PP'$ be any homologous points, and $PM$ the perpendicular from $P$ on the vanishing line of the figure generated by $P$, then $OP/PM \propto OP'$, $O$ being the c. of h.*

Let $OP$ cut the vanishing line in $I$ and the a. of h. in $L$. Then, since $I$ corresponds to the point $\Omega'$ at infinity upon $OP$, we have

$$(OI, PL) = (O\Omega', P'L).$$

Hence

$$\frac{OP}{PI} + \frac{OL}{LI} = \frac{OP'}{P'\Omega'} + \frac{OL}{L\Omega'} = \frac{OP'}{OL},$$

i.e. $OP : OP' :: PI : LI :: PM : h$,

where $h$ is the perpendicular distance between the vanishing line and the a. of h.

# Homology.

Hence $OP : PM :: OP' : h$.

**Ex.** *Prove the SP : PM property of a focus.*
Form a homologue of the conic, taking $S$ as the c. of h. and the corresponding directrix as vanishing line. Then $SP \div PM \propto SP'$. But by § 12 the locus of $P'$ is a circle with centre $S$. Hence $SP \div PM$ is constant.

**20.** *In two homologous figures, if $(X, p)$ and $(X, q)$ denote the perpendiculars from the variable point $X$ on two given lines $p$ and $q$, and if $(X', p')$ and $(X', q')$ denote the perpendiculars from the corresponding point $X'$ in the homologous figure on the corresponding lines $p'$ and $q'$, then $\dfrac{(X, p)}{(X, q)} \div \dfrac{(X', p')}{(X', q')}$ is constant.*

For take another point $Y$, and let $XY$ cut the lines $p$ and $q$ in $A$ and $B$. Then $X'Y'$ will cut $p'$ and $q'$ in the homologous points $A'$ and $B'$. Hence, since homologous figures are projective, we have

$$(AB, XY) = (A'B', X'Y'),$$

i.e. $AX/AY \div XB/YB = A'X'/A'Y' \div X'B'/Y'B'$,

i.e. $(X, p)/(Y, p) \div (X, q)/(Y, q)$
$\qquad = (X', p')/(Y', p') \div (X', q')/(Y', q')$.

Hence $(X, p)/(X, q) \div (X', p')/(X', q')$ is constant.

**Ex. 1.** *If $X$ and $Y$ be fixed and $p$ vary, then*
$\qquad (X, p)/(Y, p) \div (X', p')/(Y', p')$ *is constant.*

**Ex. 2.** *If $i$ be the vanishing line of the unaccented figure, then*
$\qquad (X, p)/(X, i) \div (X', p')$ *is constant.*
Take $q'$ at infinity; then $(X', q') = (Y', q')$.

**Ex. 3.** *If $i$ and $j'$ be the vanishing lines, then*
$\qquad (X, i) . (X', j')$ *is constant.*
Take $p$ and $q'$ at infinity.

**Ex. 4.** $OX/(X, p) \div OX'/(X', p')$ *is constant.*

Take $q$ and $q'$ as the axis of homology $a$, and notice that
$$OX/(X, a) \div OX'/(X', a) \text{ is constant,}$$
since $(OU, XX')$ is constant.

**Ex. 5.** $OX/(X, i) \div OX'$ *is constant.*

# MISCELLANEOUS EXAMPLES.

1. GENERALISE by projection and reciprocation the theorems—(1) 'The director circles of all conics inscribed in the same quadrilateral are coaxal,' (2) 'The locus of the centre of an equilateral hyperbola which passes through three given points is a circle.'

2. The portion of a common tangent to two circles $a$ and $\beta$ between the points of contact is the diameter of the circle $\gamma$. If the common chord of $\gamma$ and $a$ meets that of $\gamma$ and $\beta$ in $R$, show that $R$ is the pole for $\gamma$ of the line of centres of $a$ and $\beta$.

3. Generalise by projection the theorem—'The straight lines which connect either directly or transversely the extremities of parallel diameters of two circles intersect on their line of centres.'

4. A pair of right lines through a fixed point $O$ meet a conic in $PQ$, $P'Q'$; show that if $PP'$ passes through a fixed point, then $QQ'$ also passes through a fixed point.

5. Generalise by projection and reciprocation the theorem—'A diameter of a rectangular hyperbola and the tangent at either of its extremities are equally inclined to either asymptote.'

6. If $P$, $Q$ denote any pair of diametrically opposite points on the circumference of a given circle, and $QY$ the perpendicular from $Q$ upon the polar of $P$ with respect to another given circle whose centre is $C$, show that $QY \cdot CP$ is constant.
What does the theorem become when the circles are orthogonal?

7. Through a given point $O$ draw a line cutting the sides $BC, CA, AB$ of a triangle $ABC$ in points $A', B', C'$, such that $(OA', B'C')$ shall be harmonic.

8. Given the centre of a conic and three tangents, find the point of contact of any one of them.

9. Two similar and similarly situated conics have a common focus which is not a centre of similitude. Prove that a parabola can be described touching the common chord and the common tangents of the conics, and having its focus at their common focus.

10. Generalise by projection the theorem—'One circle can be described so as to pass through the four vertices of a square and another so as to touch its four sides, the centre of each circle being the intersection of diagonals.'

11. Two conics touch at $A$, and intersect at $B$ and $C$. Through $O$, the point where $BC$ meets the tangent at $A$, is drawn a chord $OPP'$ of the one conic, and $AP$, $AP'$ produced if necessary meet the second conic in $Q$ and $Q'$. Prove that $Q$, $Q'$ and $O$ are collinear.

12. $ABCD$ is a rectangle, and $(AC, PQ)$, $(BD, XY)$ are harmonic ranges; show that the points $P$, $Q$, $X$, $Y$ lie on a circle.

13. Through $O$, one of the points of intersection of two circles, the chords $POQ$ and $OP'Q'$ are drawn ($P$ and $P'$ being on one circle and $Q$ and $Q'$ on the other). Show that if $PO : OQ :: OP' : OQ'$, then $OP$ and $OP'$ generate a pencil in involution.

14. $O$ is the orthocentre of the acute-angled triangle $ABC$. Prove that the polar circles of the triangles $OBC$, $OCA$, $OAB$ are orthogonal, each to each.

15. A number of conics are inscribed in a given triangle so as to touch one of its sides at a given point. Show that their points of contact with the other two sides form two homographic divisions which are in perspective.

16. $AC$, $BD$ are conjugate diameters of a central conic, and $P$ is any point on the arc $AB$. $PA$, $PB$ meet $CD$ in $Q$, $R$ respectively. Prove that the range $(QC, DR)$ is harmonic.

17. Generalise by projection and reciprocation the proposition—'The locus of the foot of the perpendicular upon any tangent to an ellipse from a focus is a circle.'

18. $P$ is the pole of a chord which subtends a constant angle at the focus $S$ of a conic, and $SP$ intersects the chord in $Q$; find the locus of the point $R$ such that $(SR, PQ)$ is harmonic.

19. A straight line $AD$ is trisected in $B$, $C$; the connectors of $A$, $B$, $C$, $D$, and the point at infinity on $AD$ with any point $S$ meet another straight line in $A'$, $B'$, $C'$, $D'$, $E'$ respectively; show that
$$E'B' : E'D' = 3 . A'B' : A'D'.$$

20. From any point $Q$ on a fixed tangent $BQ$ to a circle $AA'B$, straight lines are drawn to $A$, $A'$, the extremities of a fixed diameter parallel to $BQ$, meeting the circle again in $P$, $P'$ respectively; show that the locus of the intersection of $A'P$, $AP'$ is a parabola of which $B$ is the vertex.

21. Two conics $a$ and $\beta$ intersect in the points $A$, $B$, $C$, $D$; show that if the pole of $AB$ with regard to $a$ lies on $\beta$, then the pole of $CD$ with regard to $a$ lies on $\beta$.

If the vertex of a parabola is the pole of one of its chords of inter-

# Miscellaneous Examples. 301

section with a circle, then another common chord is a diameter of the circle.

22. 'If a circle be drawn through the foci $S$, $H$ of two confocal ellipses, cutting the ellipses in $P$ and $Q$, the tangents to the ellipses at $P$ and $Q$ will intersect on the circumference of the circle.'

Generalise this theorem (1) by projection, (2) by reciprocation with respect to the point $S$, (3) by reciprocation with respect to any point in the plane.

23. 'If two circles of varying magnitude intersect on the side $BC$ of a given triangle $ABC$ and touch $AB$, $AC$ at $B$ and $C$ respectively; then the locus of $O$, their other point of intersection, is the circum-circle of the triangle; and the circle on which their centres and the point $O$ lie, always passes through a fixed point.'

Obtain by projection the corresponding theorem when the two circles are replaced (1) by conics, (2) by similar and similarly situated conics.

24. Two ranges are in perspective, and the centre of perspective $S$ is equidistant from the axes of the ranges. The axes are turned about their meet $O$ until they coincide. Show that if $S$ does not coincide with $O$, an involution is produced; and find the centre and double points.

25. 'If a circle touches two given circles, the connector of its points of contact passes through a centre of similitude of the given circles.'

Reciprocate this proposition with respect to a limiting point.

26. The pairs of points $AB$, $CD$ form a harmonic range. Prove that, if $X$ is any other point on the same axis, then the anharmonic ratios $(AB, CX)$ and $(AB, DX)$ are equal and of opposite sign.

27. The connectors of a point $D$ in the plane of the triangle $ABC$ with $B$, $C$ meet the opposite sides in $E$, $F$ respectively; show that the triangles $BDC$, $EDF$ have the same ratio as the triangles $ABC$, $AEF$.

28. $A$, $B$, $C$ are three points on a straight line; $A_1$ is the harmonic conjugate of $A$ with respect to $BC$, $B_1$ of $B$ with respect to $CA$, and $C_1$ of $C$ with respect to $AB$; show that $AA_1$, $BB_1$, $CC_1$ are three pairs of a range in involution.

29. A conic is reciprocated into a circle. Find the reciprocals of a pair of conjugate diameters.

30. Generalise by projection the theorem—'If a straight line touch a circle and from the point of contact a straight line be drawn cutting the circle, the angles which this line makes with the line touching the circle shall be equal to the angles which are in the alternate segments of the circle.'

31. The locus of the pole of a chord of a conic which subtends a right angle at a fixed point is a conic.

32. A quadrilateral $ABCD$ is circumscribed to a conic, and a fifth tangent is drawn at the point $P$; the diagonals $AC$, $BD$ meet the tangent at $P$ in $\alpha$ and $\beta$, and the points $\alpha'$, $\beta'$ are taken the harmonic conjugates of $\alpha$ and $\beta$ with respect to $A$, $C$ and $B$, $D$ respectively; show that $\alpha'$ $\beta'$, $P$ are on a straight line.

33. Through the vertex $A$ of a square $ABCD$ a straight line is drawn meeting the sides $BC$, $CD$ in $E$, $F$. If $ED$, $FB$ intersect at $G$, show that $CG$ is at right angles to $EF$.

34. Determine the envelope of a straight line which meets the sides of a triangle in $A$, $B$, $C$, so that the ratio $AB : AC$ is constant.

35. Generalise by projection the theorem—'If $OP$, $OQ$, tangents to a parabola whose focus is $S$, are cut by the circle on $OS$ as diameter in $M$ and $N$, then $MN$ will be perpendicular to the axis.'

36. Reciprocate with regard to the focus of the parabola the theorem —'The circle described on a focal radius of a parabola as diameter touches the tangent at the vertex.'

37. Two given straight lines $AB$ and $CD$ intersect in $D$, and a variable point $P$ on $CD$ is joined to the fixed points $A$, $B$ on $AB$. If a point $Q$ be taken such that the angles between $AP$ and $AQ$, and between $BQ$ and $PB$ produced are each equal to $CDA$, show that the locus of $Q$ is a straight line.

38. $M$ and $N$ are a pair of inverse points with regard to a given circle whose centre is $C$. Prove that (1) if $P$ is any point on the circle
$$PM^2 : PN^2 :: CM : CN;$$
(2) if any chord of the circle is drawn through $M$ or $N$, the product of the distances of its extremities from the straight line bisecting $MN$ at right angles is constant.

39. Points $P$, $Q$ are taken on the sides $AB$, $AC$ of a triangle respectively, such that $AP=CQ$; show that the line joining $PQ$ will envelope a parabola.

Through a given point draw a straight line to cut the equal sides $AB$, $AC$ of an isosceles triangle $BAC$ in $P$, $Q$ respectively, so that $AP$ is equal to $CQ$.

40. Given the proposition 'any point $P$ of an ellipse, the two foci, and the points of intersection of the tangent and normal at $P$ with the minor axis are concyclic,' (1) generalise it by projection, (2) reciprocate it with regard to one of the foci.

41. Generalise the following proposition (1) by reciprocating it with respect to $A$, and (2) by projection—'A fixed circle whose centre is $O$ touches a given straight line at a point $A$; the locus of the centre of a circle which moves so that it always touches the fixed circle and the fixed straight line is a parabola whose focus is $O$, and whose vertex is $A$.'

## Miscellaneous Examples. 303

42. Two circles $a$ and $\beta$ intersect a conic $\gamma$; show that the chords of intersection of $a$ and $\gamma$ meet the chords of intersection of $\beta$ and $\gamma$ in four points which lie on a circle having the same radical axis with $a$ and $\beta$.

43. Through any point $O$ in the plane of a triangle $ABC$ are drawn $OA'$, $OB'$, $OC'$ bisecting the supplements of the angles $BOC$, $COA$, $AOB$ and meeting $BC$, $CA$, $AB$ in $A'$, $B'$, $C'$ respectively; show that the six lines $OA$, $OB$, $OC$, $OA'$, $OB'$, $OC'$ form a pencil in involution.

44. Two conics $a$ and $\beta$ have double contact at $B$ and $C$, $A$ being the pole of $BC$. Tangents from a point $X$ upon $AB$ are drawn to $a$ and $\beta$ meeting $AC$ in $Y$ and $Y'$. Show that $Y$ and $Y'$ generate homographic ranges, the double points of which are $A$ and $C$.

45. A quadrangle $ABCD$ is inscribed in a parabola; through two of its vertices $C$ and $D$ straight lines are drawn parallel to the axis, meeting $DA$, $BC$ in $P$ and $Q$; show that $PQ$ is parallel to $AB$.

46. Prove that the polar reciprocal with regard to a parabola of the circle of curvature at its vertex is a rectangular hyperbola of which the circle is also the circle of curvature at a vertex.

47. The opposite vertices $AA'$, $BB'$, $CC'$ of a quadrilateral circumscribing a conic are joined to a given point $O$; $OA$ cuts the polar of $A$ in $a$, $OB$ cuts the polar of $B$ in $b$, and so on; show that a conic can be drawn through the seven points $Oaa'bb'cc'$.

48. A range on a line is projected from two different vertices on to another line. Find the double points of the projected ranges.

49. If four points $A$, $B$, $C$, $D$ be taken on the circumference of a circle, prove that the centres of the nine-point circles of the four triangles $ABC$, $BCD$, $CDA$, $DAB$ will lie on the circumference of another circle, whose radius is one-half that of the first.

50. If the orthocentre of a triangle inscribed in a parabola be on the directrix, then the polar circle of the triangle passes through the focus.

51. $A$ and $BC$ are a given pole and polar with regard to a conic; $DE$ is a given chord through $A$; $P, Q, R, \ldots$ are any number of points on the conic, and $P', Q', R', \ldots$ are the points where $EP$, $EQ$, $ER$, $\ldots$ meet $BC$. Prove that $D(PP', QQ', RR', \ldots)$ is an involution; and determine its double lines.

52. $ABCD$ is a quadrilateral circumscribing a conic $a$. $AB$, $DC$ meet in $E$, and $BC$, $AD$ in $F$, and a conic $\beta$ is drawn through the points $B$, $D$, $F$, $E$. Prove that the four tangents to $a$ at the points where the conics intersect pass two and two through the pair of points where $AC$ cuts $\beta$.

53. Two conics $a$, $\beta$ intersect in the points $A$, $B$, $C$, $D$. If the pole of $AB$ with respect to $a$ coincides with the pole of $CD$ with respect to $\beta$, prove that the pole of $CD$ with respect to $a$ will coincide with the pole of $AB$ with respect to $\beta$.

54. Three conics all pass through the same two points $A, B$. The first and second conics intersect one another in two other points $C, D$; and the pole of $AB$ with regard to the second conic lies on the first conic. The third conic touches the line joining $C, D$; and the pole of $AB$ with regard to it lies on the second conic. Show that the tangents, other than $CD$, drawn from the points $C, D$ to the third conic meet on the circumference of the first conic.

55. Given the asymptotes of a conic and another tangent, show how to construct the pair of tangents from a given point to the conic.

Given the three middle points of the sides of a given triangle, draw a straight line through a given point to bisect the triangle.

56. A conic cuts the sides of a triangle $ABC$ in the pairs of points $a_1\ a_2,\ b_1\ b_2,\ c_1\ c_2$ respectively; if $Bb_2, Cc_2$ intersect in $a_1$, and $Bb_1, Cc_1$ in $a_2$, and so on, and if $\beta_1\ \beta_2\ \beta_3\ \beta_4,\ \gamma_1\ \gamma_2\ \gamma_3\ \gamma_4$ be similarly constructed; show that the straight lines obtained by putting in various suffixes in $Aa$, $B\beta$, $C\gamma$ meet, three by three, in eight points.

57. Reciprocate the proposition that the nine-point circle of a triangle touches the inscribed circle (1) with regard to one of the angular points of the triangle, (2) with regard to the middle point of one of its sides.

58. 'If, from a point within a circle, more than two equal straight lines can be drawn to the circumference, that point is the centre of the circle.'

Generalise the above proposition (1) by reciprocation, (2) by projection.

59. $TP$ and $TQ$ are tangents of a conic and $PQ$ is bisected in $V$; also $TV$ is bisected by the curve. Show that the conic is a parabola.

60. A conic of constant eccentricity is drawn with one focus at the centre of a given circle and circumscribing a triangle self-conjugate with respect to the given circle; show that the corresponding directrices for different positions of the triangle will envelope a circle.

61. A straight line moves so as to make upon two fixed straight lines intercepts whose difference is constant; prove that it will always touch a fixed parabola, and determine the focus and directrix of the parabola.

62. By reciprocation deduce a proposition relating to the circle from the following—'The locus of a point dividing in a given ratio the ordinate $PN$ of a parabola is another parabola having the same vertex and axis.'

63. The envelope of a straight line which moves so that two fixed circles intercept on it chords of equal length is a parabola.

64. Given a conic and a pair of straight lines conjugate with regard to it, project the conic into a parabola of which the projections of the given lines shall be latus rectum and directrix.

65. An ellipse has the focus of a parabola for centre and has with it contact of the third order at its vertex. Tangents are drawn to the two conics from any point on their common tangent, and the harmonic conjugate of this latter with regard to them is taken. Prove that its envelope is the common circle of curvature of the two conics at the common vertex.

66. $ABC$, $DEF$ are two triangles inscribed in a conic. $BC$, $CA$, $AB$ are parallel respectively to $EF$, $FD$, $DE$. Prove that $AD$, $BE$, $CF$ are diameters of the conic.

67. Find the double rays of the pencils $O(ABC...)$ and $O(A'B'C'...)$, each of which is in perspective with the pencil $V(A''B''C''...)$.

68. $DD'$ is a fixed diameter of a conic and $PP'$ is a double ordinate of this diameter. A parallel through $D'$ to $DP$ meets $DP'$ in $X$. Find the locus of $X$.

69. Through a point $O$ is drawn a straight line cutting a conic in $AB$, and on $AB$ are taken points $CD$, such that
$$(1 \div OC) + (1 \div OD) = (1 \div OA) + (1 \div OB).$$
Then if $MN$ be the points of contact of tangents from $D$, and $LR$ those of tangents from $C$, show that either $LM$ and $RN$, or $LN$ and $RM$, meet in $O$.

70. Construct the conic which passes through the four points $ABCD$ and is such that $AB$ and $CD$ are conjugate lines with regard to it.

71. $AOB$ and $COD$ are two diameters of a circle and $QR$ is a chord parallel to $AB$; if $P$ be the intersection of $CQ$ and $DR$, or of $DQ$ and $CR$, and if from $P$ be drawn $PM$ parallel to $AB$ to meet $CD$ in $M$, then $OM^2 = OD^2 + PM^2$.

72. $AB$, $AC$ are two chords of an ellipse equally inclined to the tangent at $A$; show that the ratio of the chords is the duplicate of the ratio of the diameters parallel to them.

73. Construct, by means of the ruler only, a conic which shall pass through two given points and have a given self-conjugate triangle.
Also construct the pole of the connector of the given points with respect to the conic.

74. Through a fixed point $A$ any two straight lines are drawn meeting a conic in $B$, $B'$ and $C$, $C'$ respectively; parallels through $A$ to $BC'$, $B'C$ meet $B'C$, $BC'$ respectively in $D$, $E$; find the locus of $D$ and of $E$.

75. Two equal tangents $TP$ and $TQ$ of a parabola are cut in $M$ and $N$ by a third tangent; show that $TM = QN$.

76. The tangents at two points of an ellipse, whose foci are $S$, $H$, meet in $T$, and the normals at the same points meet in $O$; prove that the perpendiculars through $S$, $H$ to $ST$, $HT$ respectively divide $OT$ harmonically.
Deduce a construction for the centre of curvature at any point of the ellipse.

77. An ellipse may be regarded as the polar reciprocal of the auxiliary circle with respect to an imaginary circle of which a focus is centre. Prove this, and find the lines which correspond to the centre and the other focus of the ellipse.

78. Two conics $u$, $v$ intersect in $A$, $B$, $C$, $D$; $E$, $F$ are the poles of $CD$ with regard to the conics $u$, $v$ respectively, and $AE$, $AF$ meet $CD$ in $G$, $H$ respectively; a straight line is drawn through $A$ meeting $u$, $v$ in $P$, $Q$ respectively; show that the locus of the intersection of $PH$, $QG$ is a straight line passing through $B$ and through the intersection of $EF$, $CD$.

79. Two triangles, one inscribed in and the other escribed to a given triangle, and both in perspective with it, are in perspective.

Each of the triangles determined by the common tangents of two conics is in perspective with each of the triangles determined by the common points of the conics.

80. Two circles cut each other orthogonally; show that the distances of any point from their centres are in the same ratio as the distances of the centre of each circle from the polar of the point with respect to the other.

The directrix of a fixed conic is the polar of the corresponding focus with respect to a fixed circle; with any point on the conic as centre a variable circle is described cutting the fixed circle orthogonally; find the envelope of the polar of the focus with respect to the variable circle.

81. Obtain a construction for projecting a conic and a point within it into a parabola and its focus.

82. A conic circumscribes a triangle $ABC$, the tangents at the angular points meeting the opposite sides on the straight line $DEF$. The lines joining any point $P$ on $DEF$ to $A$, $B$, $C$ meet the conic again in $A'$, $B'$, $C'$. Show that the triangle $A'B'C'$ envelopes a fixed conic inscribed in $ABC$, and having double contact with the given conic at the points where it is met by $DEF$. Show also that the tangents at $A'$, $B'$, $C'$ to the original conic meet $B'C'$, $C'A'$, $A'B'$ in points lying on $DEF$.

83. $ABCD$ is a quadrilateral whose sides $AB$, $CD$ meet in $E$, and $AD$, $BC$ in $F$; $A$ is a fixed point, $EF$ a fixed straight line, and $B$, $C$ lie each upon one of two fixed straight lines concurrent with $EF$; find the locus of $D$.

84. All the tangents of a conic are inverted from any point. Show that the locus of the centres of all the circles into which they invert is a conic.

85. If $A$, $B$, $C$, $D$ be four collinear points, and $O$ any point whatever, prove that
$$\Sigma \left\{ OA^2 \div (AB \cdot AC \cdot AD) \right\} = 0.$$

Also show that if $A'$, $B'$, $C'$, $D'$ be four concyclic points, then $\Sigma \left\{ 1 \div (A'B' \cdot A'C' \cdot A'D') \right\} = 0$, the sign of any rectilinear segment being the same as in the preceding identity.

## Miscellaneous Examples.

86. If $O$ be the intersection of the common tangents to two conics having double contact, and if a straight line through $O$ meet the two conics in $P$, $P'$ and $Q$, $Q'$ respectively, prove that
$$PQ \cdot P'Q' \cdot (PO + P'O) = PO^2 \cdot P'Q' + P'O^2 \cdot PQ,$$
and that $\quad PQ \cdot PQ' : P'Q \cdot P'Q' :: PO^2 : P'O^2.$

87. Describe a conic to touch a given straight line at a given point and to osculate a given circle at a given point.

88. If a system of conics have a common self-conjugate triangle, any line through one of the vertices of this triangle is cut by the system in involution.

Two conics, $U$ and $U'$, touch their common tangents in $ABCD$ and $A'B'C'D'$; show that $AB$ cuts $U$, $U'$ and the other sides of the quadrilateral of tangents in six points in involution.

89. Four points $A, B, C, D$ are taken on a conic such that $AB, BC, CD$ touch a conic having double contact with it; show that $A$ and $D$ generate homographic ranges on the conic, and find the common points of the ranges.

90. The angular points $ABC$ of a triangle are joined to a point $O$ and the bisectors of the angles $BOC$, $COA$, $AOB$ meet the corresponding sides of the triangle in $a_1 a_2$, $\beta_1 \beta_2$, $\gamma_1 \gamma_2$; show that these points lie three by three on four straight lines; and that if $O$ lie on the circle circumscribing the triangle, each of the lines $a_1 \beta_2 \gamma_2$, &c. passes through the centre of a circle touching the three sides of the triangle.

91. 'If from a point $T$ on the directrix of a parabola whose vertex is $A$ tangents $TP$, $TQ$ are drawn to the curve, and $PA$, $QA$ joined and produced to cut the directrix in $M$, $N$, then will $T$ be the middle point of $MN$.'

Obtain from the above theorem by reciprocation a property of (1) a circle, (2) a rectangular hyperbola.

92. In two figures in homology $M$ and $M'$ are homologous points and $O$ is the centre of perspective. Show that $OM$ is to $MM'$ as the perpendicular from $M$ on its vanishing line is to the perpendicular from $M$ on the axis of perspective.

93. Given two points $A, B$ on a rectangular hyperbola and the polar of a given point $O$ in the line $AB$; determine the points of intersection of the curve with the straight line drawn through $O$ perpendicular to $AB$.

94. Show how to project a given quadrilateral into a quadrilateral $ABCD$ such that $AB$ is equal to $AC$, and that $D$ is the centre of gravity of the triangle $ABC$.

95. A circle has double contact with an ellipse, and lies within it. A chord of the ellipse is drawn touching the circle, and through its middle point is drawn a chord of the ellipse parallel to the minor axis. Show that the rectangle contained by the segments of this chord is

equal to the rectangle contained by the segments into which the first is divided by the point of contact.

96. $ABCDEF$ is a hexagon inscribed in one conic and circumscribing another. The connectors of its vertices with any point $O$ in its plane meet the former conic again in the vertices of a second hexagon $A'B'C'D'E'F'$. Prove that it is possible in this to inscribe another conic.

97. $ABCD$, $AB'C'D'$ are two parallelograms having a common vertex $A$ and the sides $AB$, $AD$ of the one along the same straight lines as the sides $AB'$, $AD'$ respectively of the other. Show that the lines $BD'$, $B'D$, $CC'$ are concurrent.

98. Three conics $a$, $\beta$, $\gamma$ are inscribed in the same quadrilateral. From any point, tangents $a$, $b$ are drawn to $a$, and tangents $a'$, $b'$ to $\beta$. Show that if $a$, $a'$ are conjugate lines with respect to $\gamma$, so are $b$, $b'$.

99. If three tangents to a conic can be found such that the circle circumscribing the triangle formed by them passes through a focus, the conic must be a parabola.

100. From each point on a straight line parallel to an axis of a conic is drawn a straight line perpendicular to the polar of the point; show that the locus of the foot of the perpendicular is a circle.

101. $AB$ is a diameter of a circle, and $C$ and $D$ are points on the circle. $AC$, $BD$ meet in $E$. Show that the circle about $CDE$ is orthogonal to the given circle.

102. Find the locus of the centre of a circle which divides two given segments of lines harmonically.

103. The sides $AB$, $AD$ of a parallelogram $ABCD$ are fixed in position, and $C$ moves on a fixed line; show that the diagonal $BD$ envelopes a parabola.

104. A tangent of a hyperbola whose centre is $C$ meets the asymptotes in $P$ and $Q$; show that the locus of the orthocentre of the triangle $CPQ$ is another hyperbola.

105. Through fixed points $A$ and $B$ are drawn conjugate lines for a given conic. Show that the locus of their meet is the conic through $A$, $B$ and the points of contact of tangents from $A$ and $B$.

106. $A$, $B$, $C$, $D$ are four points on a conic, and $O$ is the pole of $AB$. Show that $O(AB, CD)$ is the square of $(AB, CD)$.

107. $A$, $B$, $C$, $D$ are four points on a conic. The tangent at $A$ meets $BC$, $CD$ in $a_1$, $a_2$; the tangent at $B$ meets $CD$, $DA$ in $b_1$, $b_2$; and so on. Show that the eight points $a_1$, $a_2$, $b_1$, $b_2$, $c_1$, $c_2$, $d_1$, $d_2$ lie on a conic.

108. The centre $O$ of a conic lies on the directrix of a parabola, and a triangle can be drawn circumscribed to the parabola and self-conjugate for the conic. Show that the tangents from $O$ to the parabola are the axes of the conic.

## Miscellaneous Examples. 309

109. Two sides $AQ$, $AR$ of a triangle $AQR$ circumscribed to a given circle are given in position; the circles escribed to $AQ$ and $AR$ touch $AQ$ and $AR$ in $V$ and $U$; show that the locus of the meet of $QU$ and $RV$ is a hyperbola with $AQ$ and $AR$ as asymptotes.

110. If the chords $OP$, $OQ$ of a conic are equally inclined to a fixed line; then, if $O$ be a fixed point, $PQ$ passes through a fixed point.

111. A fixed line $l$ meets one of the system of conics through the four points $A$, $B$, $C$, $D$ in $P$ and $Q$; show that the conic touching $AB$, $CD$, $PQ$ and the tangents at $P$ and $Q$ touches a fourth fixed line.

112. Triangles can be inscribed in $a$ which are self-conjugate for $\beta$; $ABC$ is a triangle inscribed in $a$ and $A'B'C'$ is its reciprocal for $\beta$; show that the centre of homology of $ABC$ and $A'B'C'$ is on $a$.

113. Six circles of a coaxal system touch the sides of a triangle $ABC$ inscribed in any coaxal in the points $aa'$, $bb'$, $cc'$; show that these points are the opposite vertices of a quadrilateral.

114. $A$, $B$, $C$, $D$ are four points on a circle, and $A'$, $B'$, $C'$, $D'$ are the orthocentres of the triangles $BCD$, $CDA$, $DAB$, $ABC$. Show that the figures $ABCD$, $A'B'C'D'$ are superposable.

115. Any conic $a$ which divides harmonically two of the diagonals of a quadrilateral is related to any conic $\beta$ inscribed in the quadrilateral in such a way that triangles can be inscribed in $a$ which are self-conjugate for $\beta$.

116. The envelope of the axes of all conics touching four tangents of a circle is a parabola.

117. If $(AA', BB') = -1 = (AA', PQ) = (BB', PQ')$; show that $(AA', BB', QQ')$ is an involution.

118. If two conics can be drawn to divide four given segments harmonically, then an infinite number of such conics can be drawn.

119. If $(AA', BB', CC')$ be an involution, show that
$$(A'A, BC') + (B'B, CA') + (C'C, AB') = 1.$$

120. $T$ is a point on the directors of the conics $a$ and $\beta$. The reciprocal of $a$ for $\beta$ meets the polar of $T$ for $\beta$ in $Q$, $R$. Show that the angle $QTR$ is right.

121. Through the centre $O$ of a circle is drawn a conic, and $A$ and $A'$ are a pair of opposite meets of common tangents of the circle and conic; show that the bisectors of the angle $AOA'$ are the tangent and the normal at $O$.

122. A given line meets one of a series of coaxal circles in $P$, $Q$. The parabola which touches the line, the tangents at $P$, $Q$, and the radical axis has a third fixed tangent.

123. If a series of conics be inscribed in the same quadrilateral of which $A$, $A'$ is a pair of opposite vertices, and if from a fixed point $O$ tangents $OP$, $OQ$ be drawn to one of the conics, the conic through $OPQAA'$ will pass through a fourth fixed point.

124. On a tangent to a circle inscribed in a triangle $ABC$ are taken points $a$, $b$, $c$, such that the angles subtended by $Aa$, $Bb$, $Cc$ at the centre $O$ are equal; show that $Aa$, $Bb$, $Cc$ are concurrent.

125. Through two given points, four conics can be drawn for which three given pairs of lines are conjugate; and the common chord is divided harmonically by every conic through its four poles for the conics.

126. The locus of the pole of a common chord of two conics for a variable conic having double contact with the two given conics consists of a conic through the two common points on the given chord together with the join of the poles of the chord for the two conics.

127. Find the locus of the centre of a conic which passes through two given points and divides two given segments harmonically.

128. A variable conic passes through three fixed points and is such that triangles can be inscribed in it which are self-polar for a given conic. Show that it passes through a fourth fixed point.

129. If a variable conic touch three fixed lines, and be such that triangles can be drawn circumscribing it which are self-polar for a given conic, then the variable conic will have a fourth fixed tangent, and the chords of contact of the variable conic with the fixed lines pass through fixed points.

130. The directrix of a parabola which has a fixed focus and is such that triangles can be described about it which are self-polar for a given conic, passes through a fixed point.

131. A conic $U$ passes through two given points and is such that two sets of triangles can be inscribed in it, one self-polar for a fixed conic $V$ and the other self-polar for a fixed conic $W$. Show that $U$ has a fixed self-polar triangle.

132. A variable conic $U$ cuts a given conic $V$ in two given points and also touches it and is such that triangles can be inscribed in it self-polar for a given conic $W$. Show that $U$ touches another fixed conic.

133. Three parabolas are drawn, two of which pass through the four points common to two conics and the third touches their common tangents. Show that their directrices are concurrent.

134. If a system of rectangular hyperbolas have two points common, any line perpendicular to the common chord meets them in an involution.

135. The reciprocal of a circle through the centre of a rectangular hyperbola, taking the r.h. itself as base conic, is a parabola whose focus is at the centre of the r.h.

136. The reciprocal of any circle, taking any r.h. as base conic, is a conic, one of whose foci is at the centre of the r.h.; and the centre of the circle reciprocates into the corresponding directrix.

# Miscellaneous Examples. 311

137. The chords $AB$ and $A'B'$ of a conic $a$ meet in $V$. $\beta$ is the conic touching $AB$, $A'B'$ and the tangents at $A$, $B$, $A'$, $B'$. $VL$ and $VL'$ divide $AVA'$ harmonically and cut the conic $a$ in $LM$ and $L'M'$. Show that the other joins of the points $L$, $M$, $L'$, $M'$ touch $\beta$. Also any tangent of $\beta$ meets $AB$ and $A'B'$ in points which are conjugate for $a$.

138. The director circle of a conic is the conic through the circular points and the points of contact of tangents from these points to the conic.

139. Tangents to a circle at $P$ and $Q$ meet another circle in $AB$ and $CD$; show that a conic can be drawn with a focus at either limiting point of the two circles and with $PQ$ as corresponding directrix which shall pass through $ABCD$.

140. Tangents to a conic from two points $PP'$ on a confocal meet again in the opposite points $QQ'$ and $RR'$. Show that $QQ'$ lie on one confocal and $RR'$ on another; and that the tangents to the confocals at $PP'QQ'RR'$ are concurrent.

141. The centroid of the meets of a parabola and a circle is on the axis of the parabola.

142. A variable tangent of a circle meets two fixed parallel tangents in $P$ and $Q$, and a fixed line through the centre in $R$. $X$ is taken so that $(PQ, RX) = -1$. Show that the locus of $X$ is a concentric circle.

143. A triangle is reciprocated for its polar circle. Show that the reciprocal of the centroid is the radical axis of the circum-circle and the nine-point circle.

144. The reciprocal of a triangle for its centroid is a triangle having the same centroid.

145. Triangles can be circumscribed to $a$ which are self-conjugate for $\beta$. A tangent of $a$ cuts $\beta$ in $P$ and $Q$; and a conic $\gamma$ is drawn touching $\beta$ at $P$ and at $Q$. Show that triangles can be circumscribed to $a$ which are self-conjugate for $\gamma$.

146. $PP'$ is a chord of a parabola. Any tangent of the parabola cuts the tangent parallel to $PP'$ in $X$ and the tangents at $P$ and $P'$ in $R$ and $R'$; show that $RX = XR'$.

147. If the conic $a$ be its own reciprocal for the conic $\beta$, then $\beta$ is its own reciprocal for $a$.

148. Given a conic $a$ and a chord $BC$ of $a$, a conic $\beta$ can be found having double contact with $a$ at $B$ and $C$, such that $a$ is its own reciprocal for $\beta$.

149. A conic cannot be its own reciprocal for a conic having four-point contact with it.

150. If the conic $a$ be its own reciprocal for the conic $\beta$, then $a$ and $\beta$ can be projected into concentric circles, the squares of whose radii are numerically equal.

# Miscellaneous Examples.

151. Any point $P$ on a conic and the pole of the normal at $P$ are conjugate points for the director circle.

152. The pole of the normal at any point $P$ of a conic is the centre of curvature of $P$ for the confocal through $P$.

153. $ABC$ is a triangle, and $AL$, $BM$, $CN$ meet in a point, $LMN$ being points on $BC$, $CA$, $AB$. Three conics are described, one touching $BM$, $CN$ at $M$, $N$ and passing through $A$; so the others. Prove that at $A$, $B$, $C$ respectively they are touched by the same conic.

154. The lines joining four fixed points in a plane intersect in pairs in points $O_1 O_2 O_3$, and $P$ is a variable point. Prove that the harmonic conjugates of $O_1 P$, $O_2 P$, $O_3 P$ for the pairs of lines meeting in $O_1 O_2 O_3$ respectively, intersect in a point.

155. If a parabola touch the sides of a fixed triangle, the chords of contact will each pass through a fixed point.

156. The six intersections of the sides of two similar and similarly situated triangles lie on a conic, which is a circle if the perpendicular distances between the pairs of parallel sides are proportional to the sides of the triangle.

157. Two conics have double contact, $O$ being the intersection of the common tangents. From $P$ and $Q$ on the outer conic pairs of tangents are drawn to the inner, forming a quadrilateral, and $R$ is the pole of $PQ$ with respect to the inner conic. Prove that two diagonals of the quadrilateral pass through $R$, and that one of these diagonals passes through $O$.

158. A conic is drawn through the middle points of the lines joining the vertices of a fixed triangle to a variable point in its plane, and through the points in which these joining lines cut the sides of the triangle. Determine the locus of the variable point when the conic is a rectangular hyperbola; and prove that the locus of the centre of the rectangular hyperbola is a circle.

159. The feet of the normals from any point to a rectangular hyperbola form a triangle and its orthocentre.

160. $ABC$ is a triangle and $A'B'C'$ are the middle points of its sides. $O$ is the orthocentre. $AO$, $BO$, $CO$ meet the opposite sides in $DEF$. $EF$, $FD$, $DE$ meet the sides in $LMN$. Prove that $OL$ is perpendicular to $AA'$, $OM$ to $BB'$, and $ON$ to $CC'$.

161. A variable conic touches the sides $AB$, $AC$ of a triangle $ABC$ at $B$ and $C$. Prove that the points of contact of tangents from a fixed point $P$ to the conic lie on a fixed conic though $PABC$.

162. Given two tangents to a parabola and a fixed point on the chord of contact, show that a third tangent is known.

163. Tangents to a conic from two points on a confocal form a quadrilateral in which a circle can be inscribed.

## Miscellaneous Examples.

164. $AA'$, $BB'$, $CC'$ are opposite vertices of a quadrilateral formed by four tangents to a conic. Three conics pass respectively through $AA'$, $BB'$, $CC'$ and have three-point contact with the given conic at the same point $P$. Show that the poles of $AA'$, $BB'$, $CC'$ with respect to the conics through $AA'$, $BB'$, $CC'$ respectively coincide, and the four conics have another common tangent.

165. If two conics, one inscribed in and the other circumscribed to a triangle, have the orthocentre as their common centre, they are similar, and their corresponding axes are at right angles.

166. A fixed tangent is drawn to an ellipse meeting the major axis in $T$. $QQ'$ are two points on the tangent equidistant from $T$. Show that the other tangents from $Q$ and $Q'$ to the ellipse meet on a fixed straight line parallel to the major axis.

167. With a fixed point $P$ as focus a parabola is drawn touching a variable pair of conjugate diameters of a fixed conic. Prove that it has a fixed tangent parallel to the polar of $P$.

168. A conic is described having one side of a triangle for directrix, the opposite vertex for centre, and the orthocentre for focus; prove that the sides of the triangle which meet in the centre are conjugate diameters.

169. The radius of curvature in a rectangular hyperbola is equal to half the normal chord.

170. The radius of curvature in a parabola is equal to twice the intercept on the normal between the directrix and the point of intersection of the normal and the parabola.

171. Two ellipses touch at $A$ and cut at $B$ and $C$. Their common tangents, not at $A$, meet that at $A$ in $Q$ and $R$ and intersect in $P$. Prove that $BQ$ and $CR$ meet on $AP$, and so do $BR$ and $CQ$.

172. A transversal is drawn across a quadrangle so that the locus of one double point of the involution determined on it is a straight line. Show that the locus of the other is a conic circumscribing the harmonic triangle of the quadrangle.

173. $PQ$ is a chord of one conic $a$ and touches another conic $\beta$. Prove that $P$, $Q$ are conjugate for a third conic $\gamma$.

174. $XYZ$ is a triangle self-conjugate for a circle. The lines joining $XYZ$ to a point $D$ on the circle meet the circle again in $A$, $B$, $C$ respectively. Show that as $D$ varies, the centre of mean position of $ABCD$ describes the nine-point circle of $XYZ$.

175. Two conics are described touching a pair of opposite sides of a quadrilateral, having the remaining sides as chords of contact and passing through the intersection of its diagonals; show that they touch at this point.

176. With a given point $O$ as focus, four conics can be drawn having three given pairs of points conjugate; and the directrices of these conics form a quadrilateral such that the director circles of all the inscribed conics pass through $O$.

177. The line joining two points $A$ and $B$ meets two lines $OQ$, $OP$ in $Q$ and $P$. A conic is described so that $OP$ and $OQ$ are the polars of $A$ and $B$ with regard to it. Show that the locus of its centre is the line $OR$ where $R$ divides $AB$ so that $AR : RB :: QR : RP$.

178. A chord of a conic passes through a fixed point. Prove that the other chord of intersection of the conic and the circle on this chord as diameter passes through a fixed point.

179. One of the chords of intersection of a circle and a r. h. is a diameter of the circle. Prove that the opposite chord is a diameter of the r. h.

180. Tangents are drawn to a conic $a$ parallel to conjugate diameters of a conic $\beta$. Prove that they will cut on a conic $\gamma$, concentric with $a$ and homothetic with $\beta$. Also $\gamma$ will meet $a$ in points at which the tangents to $a$ are parallel to the asymptotes of $\beta$.

181. Four concyclic points are taken on a parabola. Prove that its axis bisects the diagonals of the quadrilateral formed by the tangents to the parabola at these points.

182. If four points be taken on a circle, the axes of the two parabolas through them are the asymptotes of the centre-locus of conics through them.

183. The locus of the middle point of the intercept on a variable tangent to a conic by two fixed tangents is a conic having double contact with the given one where it is met by the diameter through the intersection of the fixed tangents.

184. On two parallel straight lines fixed points $A$, $B$ are taken and lengths $AP$, $BQ$ are measured along the lines, such that $AP + BQ$ is constant. Show that $AQ$ and $BP$ cut on a fixed parabola.

185. Chords $AP$, $AQ$ of a conic are drawn through the fixed point $A$ on the conic, such that their intercept on a fixed line is bisected by a fixed point. Show that $PQ$ passes through a fixed point.

186. Three tangents are drawn to a fixed conic, so that the orthocentre of the triangle formed by them is at one of the foci; prove that the polar circle and circum-circle are fixed.

187. Given four straight lines, show that two conics can be constructed such that an assigned straight line of the four is directrix and the other three form a self-polar triangle; and that, whichever straight line be taken as directrix, the corresponding focus is one of two fixed points.

188. Parallel tangents are drawn to a given conic, and the point where one meets a given tangent is joined to the point where the

other meets another given tangent. Prove that the envelope of the joining line is a conic to which the two tangents are asymptotes.

189. With a point on the circum-circle of a triangle as focus, four conics are described circumscribing the triangle : prove that the corresponding directrices will pass each through a centre of one of the four circles touching the sides.

190. Three conics are drawn touching each pair of the sides of a triangle at the angular points where they meet the third side and passing through a common point. Show that the tangents at this common point meet the corresponding sides in three points on a straight line, and the other common tangents to each pair of conics pass respectively through these three points.

191. $ABCD$ is a quadrilateral circumscribing a conic, and through the pole $O$ of $AC$ a line is drawn meeting $CD$, $DA$, $DB$, $BC$, and $CA$ in $PQRST$ respectively. Show that $PQ$, $RS$ subtend equal angles at any point on the circle whose diameter is $OT$.

192. The normal at a fixed point $P$ of an ellipse meets the curve again in $Q$, and any other chord $PP'$ is drawn; $QP'$ and the straight line through $P$ perpendicular to $PP'$ meet in $R$; prove that the locus of $R$ is a straight line parallel to the chord of curvature of $P$ and passing through the pole of the normal at $P$.

193. Two tangents of a hyperbola $a$ are asymptotes of another conic $\beta$. Prove that if $\beta$ touch one asymptote of $a$, it touches both.

194. A conic is drawn through four fixed points $ABCD$. $BC$, $AD$ meet in $A'$; $CA$, $BD$ in $B'$; $AB$, $CD$ in $C'$; and $O$ is the centre of the conic. Prove that $\{ABCD\}$ on the conic $= \{A'B'C'O\}$ on the conic which is the locus of $O$.

195. Tangents drawn to a conic at the four points $ABCD$ form a quadrilateral whose diagonals are $aa'$, $bb'$, $cc'$ (the tangents at $ABC$ forming the triangle $abc$ and being met by the tangent at $D$ in $a'b'c'$). The middle points of the diagonals are $A'B'C'$ and the centre is $O$. Prove that $\{A'B'C'O\} = \{ABCD\}$ at any point of the conic.

196. If a right line move in a plane in any manner, the centres of curvature at any instant of the paths of all the points on it lie on a conic.

197. Defining a bicircular quartic as the envelope of a circle which moves with its centre on a fixed conic so as to cut orthogonally a fixed circle, show that it is its own inverse with respect to any one of the vertices of the common self-conjugate triangle of the fixed circle and conic, if the radius of inversion be so chosen that the fixed circle inverts into itself.

198. A quadrilateral is formed by the tangents drawn from two fixed points on the radical axis of a system of coaxal circles to any

circle of the system. Prove that the locus of one pair of opposite vertices is one conic, and of the remaining pair is another conic, and the two fixed points are the foci of both these conics.

199. Two fixed straight lines through one of the foci of a system of confocal conics meet any one of the conics in $PP'$, $QQ'$. Prove that the envelope of $PQ$ and $P'Q'$ is one parabola, and of $PQ'$, $P'Q$ is another parabola. Also the points of contact of $PQ$, $P'Q'$, $P'Q$, $PQ'$ with their respective envelopes lie on a straight line parallel to the conjugate axis of the system, which axis touches both parabolas.

200. A parallelogram with its sides in fixed directions circumscribes a circle of a coaxal system. Prove that the locus of one pair of opposite vertices is one conic and of the remaining pair is another conic, and the common tangents of these two conics are the parallels through the common points of the system to the sides of the parallelogram. Prove also that the tangents at the vertices of any such parallelogram to their respective loci meet in a point on the line of centres of the system.

201. $O$ is the centre of a conic circumscribing a triangle, and $O'$ is the pole of the triangle for this conic. Show that $O$ is the pole of the triangle for that conic which circumscribes the triangle and has its centre at $O'$.

202. $AA'$, $BB'$, $CC'$ are the three pairs of opposite vertices of a quadrilateral. A conic through $BB'$, $CC'$ and any fifth point $P$ meets $AA'$ in $X$ and $Y$. Prove that $PX$, $PY$ are the double lines of the involution $P\{AA', BB', CC'\}$.

203. If tangents be drawn to a system of conics having four common tangents, from a fixed point ($X$) on a side ($AA'$) of the self-conjugate triangle of the system, the points of contact will lie on a conic (viz. $XBB'CC'$).

204. $AA'$ $BB'$, $CC'$ are the three pairs of opposite vertices of a quadrilateral. A straight line meets $AA'$, $BB'$, $CC'$ in $LMN$. Prove that the conics $LBB'CC'$, $MCC'AA'$, $NAA'BB'$, and the conic touching the sides of the quadrilateral and also $LMN$, have a common point.

205. Three conics have double contact at the same two points, and $A$, $B$, $C$ are their centres. A straight line parallel to $ABC$ meets them in $PP'$, $QQ'$, $RR'$ respectively, and $O$ is any point on this straight line. Prove that $OP \cdot OP' \cdot BC + OQ \cdot OQ' \cdot CA + OR \cdot OR' \cdot AB = 0$.

206. In XXVIII. § 10. Ex. 4, prove that if $O'$ be this fixed point, then $CO$, $CO'$ are equally inclined to the axes, and $CO \cdot CO' = CS^2$.

207. If triangles can be inscribed in a conic $a$ and circumscribed to a conic $\beta$, the locus of the centroid of such a triangle is a conic homothetic with $a$.

208. If the conic $\beta$ be a parabola, this locus is a straight line.

## Miscellaneous Examples. 317

209. This straight line is parallel to the line joining points on the parabola where the tangents are parallel to the asymptotes of α.

210. The tangents at three points of a rectangular hyperbola form a triangle, of which the circum-circle has its centre at a vertex and passes through the centre of the hyperbola. Show that the centroid of the three points lies on the conjugate axis.

211. Show that the orthocentre of the three points in Ex. 210 is the vertex which is the centre of the circle.

212. If in Ex. 207 the conics α and β are homothetic, the centroid of the three points of contact with β of such a triangle is a fixed point.

213. If the conics α and β are coaxial, then the normals to α at the vertices of any such triangle are concurrent and also the normals to β at the points of contact of the sides; and conversely, if $PQR$ be three points on a conic such that the normals there are concurrent, a coaxial conic can be inscribed in the triangle $PQR$.

214. If the conics α and β are both parabolas, the locus of the centroid is parallel to the axis of α.

215. If α and β are parabolas with the same axis, whose latera recta are $l$ and $l'$, then $l' = 4l$.

216. Given a triangle self-conjugate for a conic, if a directrix touch a conic β inscribed in the triangle, then the corresponding focus lies on the director circle of β.

217. A conic is inscribed in a triangle self-conjugate for a rectangular hyperbola, with one focus on the hyperbola. Show that its major axis touches the hyperbola.

218. A triangle is inscribed in a conic and circumscribed to a parabola. Prove that the locus of the centre of its circumscribing circle is a straight line.

219. The following pairs of conics are related to one another as in XIV. § 2. Ex. 14—

(i) A rectangular hyperbola, and a parabola whose focus is at the centre of the r. h. and whose directrix touches the r. h.

(ii) Two rectangular hyperbolas, each passing through the centre of the other and having the asymptotes of one parallel to the axes of the other.

220. If the polar circle of three tangents to a conic passes through a focus, the orthocentre lies on the corresponding directrix.

221. If a triangle inscribed in a parabola has its orthocentre on the directrix, its polar circle passes through the focus.

222. A circle has its centre on the directrix and touches the sides of a triangle self-conjugate for a parabola. Show that it passes through the focus.

223. Triangles can be inscribed in a conic $\alpha$ so as to be self-conjugate for a conic $\beta$. A circle has double contact with $\alpha$ along a tangent to $\beta$. Show that it cuts orthogonally the director of $\beta$.

224. Two conics, in either of which triangles can be inscribed self-conjugate for a third conic, have double contact. Show that their chord of contact touches this conic.

225. From any point $P$ two tangents $PQ$, $PR$ are drawn to an ellipse: if $C$ is the centre of the ellipse, then all hyperbolas drawn through $P$ and $C$ and having their asymptotes parallel to the axes of the ellipse cut $QR$ harmonically.

226. A conic circumscribes a triangle self-conjugate for a parabola and has its centre on the parabola. Show that an asymptote touches the parabola.

227. A circle through the centre of a rectangular hyperbola cuts it in $ABCD$. Show that the circle circumscribing the triangle formed by the tangents to the r. h. at $ABC$ passes through the centre of the hyperbola, and has its centre on the hyperbola at the extremity $D'$ of the diameter through $D$; and $D'$ is the orthocentre of $ABC$.

228. Show that if $D$ be the pole of the triangle $ABC$ for a conic, then $A$, $B$, $C$ are the poles of the triangles $BCD$, $ACD$, $ABD$ respectively. *Such a quadrangle may be said to be self-conjugate for the conic.*

229. If triangles can be inscribed in $\beta$ which are self-conjugate for $\alpha$, then quadrangles can be inscribed in $\beta$ which are self-conjugate for $\alpha$; and conversely.

230. If triangles can be circumscribed to $\beta$ which are self-conjugate for $\alpha$, then quadrilaterals can be circumscribed to $\beta$ which are self-conjugate for $\alpha$; and conversely.

231. If we can describe triangles to touch a conic $\alpha$ and to be self-polar for each of two conics $\beta$ and $\gamma$, then the four intersections of $\beta$ and $\gamma$ form a self-polar quadrangle for $\alpha$.

232. If triangles can be inscribed in each of two conics $\beta$, $\gamma$ so as to be self-polar for a conic $\alpha$, then triangles self-polar for $\alpha$ can be inscribed in any conic through the intersections of $\beta$ and $\gamma$.

233. If triangles can be circumscribed to each of two conics $\beta$, $\gamma$ self-polar for a conic $\alpha$, then triangles self-polar for $\alpha$ can be circumscribed to any conic touching the common tangents of $\beta$ and $\gamma$.

234. The polars of a fixed triangle for a system of four-point conics envelope a conic touching the sides of the triangle.

235. The poles of a fixed triangle for a system of conics having four common tangents lie on a conic circumscribing the triangle.

236. If the system of four-tangent conics is a system of confocals, the locus of the poles is a rectangular hyperbola.

# Miscellaneous Examples.

237. If two conics are related as in XIV. § 2. Ex. 14, and the first passes through the centre of the second, then the second passes through the centre of the first.

238. Three tangents to a conic $a$ form a triangle. A conic $\beta$ circumscribes the triangle and passes through the centre of $a$ and the pole of the triangle with respect to $a$. Prove that its centre lies on $a$.

239. A rectangular hyperbola circumscribes a triangle and passes through the centre of one of the circles touching the sides. Show that its centre lies on this circle.

240. Hence prove Feuerbach's theorem, viz.—the nine-point circle of any triangle touches the inscribed and escribed circles.

241. Show that in Ex. 239 the poles of the triangle for these circles lie on the respective hyperbolas; and the polars of the triangle for the hyperbolas are tangents to the respective circles.

242. The nine-point circle of a triangle inscribed in a rectangular hyperbola touches the polar-circle of the triangle formed by the tangents at the vertices, at the centre of the conic.

243. The pole with respect to a parabola of the triangle formed by three tangents to it lies on the minimum ellipse circumscribing the triangle.

244. The polar in this case passes through the centre of gravity of the triangle.

245. The pole with respect to a parabola of an inscribed triangle lies on the maximum ellipse inscribed in the triangle.

246. The two conics in the last example are reciprocal with respect to a conic with its centre at this pole and having the triangle as a self-conjugate triangle.

247. Show that the polar of a triangle for a rectangular hyperbola which circumscribes it, touches the conic which touches the three sides at the vertices of the pedal triangle; and the pole of the triangle lies on the radical axis of the circum-circle and nine-point circle of the triangle.

248. A conic passes through the vertices and centroid of a fixed triangle. Show that the pole of the triangle for the conic lies on the line at infinity, and the polar touches the maximum inscribed ellipse.

249. A conic touches the sides of a triangle and passes through its centroid. Show that the polar of the triangle for this conic is a tangent to the minimum ellipse circumscribing the triangle.

250. The foci of a conic inscribed in a triangle self-conjugate for a rectangular hyperbola are conjugate points for the r. h.

251. A parabola touches the sides of a triangle $ABC$, and $S$ is its focus. The axis meets the circum-circle again in $O$. With $O$ as centre

the rectangular hyperbola is described for which the triangle is self-conjugate. Show that the axis of the parabola is an asymptote of the r.h.

252. Two parabolas touch the sides of a triangle and have their foci at the extremities of a diameter of its circum-circle. Show that their axes are the asymptotes of a rectangular hyperbola for which the triangle is self-conjugate.

253. Triangles can be inscribed in a parabola (whose latus-rectum is $l$) so as to be self-conjugate for a coaxial parabola (whose latus-rectum is $l'$). Prove that $l' = 2l$.

254. The locus of the centre of a circle of constant radius circumscribed to a triangle self-conjugate for a fixed conic is a circle concentric with the conic.

255. Given three tangents and the sum of the squares of the axes, the locus of the centre of a conic is a circle.

256. A circle of given radius is inscribed in a triangle self-conjugate for a fixed conic. Prove that the locus of its centre is a concentric homothetic conic.

257. A circle $a$ touches the sides of a triangle self-conjugate for a conic $\beta$. Show that a rectangular hyperbola having double contact with $\beta$ along a tangent to $a$ passes through the centre of the circle.

258. A circle touches a fixed straight line, and triangles can be circumscribed to it which are self-conjugate for a fixed conic. Prove that the locus of its centre is a rectangular hyperbola.

259. The orthocentre of a triangle of tangents to a rectangular hyperbola and the centre of the circle through the points of contact are conjugate points for the r.h.

260. If the centroid of a triangle inscribed in a conic lies on a concentric homothetic conic, prove that the nine-point circle cuts orthogonally a fixed circle.

261. If two circles touch respectively the sides of two triangles self-conjugate for a conic, then their centres of similitude are conjugate points for the conic.

262. If a rectangular hyperbola has double contact with a conic $a$, its centre and the pole of the chord of contact are inverse points for the director circle of $a$.

263. A circle circumscribes triangles self-conjugate for a given conic and passes through a fixed point. Prove that its centre lies on the directrix of the parabola which has double contact with the conic at the points of contact of tangents from the fixed point.

264. Triangles are circumscribed to a central conic so as to have the same orthocentre. Prove that they have the same polar circle.

## Miscellaneous Examples.

265. Two triangles are inscribed in a conic (which is not a rectangular hyperbola) so as to have the same orthocentre. Prove that they have the same polar circle.

266. Two triangles are inscribed in a conic (which is not a circle) so that their circum-circles are concentric. Prove that they are self-conjugate for a parabola.

267. Two triangles are circumscribed to a conic, so that their circum-circles are concentric. Prove that they either have the same circum-circle or are self-conjugate for a parabola.

268. A conic which is inscribed in a triangle self-conjugate for a rectangular hyperbola and has a focus at the centre of the r. h. is a parabola.

269. A conic with a focus at the centre of a rectangular hyperbola circumscribes triangles self-conjugate for the r. h. Prove that the corresponding directrix touches the r. h.

270. Triangles can be inscribed in each of two conics $a$ and $\beta$, self-conjugate for the other. Prove that the reciprocal of $a$ for $\beta$ and of $\beta$ for $a$ is the same conic $\gamma$; and $a$, $\beta$, $\gamma$ are so related that each is the envelope of lines divided harmonically by the other two and also the locus of points from which tangents to the other two form a harmonic pencil. Also any two of these conics are reciprocals for the third.

271. Two hyperbolas pass each through the centre of the other and determine a harmonic range on the line at infinity. Prove that the reciprocal of either for the other is the parabola inscribed in the quadrilateral formed by parallels through each centre to the asymptotes of the hyperbola passing through it.

272. A conic is inscribed in a given triangle and passes through its circum-centre. Show that its director circle touches the circum-circle and the nine-point circle of the triangle.

273. Find the locus of the centre of the conic in the last example.

274. The locus of the centre of a conic touching three given straight lines and passing through a given point is the conic touching the triangle formed by the middle points of the sides of the fixed triangle and such that if $D$ be the fixed point, $G$ the centroid of the triangle and $O$ the centre of the locus, then $ODG$ are collinear, and $DO = \frac{3}{4} DG$.

275. If the fixed point be the centroid of the triangle, the locus is the maximum ellipse inscribed in the triangle formed by joining the middle points of the sides.

276. A circle inscribed in a triangle self-conjugate for a hyperbola cuts the hyperbola orthogonally at a point $P$. Show that the normal at $P$ is parallel to an asymptote.

277. A circle is inscribed in a triangle self-conjugate for a conic and has its centre on its director circle. Prove that it touches the reciprocal of the director circle for the conic.

278. A circle $a$ with centre $O$ is inscribed in a triangle self-conjugate for a conic $\beta$. If $P$ and $Q$ be the points of contact of tangents to $\beta$ from $O$, then the tangents from $P$ and $Q$ to the conic which is the reciprocal for $\beta$ of its director, are also tangents to the circle $a$.

279. The six tangents to a conic from the vertices of a triangle cut again in twelve points which lie by sixes on four conics.

280. The six points in which a conic cuts the sides of a triangle can be joined in pairs by twelve other lines which are tangents by sixes to four conics.

281. If tangents are drawn to a parabola from two points $A$ and $B$, the asymptotes of the conic through $AB$ and the points of contact of the tangents from $A$ and $B$, are parallel to the tangents to the parabola from the middle point of $AB$.

282. If tangents are drawn to a parabola from $A$ and $B$, the conic through $AB$ and the points of contact will be a circle, rectangular hyperbola or parabola as $AB$ is bisected by the focus, directrix, or parabola respectively.

283. Tangents are drawn to a circle from two points on a diameter. Show that the foci of the conic touching the tangents and their chords of contact lie on the circle.

284. If tangents are drawn to a central conic from $P$ and $Q$, and $C$ be the centre and $S$ a focus, then the conic through $P$, $Q$, and the points of contact of tangents from $P$, $Q$ will be a circle if the angle $PCQ$ is bisected internally by $CS$, and if $CP \cdot CQ = CS^2$.

285. The conic in the previous example will be a rectangular hyperbola if $P$ and $Q$ are conjugate for the director circle.

286. A point and the orthocentre of the triangle formed by tangents from it to a conic and their chord of contact are conjugate points for the director circle of the conic.

287. If a conic $a$ pass through two points $A$, $B$ and the points of contact of tangents from them to a given conic, and if $\beta$ be the similarly constructed conic for two points $A'$, $B'$; then if $AB$ are conjugate for $\beta$, $A'B'$ are conjugate for $a$.

288. The reciprocal of the director circle of a conic $a$ for $a$ is confocal with $a$.

289. Along the normal to a conic at a point $O$ are taken pairs of points $PQ$ such that $OP \cdot OQ$ is equal to the square of the semi-diameter parallel to the tangent at $O$. Show that tangents to the conic from $P$ and $Q$ intersect on the circle of which a diameter is the intercept on the tangent at $O$ by the director circle.

290. The orthocentre of a triangle formed by two tangents to a conic and their chord of contact lies on the conic. Prove that the locus of the vertex of the triangle is the reciprocal of the conic for its director circle or the reciprocal for the conic of its evolute.

# Miscellaneous Examples. 323

291. The centre of the circle inscribed in a triangle formed by two tangents to an ellipse and their chord of contact lies on the conic. Prove that the locus of the vertex of the triangle is a hyperbola confocal with the ellipse, and having the equi-conjugate diameters of the ellipse for its asymptotes.

292. The centre of gravity of a triangle formed by two tangents to a conic and their chord of contact lies on the conic. Prove that the locus of the vertex of the triangle is a concentric homothetic conic.

293. From two points $BC$, tangents are drawn to a fixed conic, and the sides of the two triangles formed by these two pairs of tangents and their chords of contact touch the conic $a$. Similarly the pairs of points $CA$, $AB$ determine the conics $\beta$ and $\gamma$ respectively. Prove that if $A$ lies on $a$, then $B$ lies on $\beta$, and $C$ on $\gamma$.

294. $A'B'C'$ are the middle points of the sides of a triangle $ABC$. Prove that the conic which is concentric with the nine-point circle of $A'B'C'$ and inscribed in $A'B'C'$ has double contact with the polar circle of $ABC$ at the points where the circum-circle of $ABC$ meets the polar circle, and also has double contact with the nine-point circle of $A'B'C'$.

295. A triangle is self-conjugate for a conic. Prove that the sides of the pedal triangle touch a confocal.

296. A triangle is self-polar for a conic; show that an infinite number of triangles can be at once inscribed in the conic and circumscribed to the triangle, and vice versâ.

297. If two conics $a$ and $\beta$ are related so that the poles for $a$ of two opposite common chords lie on $\beta$, then the polars for $\beta$ of two opposite common apexes touch $a$.

298. Of all conics inscribed in a given triangle, that for which the sum of the squares of the axes is least has its centre at the orthocentre of the triangle.

299. $E$, $F$ are a pair of inverse points with respect to a circle whose centre is $A$; $B$ is the harmonic conjugate of $A$ with respect to $E$, $F$; $AP$, $BP$ and the tangent at $P$, any point on the circle, meet the polar of $E$ in $L$, $M$, $T$ respectively; show that $LT$, $TM$ subtend equal angles at $A$.

300. The connector of a pair of conjugate points with respect to a given conic passes through a fixed point and one of the pair lies on a given straight line; show that the locus of the other is a conic, and determine six points upon the locus.

THE END.

www.ingramcontent.com/pod-product-compliance
Lightning Source LLC
Chambersburg PA
CBHW030004240426
43672CB00007B/825